Springer Tracts in Electrical and Electronics Engineering

Springer Tracts in Electrical and Electronics Engineering (STEEE) publishes the latest developments in Electrical and Electronics Engineering - quickly, informally and with high quality. The intent is to cover all the main branches of electrical and electronics engineering, both theoretical and applied, including:

- Signal, Speech and Image Processing
- Speech and Audio Processing
- Image Processing
- Human-Machine Interfaces
- Digital and Analog Signal Processing
- Microwaves, RF Engineering and Optical Communications
- Electronics and Microelectronics, Instrumentation
- Electronic Circuits and Systems
- Embedded Systems
- Electronics Design and Verification
- Cyber-Physical Systems
- Electrical Power Engineering
- Power Electronics
- Photovoltaics
- Energy Grids and Networks
- Electrical Machines
- Control, Robotics, Automation
- Robotic Engineering
- Mechatronics
- Control and Systems Theory
- Automation
- Communications Engineering, Networks
- Wireless and Mobile Communication
- Internet of Things
- Computer Networks

Within the scope of the series are monographs, professional books or graduate textbooks, edited volumes as well as outstanding PhD theses and books purposely devoted to support education in electrical and electronics engineering at graduate and post-graduate levels.

Review Process

The proposal for each volume is reviewed by the main editor and/or the advisory board. The books of this series are reviewed in a single blind peer review process.

Ethics Statement for this series can be found in the Springer standard guidelines here https://www.springer.com/us/authors-editors/journal-author/journal-author-helpdesk/before-you-start/before-you-start/1330#c14214.

Vinayak Pachkawade · Koushik Guha
Editors

Micro- and Nano-Systems in 21st-Century

Designs, Developments, Applications and Perspective

Springer

Editors
Vinayak Pachkawade
VPACHKAWADE
Nagpur, Maharashtra, India

Koushik Guha
Department of Electronics
and Communication Engineering
National Institute of Technology
Silchar, Assam, India

ISSN 2731-4200 ISSN 2731-4219 (electronic)
Springer Tracts in Electrical and Electronics Engineering
ISBN 978-981-96-4658-6 ISBN 978-981-96-4659-3 (eBook)
https://doi.org/10.1007/978-981-96-4659-3

This Springer imprint is published by the registered company Springer Nature Singapore Pte Ltd.
The registered company address is: 152 Beach Road, #21-01/04 Gateway East, Singapore 189721, Singapore

Preface

The rapid evolution of micro- and nano-systems in the 21st-century has transformed the landscape of technology, driving innovations that impact nearly every aspect of human life. This book, *Micro- and Nano-Systems in 21st-Century*, explores the dynamic field of Micro-Electro-Mechanical Systems (MEMS) and Nano-Electro-Mechanical Systems (NEMS), offering readers a comprehensive understanding of their principles, applications, and future potential. As part of the esteemed *Springer Tracts in Electrical and Electronics Engineering* series, this book presents cutting-edge research and advancements that underscore the role of MEMS/NEMS in shaping modern technologies.

The book opens with Chap. 1, which lays the foundation for understanding the core mechanisms of "Actuation and Sensing in Micro/Nano System". It explores diverse transduction techniques, delving into electrostatic, piezoelectric, and optical methods for actuation, alongside capacitive and piezoresistive approaches for sensing. This chapter provides a broad overview of the fundamental principles and challenges associated with modern transducers, setting the stage for subsequent discussions.

Chapter 2 transitions into the specialized domain of "Acoustic Devices for Gas Sensing and RF Applications in MEMS/NEMS". The chapter elucidates the working principles and architectures of surface acoustic wave (SAW) and bulk acoustic wave (BAW) devices, emphasizing their role in applications ranging from precise gas detection to their critical function in 5G and 6G communication systems.

The healthcare sector benefits profoundly from MEMS, as discussed in Chap. 3, "MEMS in Twenty-First-Century Healthcare". This chapter highlights the pivotal role of MEMS in wearable devices, biosensors, lab-on-a-chip systems, and artificial implants, emphasizing how these technologies enable real-time monitoring, personalized medicine, and innovative therapeutic solutions.

Innovative microfluidic technologies are at the heart of Chap. 4, which investigates the "Analysis of Resistance and Thrombogenicity of a Composite Coating Based on Collagen and Carbon Nanotubes Using a Microfluidic Chip". By employing a microfluidic chip, the chapter demonstrates the potential for evaluating coatings under dynamic fluid flow, offering insights into improving the biocompatibility of materials used in medical implants.

The focus shifts to energy solutions in Chap. 5, “Advancements in MEMS Vibration Energy Harvesters”. The chapter examines transduction techniques, fabrication methods, and power management systems, highlighting the applications of MEMS harvesters in IoT and wearable devices. It provides a detailed analysis of overcoming limitations like narrow bandwidth and low power efficiency.

Chapter 6, “CMOS Amalgamated MEMS/NEMS: Pioneering Technology for the Modern Semiconductor Cosmos and Future World”, explores the integration of CMOS technology with MEMS/NEMS. By combining nanoscale mechanical components with CMOS circuits, the chapter presents groundbreaking advancements in performance, miniaturization, and energy efficiency, with applications in biosensors, RF MEMS, and nano-optics.

Building on this theme, Chap. 7, “CMOS-MEMS Sensors with FET-Based Transduction: A Paradigm Shift”, addresses the challenges of traditional MEMS sensors and advocates for innovative Field Effect Transistor (FET)-based architectures. This chapter explores the potential of FET-based transduction mechanisms, particularly suspended gate FETs, to streamline design and improve performance through CMOS-MEMS integration.

Finally, Chap. 8, “New and Emerging Materials for MEMS/NEMS: Properties and Characterization”, investigates the potential of advanced materials, such as nanoscale composites, metamaterials, and nano-biomaterials. These materials promise groundbreaking functionalities, including nanoscale ferroelectricity, thermal management, and shape memory, with applications extending to IoT and AI-enabled technologies.

Each chapter in this book represents the forefront of MEMS/NEMS research, reflecting the innovative approaches and interdisciplinary efforts of the global scientific community. By bringing together fundamental principles, emerging technologies, and practical applications, *Micro- and Nano-Systems in 21st-Century* serve as an essential resource for researchers, professionals, and students seeking to understand and contribute to this dynamic field.

Nagpur, India Vinayak Pachkawade
Silchar, India Koushik Guha

Contents

Part I
Transduction in Micro-Nano Devices

Chapter 1
Actuation and Sensing in Micro/Nano System

Vinayak Pachkawade

Abstract This book chapter provides a comprehensive exploration of transduction methods used in modern Micro/Nano devices, elucidating their fundamental principles, applications usage, and technological challenges. The chapter begins with a solid-state model of a transducer, offering insights into the transformation between mechanical and electrical domains. Various transduction techniques for actuation, including electrostatic, electro-thermal, electromagnetic, piezoelectric, and optical are then explained. Following this, sensing techniques, including capacitive, piezoresistive, resonant, etc., are systematically presented. The chapter concludes with an outlook on potential advancements and broader insights, emphasizing the ongoing research endeavors to refine existing techniques and explore new materials, paving the way for enhanced performance in diverse applications.

Keywords MEMS · NEMS · Sensors and Actuators · Transduction in MEMS

Introduction

Micro/Nano Systems Technology (M/NST) stands at the forefront of rapidly evolving technologies, offering unparalleled capabilities in ultra-miniaturization and seamless integration of diverse components at the micro- and nano scale. This book chapter delves into the transformative potential of M/NST across a spectrum of applications, ranging from medical and 5G/6G communication to environmental monitoring, energy systems, aerospace, consumer electronics, and automotive technologies (refer Fig. 1.1). The intrinsic ability of M/NST components to transduce energy facilitates precise quantification and measurement of physical, biological, and chemical processes through proportional electric signals.

This book chapter aims to provide a comprehensive exploration of various transduction methods employed in M/NST, with a specific focus on actuators and sensors.

V. Pachkawade (✉)
VPACHKAWADE, Nagpur, India
e-mail: vinayak@vpachkawade.com

V. Pachkawade and K. Guha (eds.), *Micro- and Nano-Systems in 21st-Century*, Springer Tracts in Electrical and Electronics Engineering,
https://doi.org/10.1007/978-981-96-4659-3_1

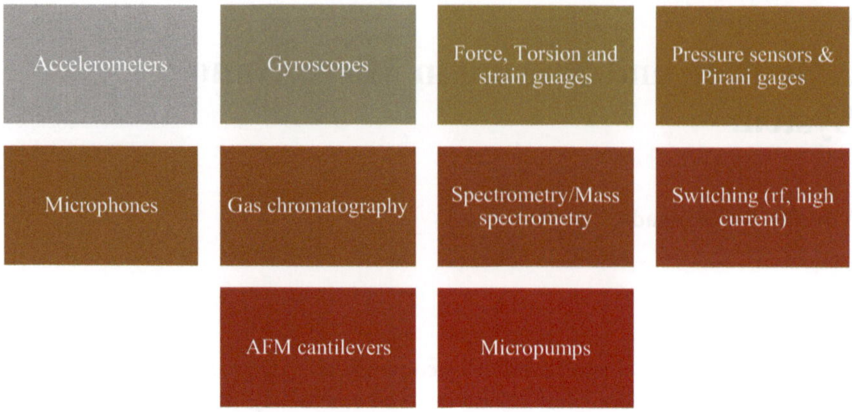

Fig. 1.1 Sample applications of M/NST

The chapter unfolds with a representative model of a solid-state transducer, offering a succinct overview of the mathematical principles and illustrating the transformation between mechanical and electrical domains. The subsequent sections meticulously scrutinize diverse transduction techniques including electrostatic, electro-thermal, electro-magnetic, and piezoelectric. Each method is elucidated with its operating principles, mathematical equations, applications, and a critical analysis of relative merits and shortcomings.

The structured presentation extends to a dedicated section on sensors, describing distinct sensing mechanisms (capacitive, piezo-resistive, resonant, and optical) along with their corresponding equations, advantages, drawbacks, and practical applications.

In conclusion, the chapter offers a forward-looking outlook, providing fundamental insights to cultivate a deeper scientific understanding among readers. The content is designed to cater to the diverse needs of researchers, practitioners, and enthusiasts in the field of M/NEMS, contributing to the ongoing discourse on the advancements and broader implications of transduction methods in this transformative domain.

Basic Transducer Model

A common solid-state model for a transducer can be represented as a one-degree-of-freedom (1-DoF) mass-spring-damper system (Fig. 1.2).

The equation of motion for such a system is given by Newton's second law of motion:

$$\sum F = Fe + Fc + Fs = Ma \tag{1.1}$$

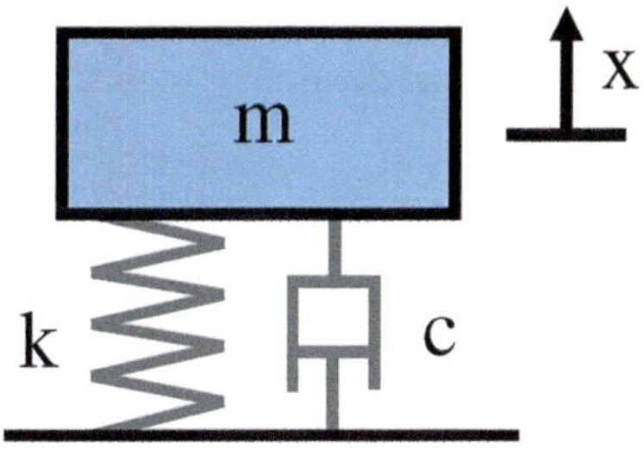

Fig. 1.2 Representative model of 1-DoF mass-spring-damper (Diez-Jimenez et al. 2021)

where:

- $\sum F$ is the net force acting on the mass
- Fe is the external force (actuation force)
- Fc is the damping force
- Fs is the restoring force
- M is the mass
- a is the acceleration

The expression for Fe can be derived based on electrostatic actuation, where the force is related to the electric field and the displacement. Assuming a parallel plate capacitor structure, the electrostatic force Fe can be expressed as:

$$Fe = \frac{1}{2}\varepsilon A \frac{V^2}{(d-x)^2} \tag{1.2}$$

where:

- ε is the permittivity
- A is the nominal electrode area
- V is the applied voltage
- d is the nominal gap distance
- x is the displacement

The restoring force Fs in a simple mechanical spring can be expressed as:

$$Fs = K_{eff}\,x$$

where:

- $K_{eff} = Km + Ke$ is the effective spring constant
- Km is the mechanical spring constant
- Ke is the electrostatic spring constant given by $-\frac{\epsilon A}{(d-x)^3}V^2$

Table 1.1 provides an analogy between the mechanical domain and its electrical domain counterpart. Each mechanical element is paired with its analogous electrical element.

Table 1.1 Analogy between mechanical and electrical domain

Mechanical domain	Electrical domain
Mass (m)	Inductance (L)
Damping (c)	Resistance (R)
Spring constant (k)	Reciprocal of capacitance (1/C)
Displacement (x)	Charge (Q)
Velocity (v)	Current (I)
Force (F)	Voltage (V)

Transduction Techniques

Figure 1.3 classifies various transduction techniques based on actuation and sensing in micro-nano systems.

(a) **Electrostatic**:

An electrostatic actuator is a device that generates mechanical motion through the application of electrostatic forces. It operates based on the principle of electrostatic attraction or repulsion between charged components. This type of actuator is commonly employed in M/NST and other precision applications due to its ability to provide precise and controlled movement at the micro- and nano scale. The fundamental principle underlying electrostatic actuators is the force of attraction or repulsion between electrically charged components. When a voltage is applied across electrodes, an electric field is created, leading to the generation of electrostatic forces. Electrostatic actuators often consist of capacitive elements, such as parallel plates or comb-drive structures. Changes in capacitance, induced by variations in the relative position of these components, result in corresponding electrostatic forces that drive

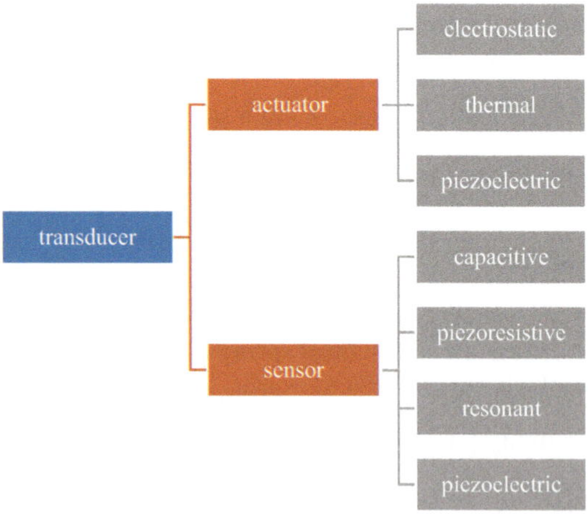

Fig. 1.3 Classification of transducers based on actuation and sensing in M/NST

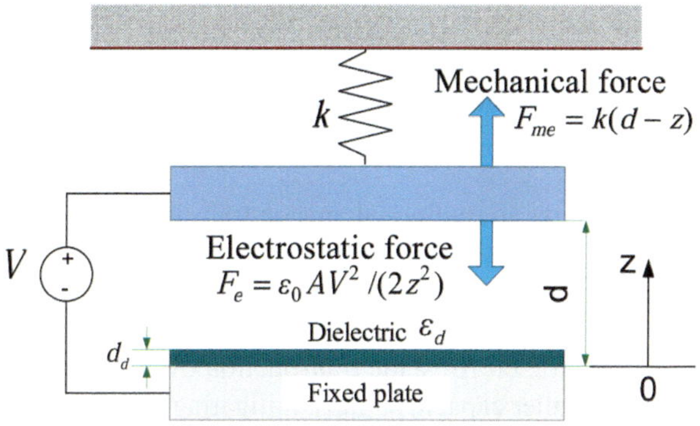

Fig. 1.4 An electrostatic actuator (Zhang et al. 2007)

mechanical motion. Figure 1.4 shows a representative schematic of an electrostatic actuator.

In a static scenario, a constant DC voltage is applied, creating a steady electrostatic force. This force causes the mechanical structure of the actuator to deflect or move to a new equilibrium position. In dynamic situations, such as resonators, both DC and AC voltages are applied. The DC component establishes the initial equilibrium position, while the AC component induces vibrations or oscillations around this position, enabling dynamic mechanical behavior.

The behavior of electrostatic actuators can be mathematically described using principles of electrostatics and mechanics. Capacitance equations, such as $C = \frac{\varepsilon A}{d}$, where ε is the permittivity, A is the nominal electrode area, and d is the capacitive air gap, are common in modeling these systems. The electrostatic force (F) acting on the actuator can be expressed, considering changes in capacitance due to motion. For example, $F = \frac{1}{2}\varepsilon A\frac{V^2}{(d-x)^2}$, where, x represents the distance, V is the applied voltage, and ε, A, and d have their usual meanings.

Electrostatic actuators offer several advantages, characterized by their simplicity in design, low power consumption, rapid response times, and seamless compatibility with electronic integration. However, these advantages come with their set of challenges, including nonlinearity in specific configurations, pull-in instability concerns, and power handling issues (Park et al. 2011; Kaajakari 2009; Agarwal et al. 2005). Fabrication poses challenges due to the necessity for deep sub-micro/nano air-gaps, making the process intricate. Despite these challenges, electrostatic actuators find diverse applications in robotics, telecommunications, healthcare, aerospace, and consumer electronics. They are integral components in RF MEMS switches, MEMS filters, oscillators, microphones, loudspeakers, mirrors, energy harvesters, and various other technological advancements.

(b) **Internal Dielectric Transduction**

Internal dielectric transduction utilizes the dielectric properties of materials to sense or generate signals with enhanced efficiency. When an electric field is applied across a dielectric material sandwiched between two conducting electrodes, the material polarizes, creating a charge separation and an electric dipole moment. This dipole moment manifests as a measurable voltage across the electrodes, the magnitude of which is influenced by the dielectric constant of the material and the applied electric field strength.

Therefore, an alternative to traditional air-gap transduction is the use of internal dielectric materials, particularly those with high dielectric constants (High-K). Filling the air-gap with such materials significantly improves transduction efficiency (Lin et al. 2006).

Advantages of dielectric electrostatic transduction over air-gap methods include the ability to achieve smaller capacitive gaps, mitigating issues like pull-in and stiction. Additionally, the high dielectric permittivity enhances driving force and capacitive sensing capabilities (Utz et al. 2018). Notably, internal dielectric transduction has demonstrated success in operating at extremely high frequencies, such as 4.51 GHz for longitudinal bulk-mode MEMS silicon, showing potential applications in high-frequency bands of the radio spectrum, such as microwave oscillators for low-power clocking in microprocessors (Lin et al. 2006).

Optimizing dielectric thickness and placement in resonators operating at high frequencies can lead to improved performance and higher resonant frequencies. Despite its advantages, internal dielectric transduction poses challenges associated with large parasitics, thin-film material processing, and integration with CMOS technology. While few recent studies have explored this transduction method, its potential to achieve enhanced performance makes it a subject of increasing interest and investigation in the realm of micro-nano systems (Lin et al. 2006).

(c) **Capacitive Sensing**

Capacitive sensing is a prevalent technology widely employed in various M/NST devices, particularly in pressure sensors where capacitive pressure sensors have displaced piezoresistive alternatives. The fundamental structure of capacitive sensing involves a parallel plate capacitance, expressed by the equation $C = \varepsilon A/d$ (also refer the section on electrostatic actuators). This equation signifies that capacitance is inversely proportional to the distance between the plates, enabling highly accurate sensing of minute displacements. Referring to Fig. 1.4, the alteration in capacitance is converted into a proportional change in voltage/current, facilitating measurement. Capacitive sensing finds application in diverse technologies, including touchscreens, accelerometers, gyroscopes, pressure sensors, and humidity sensors (Lobur and Holovatyy 2009; Qiao et al. 2018; Ruan et al. 2020; Ren et al. 2013). Noteworthy applications extend to detecting tilt (Zou et al. 2012), particles (Pachkawade and Tse 2021), biomarkers in exhaled breath, and gases (Qin et al. 2020). Capacitive sensing offers several advantages, such as high sensitivity, ultra-low power consumption, compact size, and high reliability. It is often integrated with electrostatic actuation (Rabenimanana et al. 2018), and occasionally with thermal actuation (Hajjam et al. 2009; Zhang and Hoshino 2019).

Despite its widespread use, capacitive sensing encounters challenges. Balancing high sensitivity with robustness is a primary concern, especially as miniaturization, while beneficial for many applications, can reduce capacitance and compromise sensitivity. Researchers address this challenge through optimized electrode designs, exploration of advanced dielectric materials, and innovative microfabrication techniques. Another challenge lies in cross-talk or interference within densely packed M/NST systems. To mitigate unwanted capacitance coupling, sophisticated shielding, isolation, and signal processing techniques are implemented. Temperature variations can also impact the dielectric properties of materials, affecting sensor performance. Robust calibration algorithms and environmental compensation methods are developed to enhance accuracy under changing conditions. In essence, the challenges associated with capacitive sensing are effectively tackled through a combination of material advancements, design optimizations, and signal processing strategies, solidifying its position as a versatile and reliable technology for a myriad of M/NST applications.

(d) **Thermal Actuators: Principles, Configurations, and Applications**

Thermal actuators in M/NST harness Joule heating as their fundamental principle. The application of voltage induces a current flow through the actuator material, generating heat through inherent resistance. This temperature increase prompts thermal expansion, resulting in the displacement of the actuator. The mathematical expression for the electro-thermal force F_{th} is given by $F_{th} = \alpha_T TAE(T - T_0)$, where α_T, E, A, T, and T_0 represent the coefficient of thermal expansion (CTE), Young's modulus, cross-sectional area of the material, initial, and final temperatures, respectively. Figure 1.5 shows one type of the thermal actuator (Bimorph).

Micro-thermal actuators are commonly fabricated using materials like gold, platinum, or thermally sensitive substances such as silicon/polysilicon. A thin film of thermally sensitive material is heated using an electrical current, inducing thermal expansion and actuator displacement.

In dynamic thermal actuation and sensing, an electric current induces thermal expansion for displacement, and an AC current causes alternating thermal stress,

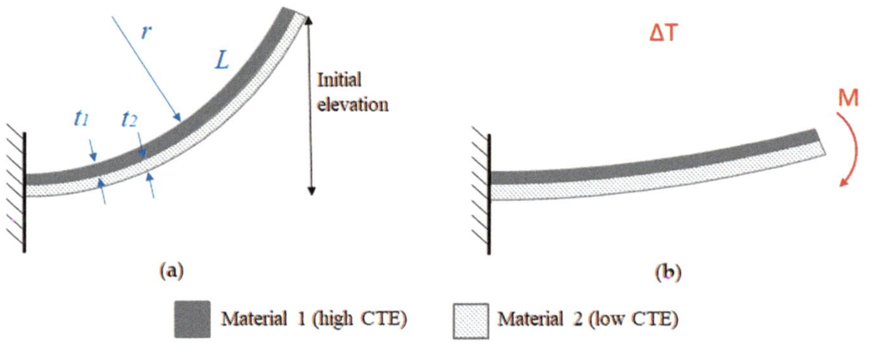

Fig. 1.5 Representative schematic of a thermal actuator, initial state (**a**) and after actuation (**b**) (Potekhina and Wang 2019)

leading to actuation in resonant mode. This mechanism, termed "Thermal-Piezo-Resistive" (TPR), finds applications in detecting ultra-fine particulate matter (PM) (Chu et al. 2018).

Various configurations of electrothermal actuators are reported, including thermal buckling actuators and chevron-shaped designs. Bimorph actuators, consisting of two bonded thin films, and unimorph actuators, with a single thin film and built-in stress gradient, offer precise and repeatable motions (Potekhina and Wang 2019).

Thermal actuators boast advantages like high efficiency, operation at low voltages, and compact size. They find utility in microgrippers, micro- and nanopositioners, scanning probes, microsatellite thermal control, micromirrors, microengines, micropumps, and RF tunable filters, among others. Fabricated using standard microfabrication techniques, they are compatible with other micro-nano devices, consuming lower power. However, challenges include slow response times, limitations in high-speed applications, and the need for precise temperature control. Multiphysics simulation aids in design, addressing challenges related to nonlinearity and ensuring uniform temperature distribution. Ongoing research explores innovative materials, design modifications, and fabrication techniques to enhance the efficiency, reliability, and scalability of electro-thermal transduction in micro-nano devices, opening avenues for applications in various fields, including micromanipulation, micro-assembly, and optical systems.

(e) **Thermal Sensing**

M/NST thermal sensors function by converting temperature changes into electrical signals, enabling precise and real-time temperature monitoring and control (Zhang and Hoshino 2019). Among these sensors, the thermoelectric sensor (Kao et al. 2010) stands out, operating by detecting the electromotive force generated due to temperature differences between two conductors (see Fig. 1.6a). With applications in automotive, aerospace, and industrial settings, this sensor can operate across a wide temperature range. Another variant, the thermal resistor, measures electrical resistance changes with temperature and finds use in consumer electronics like smartphones and laptops for temperature monitoring. Microbolometers, depicted in Fig. 1.6b, operate as MEMS-based thermal transducers that detect infrared radiation emitted by objects (Yu et al. 2020). By arranging them in a 2-D array, microbolometers contribute to thermography using an uncooled infrared sensor.

M/NST-based thermal transducers also include Surface Acoustic Wave (SAW) devices (Mujahid et al. 2019), relying on changes in acoustic wave propagation velocity with temperature. Resonant cantilevers (as shown in Fig. 1.6c) work based on the change in resonant frequency (Δf) with temperature (ΔT), offering sensitivity ($S = \Delta f/\Delta T$). Applications span mass flow metering, wind sensing, detection of hydrogen gas in vehicle interiors, and measuring liquid flow down to 0.05 μl/min (Wang et al. 2007).

Minimizing conductive heat losses in MEMS thermal flow sensors enhances sensitivity and reduces power consumption. Techniques involve low thermal conductivity substrates and membranes under heaters. Computational methods and experimental studies optimize geometry for improved accuracy and response.

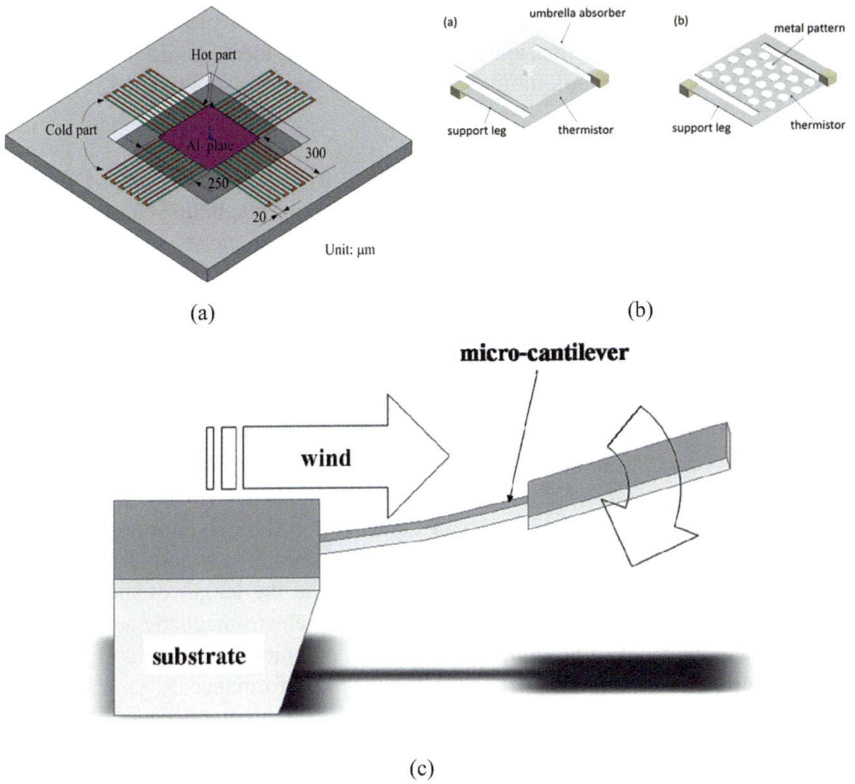

Fig. 1.6 Thermoelectric microgenerator, **a** Kao et al. (2010), infrared detection using microbolometer, **b** Yu et al. (2020), and application of cantilever as gas flow sensor, **c** Wang et al. (2007)

Low-cost M/NST thermal sensors for IoT and wearables face challenges in wafer-level processing and packaging, especially in varied pressure conditions. Analytical modeling and simulations address performance concerns under different pressures, from high vacuum to atmospheric pressure.

Ensuring high sensitivity and accuracy across a broad temperature range poses challenges. For thermoelectric sensors, optimization of conductor materials and precise calibration address these concerns. Thermal resistors use thin-film deposition with quality control to maintain consistent properties. Microbolometers require improvements in design and materials for enhanced performance, especially in applications requiring high spatial resolution. SAW devices and resonant cantilevers address challenges in maintaining high sensitivity and minimizing cross-sensitivity to external factors through material choices and design modifications.

Applications of M/NST thermal sensors, from mass flow metering to gas detection, demand rigorous testing and compensation algorithms for varying environmental conditions and dynamic operational requirements. Ongoing research focuses on integrating multiple sensing elements and advanced signal processing to enhance accuracy, reduce response time, and compensate for non-ideal conditions, paving the way for advancements in automotive, consumer electronics, and beyond.

(f) **Electromagnetic Actuator**

Electromagnetic actuators play a crucial role in Micro/Nano Systems, providing controlled and precise actuation through the interaction of magnetic fields and electric currents. These actuators utilize electromagnetic forces to induce mechanical motion, enabling various applications in micro-nano devices.

The fundamental principle behind electromagnetic actuators involves the interaction of magnetic fields with electric currents. When a current flows through a conductor in the presence of a magnetic field, a force is exerted on the conductor. This force, known as the Lorentz force, induces mechanical displacement in the actuator. In micro-nano systems, electromagnetic actuators are commonly designed as coils or conductive structures interacting with external magnetic fields or other magnetic components. Figure 1.7 shows an electromagnetic actuator.

The force (F) exerted on a conductor carrying a current (I) in a magnetic field (B) can be expressed by the Lorentz force equation: $F = BIL$, where, F is the force, B is the magnetic flux density, I is the current, L is the length of the conductor perpendicular to the magnetic field. In the context of electromagnetic actuators, this force leads to mechanical motion, and the design parameters such as coil geometry and magnetic field strength influence the actuator's performance.

Electromagnetic actuators in micro-nano systems can take various forms, including solenoids, micro-coils, and inductor-based designs. These configurations may involve single or multiple coils, depending on the specific application requirements. The choice of materials, coil dimensions, and magnetic components impacts the actuator's efficiency and response. Electromagnetic actuators find applications in micro-nano devices such as sensors, switches, valves, and resonators. They are employed in micro-scale systems for precise positioning, actuation, and manipulation of components. For example, electromagnetic microactuators can be utilized in micromirrors for optical systems, microvalves for fluid control, or resonant devices for sensing applications. While electromagnetic actuators offer precise control and versatility, challenges include power consumption, miniaturization, and integration with other components. Efficient electromagnetic actuation in the micro and nano domains often requires optimization of coil design, material selection, and external magnetic field control.

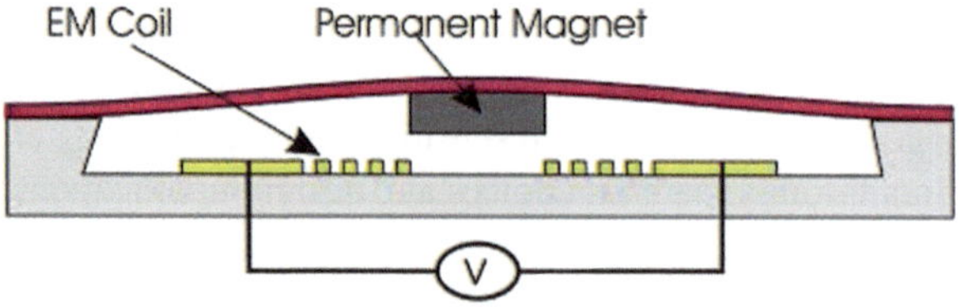

Fig. 1.7 An electromagnetic actuator Yunas et al. (2020)

(g) **Electromagnetic Sensing**

Micro-nano electromagnetic sensors operate on the principle of electromagnetic induction, where changes in electromagnetic fields induce electrical signals for processing and analysis. MEMS electromagnetic sensors typically comprise a magnetic core generating a magnetic field and a coil detecting changes in that field. When an external magnetic field is applied, it induces an electric current in the coil, generating a measurable voltage Fig. 1.8a.

One example is the magnetic field sensor, widely used to measure the strength and direction of magnetic fields, essential in navigation, robotics (Fig. 1.8b), and automotive systems. Hall Effect Sensors, depicted in Fig. 1.8a, utilize the Hall Effect to measure magnetic fields. When a conductor experiences a magnetic field perpendicular to the current flow, it generates a voltage. Hall Effect sensors are prevalent in position, speed, and current sensing applications (Lee et al. 2021). Other types include Magneto-Optical Sensors, employing the Faraday Effect for non-destructive testing and magnetic field imaging. Magneto-resistive Sensors detect weaker magnetic fields by measuring changes in resistance caused by the field, commonly used in compasses, position sensing, and current sensing. Eddy Current Sensors measure changes in magnetic fields through electromagnetic induction, commonly applied in

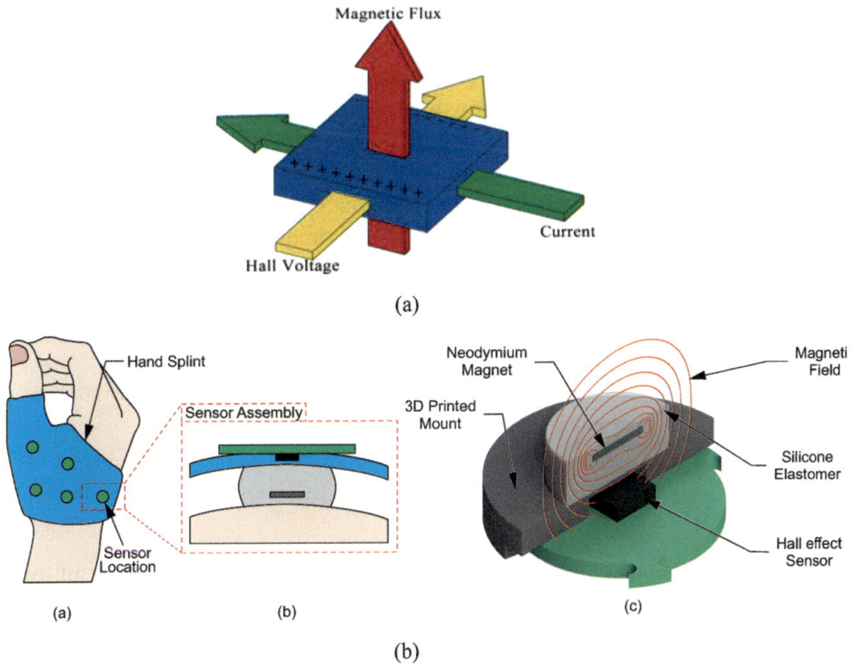

Fig. 1.8 MEMS sensors principle **a** Lee et al. (2021) and an application of magnetic field sensing, **b** Jones et al. (2020)

non-destructive testing and proximity sensing. Electromagnetic transduction finds application in human respiratory monitoring (Diez-Jimenez et al. 2021).

However, electromagnetic sensing in micro-nano devices faces challenges. Achieving high sensitivity and accuracy in detecting changes in magnetic fields is fundamental. Sensors like Hall Effect Sensors need to minimize cross-sensitivity and ensure optimal performance in varying environmental conditions, relying on advanced signal processing and calibration techniques. Magneto-Optical Sensors face challenges related to optical component design and robust magnetic field imaging, driving ongoing research for enhanced performance in non-destructive testing applications. Magneto-resistive Sensors tackle challenges in detecting weaker magnetic fields and maintaining stability amidst external interference, involving material optimizations and signal conditioning techniques. Similarly, Eddy Current Sensors encounter challenges in signal accuracy and interference, leading to the development of advanced algorithms for signal processing and proximity sensing.

The emerging application of electromagnetic transduction in human respiratory monitoring introduces challenges such as ensuring sensor comfort, accuracy, and reliability over extended periods. Addressing these challenges involves miniaturization and the use of biocompatible materials, making the sensors suitable for continuous and unobtrusive monitoring.

(h) **Piezoelectric Actuators**

Piezoelectric materials exhibit a unique property known as the piezoelectric effect, wherein they generate an electric charge in response to applied mechanical stress and, conversely, undergo mechanical deformation when subjected to an electric field. This bidirectional transduction ability allows piezoelectric transducers to serve as both sensors and actuators. Unlike their electrostatic counterparts, piezoelectric transduction offers a more direct and linear relationship in both actuation and sensing. The direct piezoelectric effect occurs when a piezoelectric material converts mechanical strain into electric polarization, while the inverse piezoelectric effect involves the conversion of electric polarization into proportional strain. The governing equations that describe the relationship between strain (ε), electric displacement (D), stress (S), and electric field (E) are given by:

$$\varepsilon = s_E S + d^T E$$

$$D = dS + \varepsilon_0 \varepsilon_r E$$

Here, s_E, d, and ϵ_r represent the compliance, coupling matrix, and relative permittivity of the material, respectively. The superscript T denotes the transpose operation. These equations can be formulated in either the strain–charge (d-form) or stress–charge (e-form), and they can be interconverted. The first part of the equation expresses that the stress field and electric field induce strain, while the second part indicates that the electric polarization is generated by the stress field and electric field. The second term in the first part of the equation represents the electric contribution to the strain.

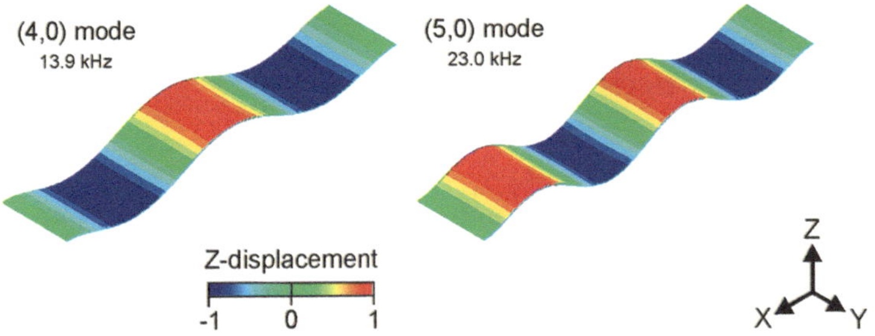

Fig. 1.9 Piezoelectric actuator used in nano-positioning applications (Ruiz-Díez et al. 2021)

This mathematical framework provides a foundation for modeling and understanding the piezoelectric coupling phenomenon in micro-nano devices and systems.

Piezoelectric actuators find applications in micro/nano positioning systems (Fig. 1.9), scanning probe microscopy, optical components tuning, inkjet printers, and microvalves, among others. Their ability to provide precise motion at the micro and nano scales makes them valuable in various fields. Despite their advantages, piezoelectric actuators face challenges such as hysteresis, nonlinearity, and the generation of heat during operation. These challenges require careful consideration in the design and control of piezoelectric-based systems.

i. Piezo-acoustic Transduction:

Piezo-acoustic transduction refers to the process of converting acoustic signals (sound waves) into electrical signals or vice versa using piezoelectric materials. When an acoustic signal, such as a sound wave, reaches a piezoelectric material, it causes mechanical vibrations or stress in the material. The piezoelectric material responds to the mechanical stress by generating an electric charge. This phenomenon is known as the direct piezoelectric effect. The charge produced is proportional to the applied stress. The generated electric charge can be captured and converted into an electrical signal. This is useful in applications such as piezoelectric microphones, where the sound waves cause the piezoelectric material to produce an electrical signal representing the original sound. Conversely, the reverse process can also occur: When an electrical signal is applied to a piezoelectric material, it generates an electric field within the material. The electric field causes the piezoelectric material to undergo mechanical deformation or vibrations. This is known as the inverse piezoelectric effect. The mechanical vibrations result in the emission of acoustic waves, producing sound. This is utilized in applications such as piezoelectric speakers or buzzers.

Piezoacoustic transduction finds diverse practical applications across various technological domains. Piezoelectric microphones utilize the transduction process to convert incoming sound waves into electrical signals, facilitating the recording and amplification of audio. In the realm of audio output, piezoelectric speakers (Fig. 1.10)

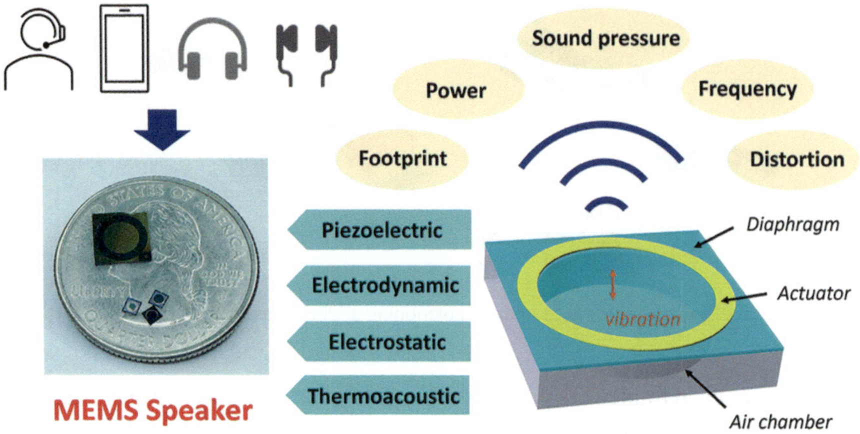

Fig. 1.10 MEMS speaker (Wang et al. 2021)

play a crucial role by transforming electrical signals into sound waves. The technology extends beyond audible frequencies to ultrasonics, where ultrasonic transducers are employed for generating and detecting ultrasonic waves. This is particularly valuable in fields such as medical imaging, distance measurement, and industrial cleaning. Additionally, piezoacoustic transduction is integral to the functionality of vibration sensors, which play a crucial role in detecting vibrations within structures or machinery. By converting these vibrations into electrical signals, these sensors enable effective monitoring and control processes. The broad frequency range, efficiency, and reliability of piezoacoustic transduction contribute to its widespread adoption and significance in numerous technological applications.

Piezoacoustic transduction harnesses the coupling between piezoelectricity and acoustics, enabling the use of piezoelectric materials as acoustic transducers in various applications. Two prominent types are Bulk Acoustic Wave (BAW) and Surface Acoustic Wave (SAW) transducers, each with distinct structures and functions.

ii. Bulk Acoustic Wave (BAW) Transducers:

BAW transducers consist of a thin piezoelectric layer sandwiched between two electrodes, with a substrate providing support. Applying voltage to the electrodes induces expansion and contraction of the piezoelectric layer, generating acoustic waves in the substrate. These waves create a standing wave resonating at a specific frequency determined by the piezoelectric layer's thickness, substrate properties, and electrode pattern. BAW transducers offer high stability, low energy consumption, and high-quality factor (Q), making them ideal for signal filtering and amplification in wireless communication systems. Additionally, they find applications in pressure and temperature sensors.

iii. Surface Acoustic Wave (SAW) Transducers:

SAW transducers employ electrodes on the substrate's surface, and a thin film of piezoelectric material is deposited on top. An applied electrical signal causes expansion and contraction, producing surface acoustic waves that propagate along the substrate's surface. Reflection between electrodes forms a standing wave resonating at a specific frequency. SAW transducers share advantages like stability and low energy consumption with BAW transducers and are widely used in sensing applications.

iv. Harmonic Electrical Excitation:

Piezoelectric materials can also be excited harmonically with electrical stimuli in the kHz to MHz range. This excitation induces structural vibrations in the material, generating acoustic waves in the surrounding fluid medium. This principle is utilized in devices like Piezoelectric Micromachined Ultrasound Transducers (PMUT), finding applications in hydrophones, phased array microphones, speakers, bio-imaging, ultrasound, and more.

v. Applications and Challenges:

Piezoacoustic transduction plays a vital role in various devices, including microphones, speakers, mirrors, gyroscopes, accelerometers, micro-grippers, gas sensors, RF filters, oscillators, and ultrasound medical imaging. However, challenges such as hysteresis, material quality limitations at smaller thicknesses, and scaling-related issues need addressing. Researchers are actively exploring advanced materials, fabrication techniques, and control methodologies to enhance the performance and reliability of miniaturized piezoelectric devices.

Despite manufacturing challenges, some commercial applications, such as FBAR filters in smartphones, are already in use. Commonly employed thin films for piezoelectricity include Lead Zirconate Titanate (PZT), Aluminum Nitride (AlN), Lithium Niobate (LiNbO3), Quartz, Polyvinylidene Fluoride (PVDF), and Zinc Oxide (ZnO), with wafer-scale film thickness around 200 nm. The ongoing exploration of novel materials and integration with other technologies continues to drive advancements in piezo-acoustic transduction, enabling diverse applications in MEMS for sensing, actuation, and energy harvesting.

(i) **Resonant Sensing**

Resonant sensing stands as a powerful technique harnessing the vibrational properties of solid-state structures to detect alterations in physical parameters, spanning mass, position, strain, pressure, electric field, temperature, acceleration/tilt, etc. (Zhang and Kim 2005; Chen et al. 2020; Wojciechowski et al. 2004; Zega et al. 2014; Kim et al. 2012). Widely explored, this technology finds applications in diverse fields, from chemical sensing to medical diagnostics and environmental monitoring. A commonly employed resonant sensor in the realm of Micro/Nano-Electro-Mechanical Systems (M/NEMS) is the resonant poly/silicon beam sensor. This thin, flexible beam resonates at its natural frequency, responding to applied loads with a shift in resonant frequency. The interaction is captured through the charge–voltage relationship in an electrostatic capacitor ($Q = CV$, $V = Vac + Vdc$).

The ensuing current through the structure, approximated as $i \approx C_0 (\partial Vac/\partial t) + \eta (\partial x/\partial t)$, embodies both the static capacitor current and the motion-induced current. This dynamic is modeled electrically, revealing a transformer-driven input force and an output transformer translating mechanical velocities into electric current. The equivalent circuit incorporates parameters such as η_{in} and η_{out} (electromechanical coupling coefficients), C_{pad} (shunt capacitor), and others, forming a sophisticated representation of resonant transduction. Key parameters such as L_m, C_m, R_m, and η play pivotal roles in characterizing inductance, capacitance, motional resistance, and electromechanical coupling factor, respectively.

From this intricate model, crucial insights emerge, including the series resonant frequency (ω_s) and quality factor (Q). Understanding these parameters is vital for achieving high sensitivity, a core challenge in resonant sensing. Maintaining stability amidst external factors like temperature variations or mechanical disturbances requires advanced sensor designs and materials resilient to environmental changes.

Stability concerns, dynamic adjustments, and precise control over resonance frequencies demand ongoing research and innovation. The resonant poly/silicon beam sensor, emblematic of this technology, exemplifies the delicate balance between size reduction and preserving resonant characteristics. As resonant sensing integrates into various applications, addressing challenges becomes paramount for unlocking the full potential of this versatile and robust technology. Ongoing research endeavors focus on exploring advanced materials, refining fabrication techniques, and implementing innovative control strategies to elevate the sensitivity, stability, and adaptability of resonant sensing technologies.

(j) Piezo-resistive Sensing

Piezo-resistivity characterizes the alteration in a material's conductivity under applied stress, particularly observed in semiconductors due to the strain-induced modifications in the material's band structure. This phenomenon influences carrier (electron) mobility and number density, leading to the manifestation of piezo-resistance. The manufacturing of piezo-resistors is straightforward through common semiconductor processes, offering inherent immunity to electromagnetic interference and seamless integration with electronic interfaces. Notably, pressure sensors leveraging the piezo-resistive effect were among the pioneering M/NST devices to undergo mass production.

In the configurations depicted in Fig. 1.11, transducers are designed for detecting and measuring physical quantities or bio/chemical traces by capturing changes in resistance. When the structure undergoes movement, embedded piezo-resistors experience alterations, and the change in resistance (ΔR) is directly proportional to the applied strain (ε), expressed as $\Delta R = G\varepsilon R$, where G is the gauge factor, and R is the nominal resistance. Subsequently, this change in resistance is translated into a proportionate voltage difference through a Wheatstone bridge circuit.

In the realm of M/NEMS, semiconductors like poly/silicon often serve as structural materials. The sensor's output, quantified as $\Delta R/R$ (normalized), considers the change in resistance (ΔR) and the nominal resistance (R) of the material (poly/silicon/

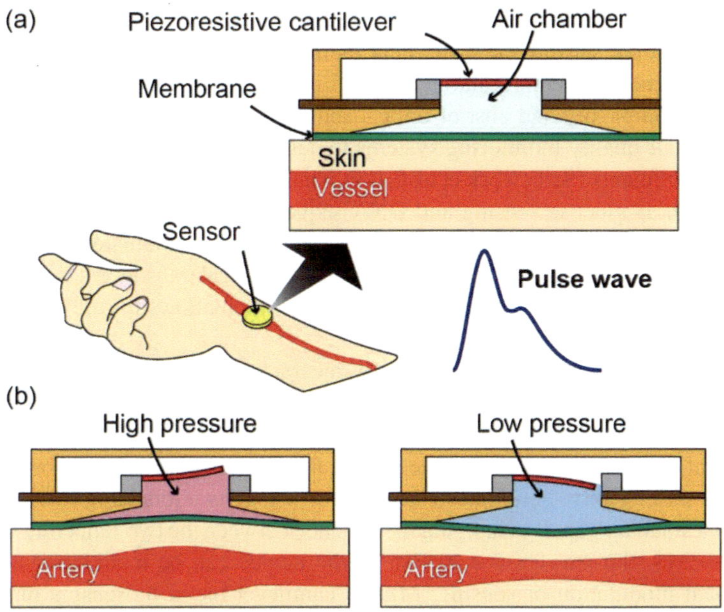

Fig. 1.11 Research depicting usage of piezo-resisitive readout (Nguyen et al. 2020)

germanium) used as a resistor. Common materials for piezo-resistive sensors also include specific polymers. Applications ranging from pressure sensors to humidity sensors, force sensors, and strain gauges exemplify the versatility of piezo-resistive effects.

Despite its widespread use, the piezo-resistive effect poses challenges. Achieving high sensitivity while minimizing cross-sensitivity to environmental factors, such as temperature variations, stands as a primary challenge. Researchers address this by utilizing advanced materials with optimized piezo-resistive coefficients and carefully engineering sensor structures. The need for precise calibration to accommodate variations in material properties and fabrication processes further prompts the development of sophisticated techniques and compensation algorithms.

Microfabrication processes, crucial for achieving miniaturization and enhancing sensor performance, introduce challenges related to process variability. Robust packaging solutions are employed to shield sensors from environmental influences. Ongoing research explores innovative materials and design strategies to enhance the reliability and accuracy of piezo-resistive sensing in MEMS applications. Addressing these challenges necessitates a multidisciplinary approach, integrating materials science, sensor design, microfabrication, and signal processing techniques to propel the capabilities of piezo-resistive sensors in MEMS applications.

(k) **Optical Sensing**

MEMS optical sensors can detect various optical parameters, including intensity, wavelength, polarization, and phase, and convert them into proportionate electrical signals. One of the key advantages of MEMS optical sensors is their ability to integrate multiple sensing elements into a single device, enabling the detection of multiple optical parameters simultaneously (Li et al. 2018). This integration can also reduce the complexity and cost of optical sensing systems. MEMS optical sensors are used in air quality monitoring systems to detect and measure particulate matter and other pollutants (Schieweck et al. 2018). They are also used in biomedical applications such as glucose sensing and DNA sequencing (Kang et al. 2009). MEMS optical transducers can be used as optical switches, filters, modulators, and sensors, among others. One of the key advantages of MEMS optical transducers is their miniaturization, low power, and integration capability, which allows for the development of compact and high-performance devices.

Various types of MEMS optical transducers have been proposed and fabricated, including MEMS-based Fabry–Perot interferometers, micro-ring resonators, photonic crystal cavities, and microcantilevers. These devices rely on the mechanical movement of microstructures to achieve optical modulation or sensing. For example, a MEMS-based Fabry–Perot interferometer consists of two reflecting surfaces separated by a small air gap. By changing the distance between the two reflecting surfaces using MEMS, the interference pattern of the device can be modulated, allowing for optical modulation or sensing. MEMS cantilever-based biosensors utilize the mechanical deflection of a cantilever due to the binding of a biomolecule to its surface (Fig. 1.12). The change in deflection can be detected optically, allowing for label-free detection of the biomolecule. Optical MEMS has become prevalent because of its high performance (high displacement accuracy) and resistance to electromagnetic interference. An example application is an accelerometer using Fabry–Pérot interfere technology, optical Fabry–Pérot (FP) pressure sensor for wind pressure measurement.

Summary and Conclusion

In conclusion, the world of transduction methods in micro and nano systems is diverse, providing a range of tools for sensing and actuation, each with its unique principles and applications. Electrostatic transduction relies on charged entities for actuation and sensing in micro/nano systems. Electromagnetic transduction uses electromagnetic induction principles for micro/nano electromagnetic sensors, measuring parameters like strength and direction. Piezoelectric transduction converts mechanical stress into electric charge and vice versa, offering linear actuation and sensing. Piezoacoustic transduction combines piezoelectricity and acoustics in bulk and surface acoustic wave transducers. Capacitive sensing, found in pressure sensors and touchscreens, relies on capacitance changes due to physical alterations. Thermal flow transduction in MEMS thermal sensors converts temperature changes into electrical signals for applications like mass flow metering and gas detection. Resonant

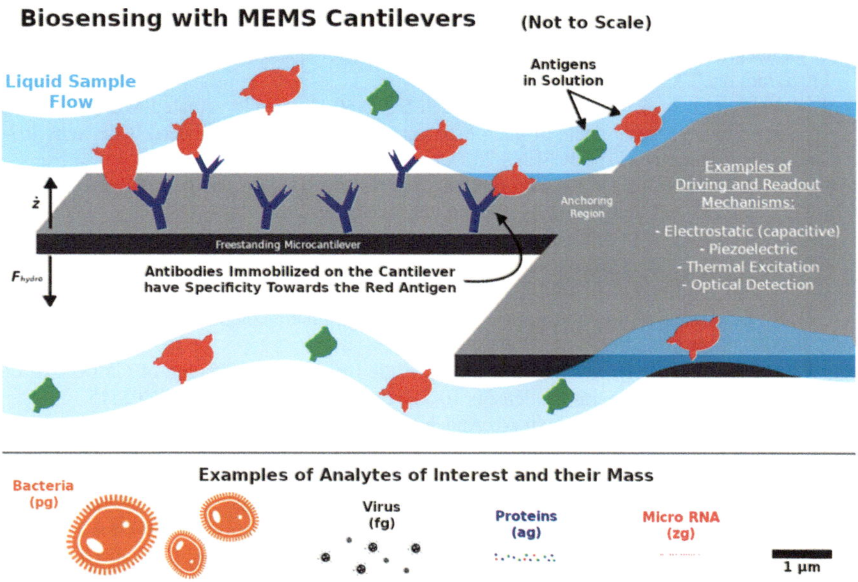

Fig. 1.12 Example of research where use of optical detection is shown (Mouro et al. 2021)

sensing utilizes the resonant properties of solid-state structures for diverse applications, facing challenges in maintaining high-quality factors. Piezo-resistive transduction observes changes in conductivity under applied stress, with widespread use in pressure and force sensing. In the scientific context, these methods drive research in micro and nano systems, offering solutions for healthcare, communication, and environmental monitoring. The pursuit of higher sensitivity, reliability, and miniaturization through a multidisciplinary approach continues to advance these technologies, promising to revolutionize our interaction with the micro and nano scales.

Competing Interests: The author has no conflicts of interest to declare that are relevant to the context of this chapter.

References

Agarwal M et al (2005) Non-linearity cancellation in MEMS resonators for improved power-handling. In: Technical digest—international electron devices meeting IEDM. https://doi.org/10.1109/iedm.2005.1609330

Chen D et al (2020) Ultrasensitive resonant electrometry utilizing micromechanical oscillators. Phys Rev Appl. https://doi.org/10.1103/PhysRevApplied.14.014001

Chu CC, Dey S, Liu TY, Chen CC, Li SS (2018) Thermal-piezoresistive SOI-MEMS oscillators based on a fully differential mechanically coupled resonator array for mass sensing applications. J Microelectromech Syst 27(1):59–72. https://doi.org/10.1109/JMEMS.2017.2778307

Diez-Jimenez E, C. Alén-Cordero C, Alcover-Sánchez R, Corral-Abad E (2021) Modelling and test of an integrated magnetic spring-eddy current damper for space applications. Actuators 10(1). https://doi.org/10.3390/act10010008

Hajjam A, Rahafrooz A, Wilson JC, Pourkamali S (2009) Thermally actuated MEMS resonant sensors for mass measurement of micro/nanoscale aerosol particles. Proceed IEEE Sens 707–710. https://doi.org/10.1109/ICSENS.2009.5398557

Jones D et al (2020) Design and evaluation of magnetic hall effect tactile sensors for use in sensorized splints. Sensors (Switzerland) 20(4). https://doi.org/10.3390/s20041123

Kaajakari V (2009) MEMS tutorial: nonlinearity in micromechanical resonators. MEMS Material-Ville Kaajakari

Kang K, Nilsen-Hamilton M, Shrotriya P (2009) Novel differential surface stress sensor for detection of DNA hybridization. In: Proceedings of the ASME summer bioengineering conference 2009, SBC2009. https://doi.org/10.1115/SBC2009-206838

Kao PH, Shih PJ, Dai CL, Liu MC (2010) Fabrication and characterization of CMOS-MEMS thermoelectric micro generators. Sensors 10(2). https://doi.org/10.3390/s100201315

Kim M, Avila JC, Avila JC, Kapandji AI, Andaluz Ortiz GM (2012) Materials and device design for piezoelectric mechanical vibration energy harvesters. Energy

Lee CY, Lin YY, Kuo CK, Fu LM (2021) Design and application of mems-based hall sensor array for magnetic field mapping. Micromachines (Basel) 12(3). https://doi.org/10.3390/mi12030299

Li CR et al (2018) Integrating temperature, humidity, and optical aerosol sensors for a wireless module for three-dimensional space monitoring. In: 2018 IEEE sensors applications symposium, SAS 2018—proceedings, vol 2018, pp 1–4. https://doi.org/10.1109/SAS.2018.8336731

Lin Y-W, Li S-S, Xie Y, Ren Z, Nguyen (2006) Vibrating micromechanical resonators with solid dielectric capacitive transducer gaps. https://doi.org/10.1109/freq.2005.1573914.

Lobur M, Holovatyy A (2009) Overview and analysis of readout circuits for capacitive sensing in MEMS gyroscopes (MEMS angular velocity sensors). In: 2009 proceedings of the 5th international conference on perspective technologies and methods in MEMS design, MEMSTECH 2009

Mouro J, Pinto R, Paoletti P, Tiribilli B (2021) Microcantilever: dynamical response for mass sensing and fluid characterization. Sensors (Switzerland), 21(1). https://doi.org/10.3390/s21010115

Mujahid A, Afzal A, Dickert FL (2019) An overview of high frequency acoustic sensors—QCMs, SAWs and FBARs—chemical and biochemical applications. Sensors (Switzerland) 19(20). https://doi.org/10.3390/s19204395

Nguyen TV, Mizuki Y, Tsukagoshi T, Takahata T, Ichiki M, Shimoyama I (2020) MEMS-based pulse wave sensor utilizing a piezoresistive cantilever. Sensors (Switzerland) 20(4). https://doi.org/10.3390/s20041052

Pachkawade V, Tse Z (2022) MEMS sensor for detection and measurement of ultra-fine particles: a review. http://arxiv.org/abs/2112.09837

Park HW, Kim YK, Jeong HG, Song JW, Kim JM (2011) Feed-through capacitance reduction for a micro-resonator with push-pull configuration based on electrical characteristic analysis of resonator with direct drive. Sens Actuators A Phys. https://doi.org/10.1016/j.sna.2011.05.032

Potekhina A, Wang C (2019) Review of electrothermal actuators and applications. Actuators 8(4). https://doi.org/10.3390/ACT8040069

Qiao Z, Boom BA, Annema AJ, Wiegerink RJ, Nauta B (2018) On frequency-based interface circuits for capacitive MEMS accelerometers. Micromachines (Basel). https://doi.org/10.3390/mi9100488

Qin X et al (2020) A paper based milli-cantilever sensor for detecting hydrocarbon gases via smartphone camera. Anal Chem 92(12):8480–8486. https://doi.org/10.1021/acs.analchem.0c01240

Rabenimanana T, Walter V, Kacem N, Le Moal P, Lardiès J (2018) Nonlinear analytical model of two weakly coupled mems cantilevers for mass sensing using electrostatic actuation. Proc West Mark Ed Assoc Conf. https://doi.org/10.3390/proceedings2131084

Ren S, Yuan W, Qiao D, Deng J, Sun X (2013) Pressure sensor with integrated resonator operating at atmospheric pressure. Sensors (Switzerland). https://doi.org/10.3390/s131217006
Ruan Z, Ding X, Qin Z, Jia J, Li H (2020) Automatic mode-matching method for mems disk resonator gyroscopes based on virtual coriolis force. Micromachines (Basel). https://doi.org/10.3390/mi11020210
Ruiz-Díez V, Hernando-García J, Toledo J, Ababneh A, Seidel H, Sánchez-Rojas JL (2021) Piezoelectric mems linear motor for nanopositioning applications. Actuators 10(2). https://doi.org/10.3390/act10020036
Schieweck A et al (2018) Smart homes and the control of indoor air quality. Renew Sustain Energy Rev 94:705–718. https://doi.org/10.1016/J.RSER.2018.05.057
Utz A, Walk C, Stanitzki A, Mokhtari M, Kraft M, Kokozinski R (2018) A high-precision and high-bandwidth mems-based capacitive accelerometer. IEEE Sens J. https://doi.org/10.1109/JSEN.2018.2849873
Wang YH, Lee CY, Chiang CM (2007) A MEMS-based air flow sensor with a free-standing microcantilever structure. Sensors 7(10). https://doi.org/10.3390/s7102389
Wang H, Ma Y, Zheng Q, Cao K, Lu Y, Xie H (2021) Review of recent development of MEMS speakers. Micromachines 12(10). https://doi.org/10.3390/mi12101257
Wojciechowski KE, Baser BE, Pisano AP (2004) A MEMS resonant strain sensor operated in air. In: Proceedings of the IEEE international conference on micro electro mechanical systems (MEMS). https://doi.org/10.1109/mems.2004.1290718
Yu L, Guo Y, Zhu H, Luo M, Han P, Ji X (2020) Low-cost microbolometer type infrared detectors. Micromachines 11(9). https://doi.org/10.3390/MI11090800
Yunas J et al (2020) Polymer-based MEMS electromagnetic actuator for biomedical application: a review. Polymers 12(5). https://doi.org/10.3390/POLYM12051184
Zega V, Comi C, Corigliano A, Valzasina C (2014) Integrated structure for a resonant micro-gyroscope and accelerometer. Frattura Ed Integrita Strutturale. https://doi.org/10.3221/IGF-ESIS.29.29
Zhang WM, Meng G, Chen D (2007) Stability, nonlinearity and reliability of electrostatically actuated MEMS devices. Sensors 7(5). https://doi.org/10.3390/s7050760
Zhang JXJ, Hoshino K (2019) Mechanical transducers: cantilevers, acoustic wave sensors, and thermal sensors. Mol Sens Nanodev. https://doi.org/10.1016/b978-0-12-814862-4.00006-5
Zhang H, Kim ES (2005) Micromachined acoustic resonant mass sensor. J Microelectromech Syst. https://doi.org/10.1109/JMEMS.2005.845405
Zou X, Thiruvenkatanathan P, Seshia AA (2012) Micro-electro-mechanical resonant tilt sensor. In: 2012 IEEE international frequency control symposium, IFCS 2012, proceedings. https://doi.org/10.1109/FCS.2012.6243702

Chapter 2
Acoustic Devices for Gas Sensing and RF Applications in MEMS/NEMS

Soumi Mukherjee, Alok J. Verma, Shobha Shukla, and Sumit Saxena

Abstract The chapter provides in-depth discussions about the physics, device architectures, and modes of operation, providing a comprehensive understanding of how the MEMS/NEMS-based acoustic devices work. The chapter explains the fundamental concepts of acoustic and then discusses the design and operation of various acoustic devices, such as surface acoustic wave (SAW) devices, bulk acoustic wave (BAW) devices, film bulk acoustic resonators (FBARs), and photoacoustic devices. The chapter emphasizes the importance of materials and manufacturing procedures for device performance and provides insights into acoustic device material choices. Moving on, the chapter expands into the relevance of acoustic devices in gas sensing applications, illustrating how these devices use their unique features to detect gases precisely and quickly in a variety of situations. Furthermore, the chapter explains how acoustic devices play an important part in 5G/6G communication systems, where they act as oscillators and filters to improve signal processing, filtering, and frequency management.

Keywords MEMS · NEMS · Acoustic · SAW · BAW · FBARs · Photoacoustic · Gas sensors · Resonators · Oscillators

S. Mukherjee · S. Shukla · S. Saxena (✉)
Nanostructures Engineering and Modeling Laboratory, Department of Metallurgical Engineering and Materials Science, Indian Institute of Technology Bombay, Mumbai, MH, India
e-mail: sumit.saxena@iitb.ac.in

S. Mukherjee
e-mail: 214110017@iitb.ac.in

A. J. Verma
Society for Applied Microwave Electronics Engineering and Research (SAMEER), IIT Campus, Powai, Mumbai, India

S. Shukla · S. Saxena
Water Innovation Center: Technology Research and Education, Department of Metallurgical Engineering and Materials Science, Indian Institute of Technology Bombay, Mumbai, MH, India

V. Pachkawade and K. Guha (eds.), *Micro- and Nano-Systems in 21st-Century*, Springer Tracts in Electrical and Electronics Engineering,
https://doi.org/10.1007/978-981-96-4659-3_2

List of Abbreviations

SONAR	Sound Navigation and Ranging
RF	Radio Frequency
SAW	Surface Acoustic Waves
BAW	Bulk Acoustic Waves
QCM	Quartz Crystal Microbalance
QCM-D	Quartz Crystal Microbalance with Dissipation
FBRAR	Film-based Bulk Acoustic Resonators
TC-FBAR	Temperature-Compensated Film Bulk Acoustic Wave Resonator
OBAR	Overmoded Bulk Acoustic Resonator
FOM	Figure-Of-Merit
PZT	Lead Zirconate Titanate
CMOS	Complementary Metal-oxide Semiconductor
RTA	Rapid Thermal Annealing
FWHM	Full-Width at Half-Maximum
ScAlN	Scandium-doped Aluminum Nitride
LD ScAlN	Lightly-doped Scandium Aluminum Nitride
CNT	Carbon Nano Tubes
SWCNTs	Single-Wall Carbon Nanotubes
MWCNTs	Multi-Wall Carbon nanotubes
ITO	Indium Tin Oxide
PA	Photo Acoustic
PAS	Photo Acoustic Spectroscopy
VOC	Volatile Organic Compounds
PEI	Poly(ethylene imine)
PSS	Poly(sodium 4-styrenesulfonate)
PDDA	Poly(diallyldimethylammonium
PAA	Poly(acrylic acid)
PVA	Poly(vinyl alcohol)
PANI	Polyaniline
FPGA	Field-Programmable Gate Array
RH	Relative Humidity
DFB	Distributed Feedback Lasers
IoT	Internet of Things
RFFE	Radiofrequency Front-Ends
EM	Electromagnetic
IL	Insertion Loss
LFE	Lateral Field-Excited
CLMRs	Cross-Sectional Lame' Mode Resonators

Introduction

Acoustics is the scientific exploration of sound and its behavior across various media, encompassing the creation, propagation, transmission, reception, and effects of sound waves. It delves into the origins of sound, its journey through diverse materials, and its interactions with the surroundings, encompassing reflection, absorption, and diffusion. Acoustic devices, serving as architects of sound modification, translate these principles into tangible technologies. From microphones capturing diverse voices to loudspeakers resonating with melodies and ultrasound revealing uncharted territories, acoustic equipment spans the ordinary to the extraordinary, leaving an enduring impact on modern life.

In acoustic engineering, ultrasound plays a crucial role, operating beyond audible frequencies, surpassing the 20 kHz human hearing limit. The heart of ultrasound technology lies in piezoelectric transducers, adept at converting electrical energy into acoustic waves. Constructed from crystalline solids with a unique interplay of mechanical strain and electrical polarization, these transducers seamlessly receive ultrasonic waves and translate them back into electronic signals—a cyclical process that unfolds harmoniously (Rossing 2007).

The chronicles of ultrasound trace back to the echoes of World War I, where the first underwater sound navigation and ranging system, christened SONAR, pioneered large-scale ultrasound implementation. Fast forward to the 1930s, and ultrasound ventured beyond military applications, stepping into the realms of physical therapy with promises of deep tissue heating. As the decades waltzed by, ultrasound metamorphosed into a medical maestro, donning diagnostic and therapeutic roles (Lurton 2002). Gynecology and cardiology welcomed it into the fold of imaging, while its capabilities are extended to drug delivery and disease treatment (Woo 2002). Beyond the realms of medicine, ultrasound found its place in the industrial stage, from precision cleaning and welding to facilitating non-destructive testing.

The journey of acoustic devices is a harmonious blend of science, innovation, and practicality. In our exploration of their role in gas sensing and RF applications within MEMS/NEMS, we uncover the technology layers utilizing sound's essence to push boundaries. This evolution converges scientific principles, innovation, and functionality, delving into basic acoustic principles and phenomena. The focus then shifts to acoustic devices, covering introductions, material selection, fabrication intricacies, and applications. Specialized sections explore gas sensing intricacies and extend to 5G/6G communications, addressing oscillators, filters, and challenges. The concluding section synthesizes the exploration, weaving together the diverse facets of acoustic devices into a cohesive technological narrative.

Fundamentals of Acoustic Waves

History of Acoustics

Ancient Greek philosopher Pythagoras (570–495 BC) is credited with realizing that sound travels in wave-like patterns. Driven by curiosity about pleasing musical combinations, he discovered that harmonious tones occurred when the lengths of vibrating strings were expressed as ratios of whole numbers (e.g., 2–3, 3–4). The exploration of sound, dating back to the sixth century, began with Pythagoras, and around 2000 years later, Galileo fully developed the study by demonstrating the connection between the frequency of a wave and the pitch it produces (Zhang 2022).

In the sixteenth century, French mathematician Marin Mersenne (1588–1648) conducted groundbreaking experiments to determine the speed of sound in air, marking a significant milestone in the understanding of sound propagation. This era saw the true beginning of the study of the fundamental principles governing the transmission of sound (Alton Everest 2001).

Sir Isaac Newton (1642–1727) laid the foundation of acoustics by mathematically formulating the relationship between the properties of a material and the speed of sound waves traveling through it. His deduction of the formula for wave velocity in solids played a pivotal role in advancing physical acoustics and contributed to the broader understanding of wave propagation in various materials (Rossing 2007).

The term 'acoustics' is believed to have been coined by French physicist Joseph Sauveur (1653–1716), whose contribution went beyond naming. Sauveur's work laid the foundation for a structured and scientific approach to acoustics, delving into mathematical representations of musical intervals and the relationship between vibrations and musical harmony. This marked significant progress in quantifying and analyzing sound intricacies, exemplifying the essence of the newly coined term. In the aftermath of the eighteenth century, continuum physics, or field theory, was formally established, acquiring a well-defined mathematical framework. The wave equation emerged, including its application to sound propagation through air. This period witnessed substantial advancements in acoustics, marked by an increasing reliance on rigorous mathematical and physical principles. Euler (1707–1783), Lagrange (1736–1813), and d'Alembert (1717–1783) were pivotal figures contributing significantly to this notable development (Rossing 2007; Kinsler 2000).

The nineteenth century witnessed a transformative period in acoustics. Hermann von Helmholtz (1821–1894) made groundbreaking contributions by devising resonators and using mathematical techniques to analyze complex sounds. He introduced the concept of the harmonic series as the basis for understanding complex waveforms and the perception of timbre in musical sounds. In 1885, Lord Rayleigh predicted the existence of surface sound waves on solids, known as Rayleigh waves. This discovery laid the foundation for studying surface acoustic waves, contributing to the later development of resonators, filters, and oscillators (Abigai 2009).

The twentieth century marked significant technological advancements in acoustics, with a surge in studies on underwater acoustics and architectural acoustics. The tragic sinking of the Titanic in 1912 accelerated efforts to develop systems for visualizing underwater structures, and the First World War spurred the development of high-frequency sound wave technologies for submarine detection. A notable outcome was the U.S. Navy's development of SONAR (Sound Navigation and Ranging) in the 1930s. American physicist Wallace Sabine pioneered the systematic application of scientific principles to architectural acoustics, founding the field. His work transformed architectural design to prioritize acoustic comfort and quality, enabling manipulation of room dimensions, surface materials, and interior features for desired acoustic effects (Lurton 2002; Auld 1973).

Sound Propagation: Development of Wave Theory

The inception of wave theory is marked by the understanding of how waves travel through solids. In 1831, Siméon Denis Poisson proposed that an unbounded, uniform, and isotropic elastic solid can convey two types of waves at different speeds. The faster one corresponds to the longitudinal wave, and the slower one corresponds to the transverse wave. In 1849, Sir George Stokes formulated the velocity components for these waves, revealing that longitudinal waves consist of particles polarizing and vibrating in an irrotational dilation manner. The equation for the velocity of longitudinal wave is given by

$$v_l = \sqrt{(\lambda_L + 2\mu_L)}$$

As for the transverse part, it is composed of particles moving in an equivoluminal rotation motion. The wave travels with a velocity given by

$$v_t = \sqrt{\mu_L/\rho}$$

where λ_L and μ_L are elastic constants (known today as the Lamé constants) of the medium, and ρ its density.

This theory demonstrated that waves propagate in a bulk medium through dilation and distortions. However, it overlooked the presence of boundaries. While effective in explaining phenomena within the medium, challenges arose when applying it to seismic activity, as observations and recordings of seismic vibrations were conducted on the Earth's surface.

Lord Rayleigh, or John William Strutt, made significant contributions to the understanding of sound and wave propagation in his seminal work "The Theory of Sound," published in two volumes in 1877 and 1878. This work is a cornerstone in the mathematical physics of wave propagation and vibration, offering a comprehensive treatment that integrates experimental observations and mathematical analyses. Rayleigh initiated his study with simple harmonic motion, focusing on the oscillation of a pendulum. The motion of a pendulum can be described by a second-order linear differential equation, such as:

$$\frac{d^2\theta}{dt^2} = \frac{g}{l}\text{Sin}\theta = 0,$$

where, θ is the angular displacement, t is time, g is the acceleration due to gravity, and l is the length of the pendulum.

Rayleigh's mathematical exploration extended to the realm of vibrating strings, where he employed the wave equation to model the transverse motion of a string under tension. The dynamic behavior of the string is described by the wave equation expressed as.

$$\frac{d^2u}{dt^2} = c^2\frac{d^2u}{dx^2},$$

Here, u represents the transverse displacement of the string, t denotes time, x signifies the spatial coordinate along the string, and c is the speed at which waves propagate through the string. This equation elegantly describes the interplay between time, space, and displacement, offering a mathematical framework for understanding the intricate vibrations and harmonic phenomena intrinsic to vibrating strings.

Rayleigh's scientific journey took a significant turn as he explored fluid dynamics, specifically investigating the motion of air to understand the complexities of sound wave propagation. Concentrating on subtle changes in pressure and density within a compressible fluid, Rayleigh aimed to reveal the fundamental principles governing sound transmission through the air. Through meticulous derivation, he formulated the linearized equations of motion for compressible fluids, ultimately unveiling the wave equation for sound. The resulting equation, expressed as

$$\frac{d^2p}{dt^2} = c^2\frac{d^2u}{dx^2}$$

encapsulates the dynamics of sound waves in air. In this formulation, p symbolizes the sound pressure, t signifies time, x denotes spatial coordinates, and c represents the speed at which sound propagates through the fluid. This mathematical representation forms the backbone of acoustics, offering a sophisticated tool to comprehend the intricate interplay between time, space, and pressure in the transmission of sound through fluid mediums. Rayleigh's contributions in this realm laid the groundwork for modern acoustics and our understanding of how sound waves traverse through the air (Strutt 2011).

Types of Acoustic Waves

Acoustic waves are a subset of mechanical waves that involve the propagation of energy through a medium by means of particle oscillations or vibrations. They are commonly associated with sound, and there are several types of acoustic waves based on their characteristics and behavior:

Acoustic waves, a subset of mechanical waves, involve energy propagation through a medium via particle oscillations. They include (Bean 2009; Auld 1973; Strutt 2011; Fogel et al. 2016).

1. **Longitudinal (Compression) Waves**
   - Involve parallel particle oscillations.
   - Mechanism for sound transmission in gases, liquids, and solids.
2. **Transverse Waves**
   - Less common, primarily within solids.
   - Entail particle oscillations perpendicular to the wave, forming crests and troughs.
3. **Rayleigh (R) Waves**
   - Traverse a substrate's surface, with polarization perpendicular to the surface.
   - Induce elliptical particle movement, influenced by substrate material and crystal orientation.
4. **Love (L) Waves**
   - Propagate horizontally along the Earth's surface.
   - Display lateral particle motion perpendicular to propagation, contributing to seismic ground shaking.
5. **Shear (S) Waves**
   - Transverse waves traversing solids, causing shearing or twisting.
   - Generate "secondary" seismic effects during earthquakes.
6. **Compression (P) Waves**
   - Longitudinal waves propagating through solids.
   - Feature particle oscillations parallel to the wave, causing compressions and rarefactions.

Figure 2.1 shows a diagrammatic representation of the various types of acoustic waves. These acoustic waves play essential roles in fields such as communication, seismology, medical imaging (ultrasound), and non-destructive testing. Understanding their characteristics and behavior is fundamental for designing devices, diagnosing problems, and exploring the properties of materials.

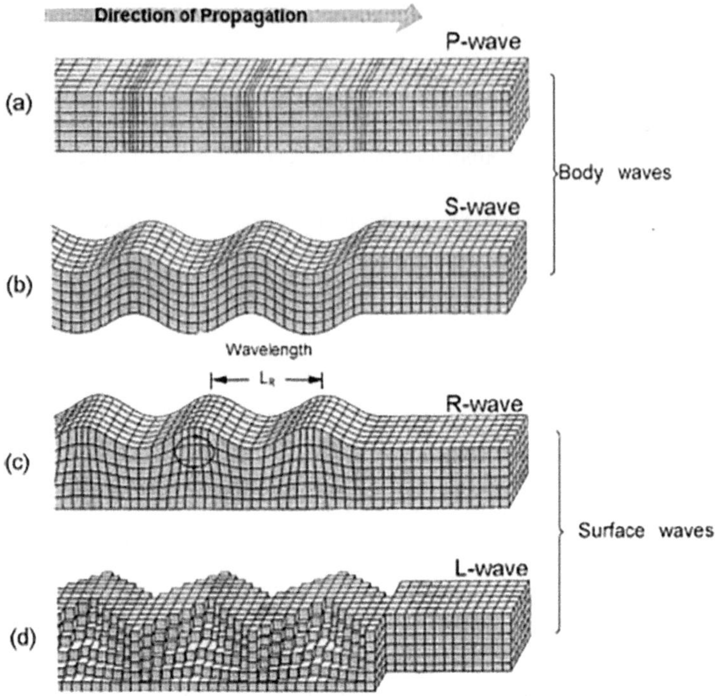

Fig. 2.1 Types of acoustic waves (Athanasopoulos et al. 2000) (reprinted with permission)

Acoustic Devices

Acoustic wave sensors function by observing alterations in the physical characteristics of an acoustic wave when exposed to a measurable quantity, known as the measurand. These sensors often utilize piezoelectric materials, which are crystalline solids lacking a center of inversion symmetry. These materials exhibit a strong connection between mechanical strain and electrical polarization. In the context of acoustic sensors, piezoelectric materials are instrumental in generating acoustic waves within solid substances. This is achieved by applying precisely tailored electric fields. The subsequent detection of acoustic waves occurs through the measurement of the charge generated as a result of induced mechanical deformation in these materials (Fogel et al. 2016).

The overarching classification of acoustic sensors extends to three main categories: surface acoustic wave (SAW) sensors, bulk acoustic wave (BAW) sensors, and unique photoacoustic sensors. SAW sensors predominantly operate on the surface of materials, BAW sensors delve into the bulk or interior of the material, while photoacoustic sensors offer a distinctive approach by responding to light-induced acoustic phenomena (Fig. 2.2). This classification serves as a cornerstone in comprehending the diverse methodologies employed in harnessing and detecting acoustic

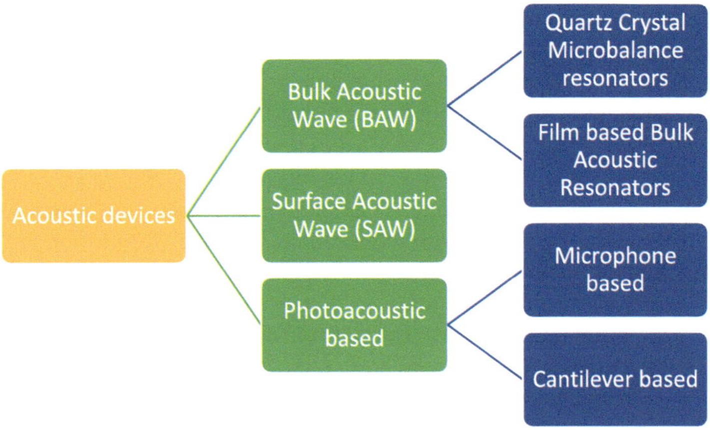

Fig. 2.2 Classification of acoustic devices

waves, paving the way for a rich tapestry of applications in sensing and measurement technologies (Ali and Prasad 2020; Razansky and Manohar 2016).

Quartz Crystal Microbalance Sensors

The Quartz Crystal Microbalance (QCM) is a significant mass sensor based on BAW resonators. In 1959, Sauerbrey, a German physicist, and his research team explored the correlation between mass absorption on a quartz surface and the frequency offset during the gas phase. Their discoveries laid the groundwork for extensive QCM research across various scientific disciplines (Alanazi et al. 2023).

A piezoelectric quartz crystal resonator is a meticulously cut slice derived from a natural or synthetic quartz crystal. The pristine form of quartz crystal is depicted in Fig. 2.3a. Illustratively represented in Fig. 2.3b, a quartz crystal microbalance (QCM) comprises a slender quartz disk with electrodes deposited on its surface (O'Sullivan and Guilbault 1999).

The application of voltage to a quartz crystal induces oscillation at a defined frequency. The alteration in mass on the quartz surface correlates directly with the frequency change of the oscillating crystal, as indicated by the Sauerbrey Eq. (2.1).

$$\nabla m = -C.\nabla f \tag{2.1}$$

The change in Quartz Crystal Microbalance (QCM) frequency indicates adsorbed analyte mass, measured in nanograms per square centimeter. The Sauerbrey equation suits elastic substances but is inadequate for inelastic materials like cells and polymers due to energy loss. It becomes invalid for soft or viscoelastic films, causing mass underestimation beyond a 2% change. Quartz Crystal Microbalance with Dissipation

(QCM-D) was developed to assess energy dissipation in biological samples (Huang et al. 2017; Dixon 2008).

A shear mode QCM is made of a thin quartz disk with coated electrodes. The quartz crystal plate is precisely cut into either AT or BT orientation, ensuring the acoustic wave propagates perpendicular to the crystal surface (Fig. 2.4). The resonant frequency depends on cutting angles, commonly using the AT cut at 35° 15 from the Z axis. The AT-cut crystal exhibits negligible temperature dependence at 25 °C, minimizing frequency changes with temperature variations. Deposition of mass on the QCM surface results in a frequency decrease (O'Sullivan and Guilbault 1999).

The sensitivity of the QCM to mass is contingent on the thickness of the crystal, influencing its resonant frequency. Thinner QCMs exhibit heightened frequencies and sensitivities, albeit with increased fragility. Typically, commercially utilized QCM crystals have a diameter of 0.5 in. In the context of liquid-phase QCM sensing, optically polished crystals are favored to minimize nonspecific interactions. However, challenges emerge in this domain, stemming from frequency shifts and signal attenuation caused by variations in viscosity. Addressing these challenges necessitates the implementation of suitable algorithms to disentangle these effects (Huang et al. 2017).

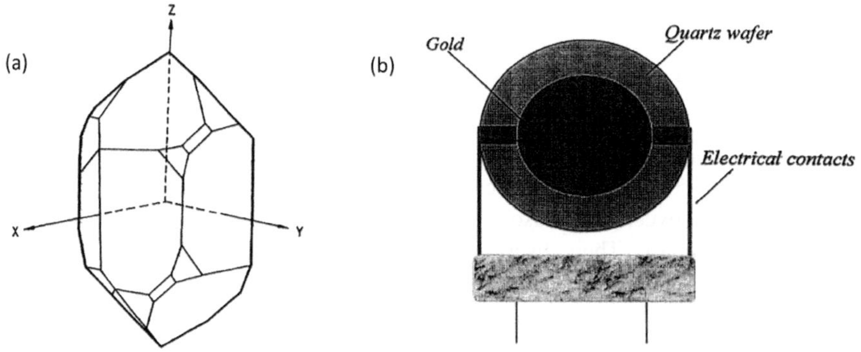

Fig. 2.3 **a** A pristine quartz crystal with axes and **b** Schematic of a quartz crystal microbalance (O'Sullivan and Guilbault 1999) (reprinted with permission)

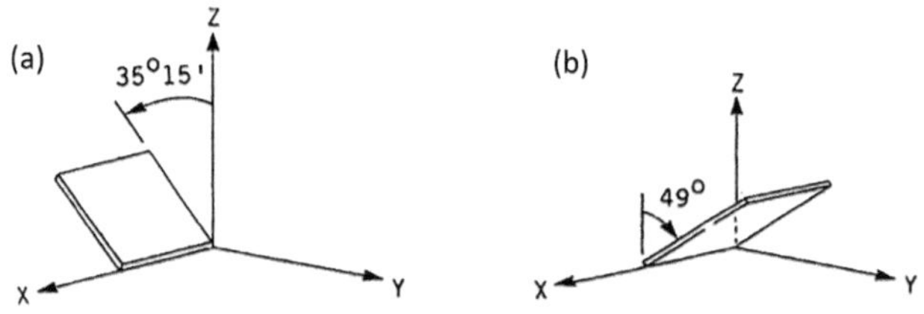

Fig. 2.4 Orientation of quartz crystal **a** AT cut **b** BT cut (O'Sullivan and Guilbault 1999) (reprinted with permission)

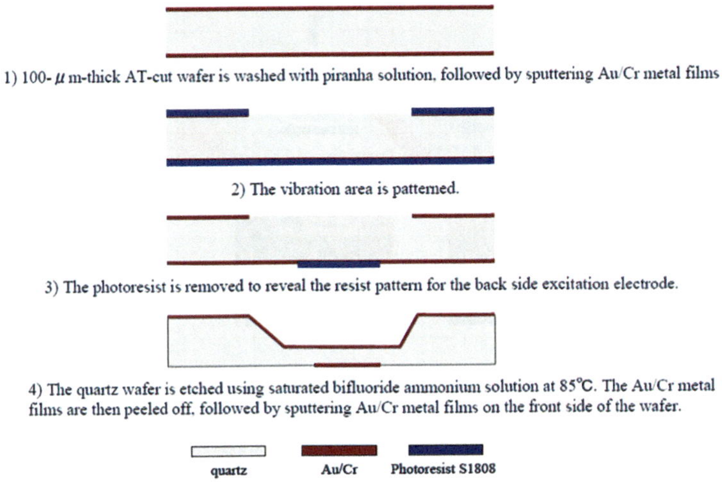

Fig. 2.5 Fabrication process flow for a QCM resonator (Liang et al. 2013) (reprinted with permission)

Materials and fabrication

The fabrication of MEMS/NEMS-based QCM sensors involves a series of intricate steps, beginning with substrate selection. High-quality quartz crystal substrates, often in the form of silicon wafers coated with silicon dioxide, provide a solid foundation for building these hybrid sensors. MEMS fabrication techniques, such as photolithography and etching, are then employed to create the micro/nanostructures that forms the mechanical resonators at the heart of the sensor. The process flow for the fabrication of a QCM resonator is shown in Fig. 2.5 (Liang et al. 2013).

Piezoelectric integration involves depositing a thin layer on resonator structures, introducing the piezoelectric effect to the MEMS/NEMS device. This effect transduces mechanical vibrations into electrical signals. Metal electrodes on the piezoelectric layer are crucial for the QCM sensor, applying the electric field needed for the piezoelectric effect. The quartz crystal resonates through the Converse Piezoelectric Effect when subjected to an alternating electric field, while the Direct Piezoelectric Effect generates an electric field within the crystal through applied mechanical strain, essential for quartz oscillators' operation (Dixon 2008; Uchino 2017).

Sensor functionalization, though not mandatory, plays a pivotal role by enabling the tailoring of the sensor's surface to cater to specific applications. This process entails coating the sensor with materials that exhibit selective interactions with target molecules, thereby amplifying the sensor's sensitivity and selectivity. In the work conducted by Wang et al., QCM sensors were functionalized for formaldehyde detection using nanoporous TiO_2 (Fig. 2.6). This involved drop-casting ethylene glycol (EG)-dispersed TiO_2 nanofibers onto the QCM electrode, followed by the functionalization of sensing polyethylenimine (PEI) on these fibers (Wang et al. 2012).

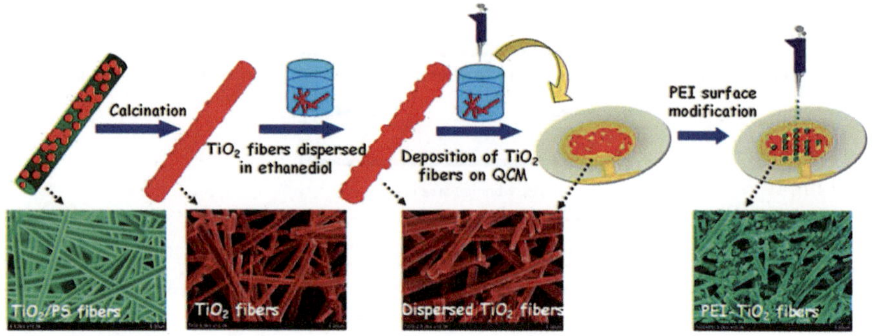

Fig. 2.6 Schematic of functionalization of QCM sensor using nanoporous TiO_2 (Wang et al. 2012) (reprinted with permission)

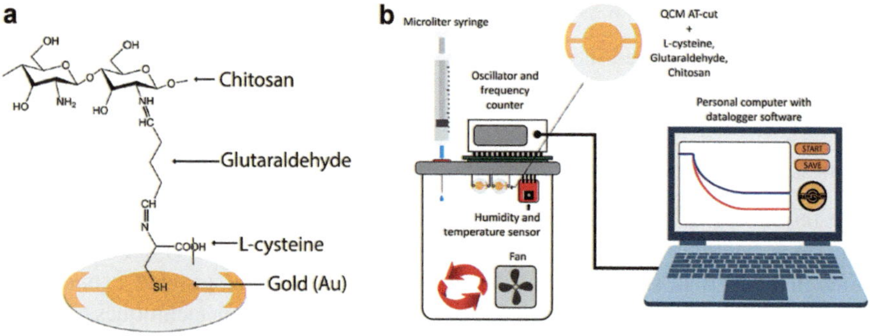

Fig. 2.7 **a** Schematic of chemical structure and active layer deposition of the chitosan-based QCM chip **b** Vapour sensing setup of the QCM system (Triyana 2018) (reprinted with permission)

Polymer composites, with their versatile properties, are a preferred material for Quartz Crystal Microbalance (QCM) sensor fabrication, offering finely adjustable characteristics. In a study by Triyana et al., a chitosan active layer immobilized onto the QCM substrate demonstrated selectivity towards alcohols. Utilizing a self-assembled monolayer of L-cysteine and glutaraldehyde as a cross-linking agent (Fig. 2.7), the QCM sensing chip exhibited optimal responsiveness to n-amyl alcohol vapor, achieving a sensitivity of approximately 4.4 Hz mg^{-1} L (Triyana 2018).

Film-Based Bulk Acoustic Resonators (FBAR)

In the late twentieth century, Film Bulk Acoustic Resonator (FBAR) devices transformed sensing technology, leveraging advancements in microfabrication and piezoelectric materials. Initially vital for communication systems' frequency stability, FBAR research evolved in the early 2000s to enhance frequency regulation through materials and design. As MEMS and NEMS technologies matured, FBAR devices became highly sensitive sensors, detecting subtle physical changes. Today, they serve in environmental monitoring, healthcare diagnostics, and industrial sensing, offering compact size, high sensitivity, and adaptability to harsh conditions. Ongoing research focuses on material and fabrication optimization for further miniaturization and improved precision in sensing technologies.

FBARs operate on the piezoelectric effect, where materials like aluminum nitride or zinc oxide generate electric charges in response to mechanical stress. Initially introduced by Sliker and Roberts using a composite crystal resonator on a bulk quartz crystal, FBARs have evolved with advancements by Lakin and Wang, incorporating a thin silicon membrane for a contemporary four-layer composite structure (Liu et al. 2020).

In its current form, an FBAR includes a membrane support layer (e.g., sputtered SiO_2) for mechanical rigidity, a nucleating metal layer (e.g., gold) for piezoelectric film growth, and a top metal film (e.g., gold or aluminum) over the piezoelectric layer. Various FBAR configurations highlight the resonator's adaptability. The FBAR equivalent circuit uses a figure-of-merit (FOM) for performance assessment, dependent on factors like film quality, material properties, thickness, topology, and harmonic mode (Liu et al. 2020; Krishnaswamy et al. 1990).

Predicting resonator response involves a one-dimensional Mason model, considering acoustic properties of each layer. Extracting an equivalent circuit includes measuring impedance near resonances, aiding in understanding FOM variations related to resonator features. The electrode configuration in Fig. 2.8, operating in the second harmonic mode, boasts the best reported FOM value of 60, showcasing the precision and utility of FBARs in oscillator and filter circuitry applications (Krishnaswamy et al. 1990).

Materials and fabrication

The first step in crafting an FBAR involves the deposition of a thin film onto a substrate. Thin film deposition methods, such as sputtering or chemical vapor deposition, are employed to ensure uniformity and precise control over the film's thickness. The thin film acts as the resonating element that responds to mechanical stress induced by acoustic waves.

Currently, three primary piezoelectric materials are employed in BAW resonators: AlN, PZT, and ZnO. Despite PZT's high piezoelectric material-coupling coefficient (kt_2) and large dielectric constant, its intrinsic drawbacks include significant loss, low sound speed, and challenges in maintaining uniform composition during preparation. ZnO, while possessing a larger kt_2 than AlN, exhibits a much smaller longitudinal sound velocity. Additionally, the presence of zinc in ZnO processing reduces carrier lifetime and is incompatible with CMOS technology (Ali and Prasad 2020; Uchino 2017).

Various techniques, including magnetron sputtering, molecular beam epitaxy, pulsed laser deposition, and metal organic chemical vapor deposition, can produce Aluminum Nitride (AlN) thin films crucial for FBAR fabrication. Molecular beam epitaxy faces challenges like low deposition rates and high costs, while pulsed laser

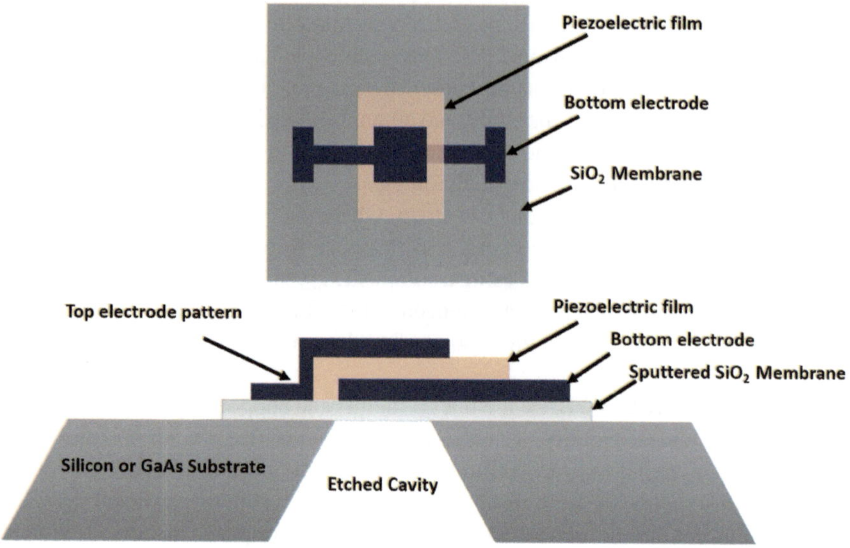

Fig. 2.8 FRAR configuration with bottom electrode sandwiched between piezoelectric film and supporting membrane (adapted from Krishnaswamy et al. (1990))

deposition is unsuitable for large-area film deposition. Metal organic chemical vapor deposition carries the risk of inducing significant residual stress due to its high deposition temperature. Currently, magnetron sputtering is the predominant method for producing piezoelectric AlN films, operating at lower temperatures (around 200–300 °C) and ensuring compatibility with CMOS technology. This technique offers advantages such as high efficiency, cost-effectiveness, and the ability to cover large deposition areas. An additional strategy to enhance AlN film quality involves implementing rapid thermal annealing (RTA) following the sputtering process (Liu et al. 2020; Xu et al. 2001).

Liu et al. found that FBAR device characteristics depend on the thickness and quality of the AlN film, as well as the thickness and materials of the top electrode. Decreasing the thickness of the AlN film or the top electrode raises the resonance frequency, offering potential for precise FBAR device tuning. Additionally, increasing the AlN film thickness or using higher Radio Frequency (RF) power enhances c-axis orientation and reduces the rocking curve Full-Width at Half-Maximum (FWHM) (Figs. 2.9 and 2.10). However, these modifications also lead to larger grain size and elevated surface roughness (Liu et al. 2009).

Doped aluminum nitride, specifically scandium-doped aluminum nitride (ScAlN), exhibits enhanced.

performance in solidly mounted Film Bulk Acoustic Resonators (FBARs) compared to pristine AlN. Resonators utilizing ScAlN thin films as the piezoelectric layer show a substantial increase in the electromechanical coupling factor, rising from approximately 3% to about 12% compared to pure AlN. However, the reduced

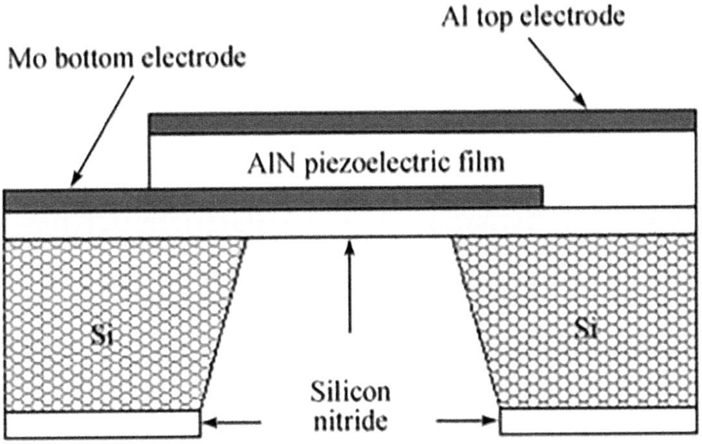

Fig. 2.9 Cross-sectional view of FBAR (Liu et al. 2009) (reprinted with permission)

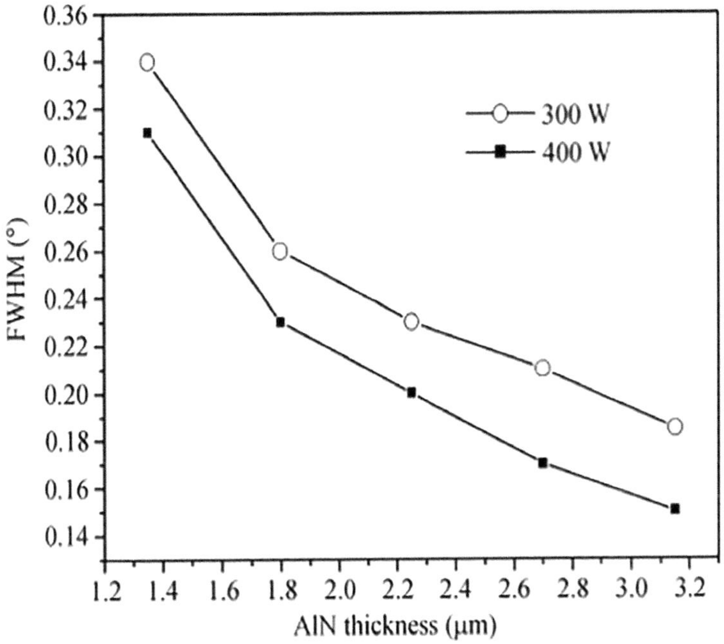

Fig. 2.10 The correlation between the thickness of AlN films and the Full-Width at Half-Maximum (FWHM) of the (002) peak in AlN films grown on Mo bottom electrodes under varying Radio Frequency (RF) powers (Liu et al. 2009) (reprinted with permission)

stiffness of ScAlN leads to a decline in the quality factor, decreasing from around 300 to about 100 due to heightened damping losses within the piezoelectric material (Schneider et al. 2017).

Beyond the influence of piezoelectric materials, electrode material selection is pivotal. An ideal electrode material should possess high acoustic impedance for improved electromechanical coupling, low resistivity to minimize series resistance (Rs), and a low density to optimize high-frequency performance by reducing mass loading. The type and quality of the electrode directly impact the piezoelectric material, requiring favorable characteristics such as good orientation, low surface roughness, and minimal lattice distortions. Research comparing various electrode materials (Pt, Mo, W, Ru, Ir, Ta, Al/Au/Cu, CNT, graphene, and ITO) identifies Mo as the preferred choice due to its balanced acoustic impedance, density, and resistivity (Lee et al. 2004; Lee et al. 2002).

Fabricating FBRARs poses challenges in controlling film thickness, ensuring uniformity, and addressing stress and strain. Researchers explore innovative solutions, including advanced nanofabrication techniques and novel materials, to enhance FBAR sensor performance. The future of FBAR fabrication holds promise, focusing on new piezoelectric materials, refined techniques, and expanded applications. Anticipated advances in nanotechnology and materials science are set to transform FBAR technology, reshaping the sensing technology landscape with unparalleled precision. The meticulous fusion of materials science, microfabrication techniques, and engineering precision positions FBARs at the forefront of sensing technology advancements.

Surface Acoustic Wave (SAW)-Based Devices

The integration of surface acoustic wave (SAW) devices offers a groundbreaking solution, addressing both sensing and actuation within a unified platform. The physics governing SAWs prove advantageous for MEMS devices, leading to versatile SAW sensors used in applications beyond traditional filters and oscillators. SAW devices exhibit adaptability, serving as exceptional actuators and sensors responsive to diverse environmental stimuli. Their applications span mechanical, chemical, electrical, physical science, and biological fields, attesting to their utility across diverse domains. Figure 2.11 illustrates the varied applications of SAW devices, emphasizing their exceptional capability to detect even the slightest disturbances on surfaces, making them indispensable in scenarios requiring precision and sensitivity.

When tailoring a Surface Acoustic Wave (SAW) sensor for a precise application, a profound understanding of the various types of waves that can be generated within a substrate becomes imperative. The unique characteristics of Rayleigh waves, Shear waves, Lamb waves, and Love waves make them integral to the design and functionality of SAW devices.

In SAW devices, the actuation process begins with applying an electrical signal to interdigitated electrodes (IDTs) on a piezoelectric substrate, converting it into

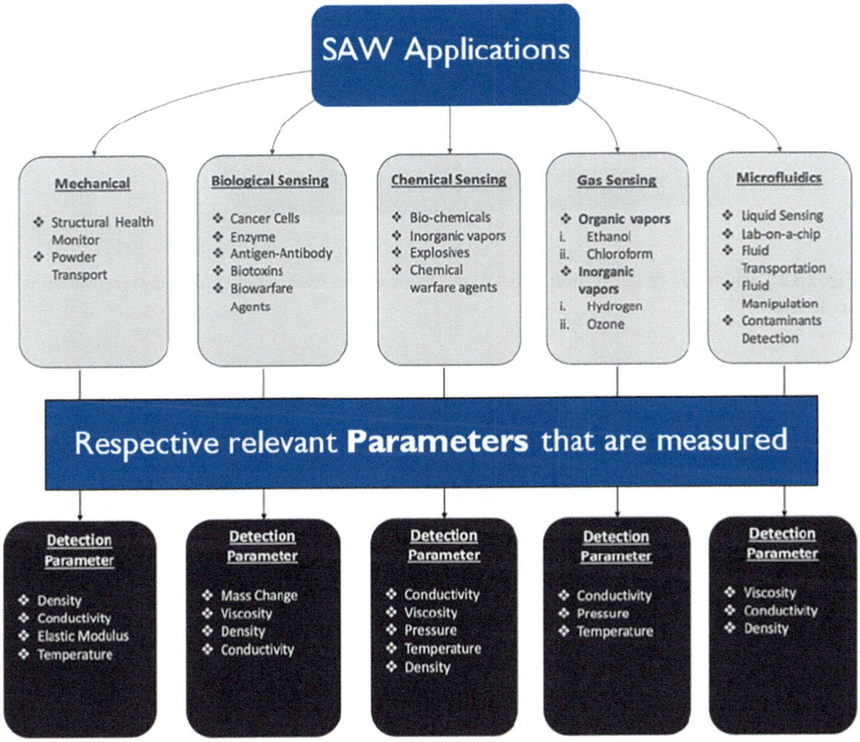

Fig. 2.11 Applications of SAW devices (Mandal and Banerjee 2022) (reprinted with permission)

mechanical energy and generating surface acoustic waves (Fig. 2.12). These waves, with a dual role in sensing and actuation, are crucial for mechanoelectrical coupling in the reverse acoustic sensing process. To suit various applications, IDTs can be configured in different ways, emphasizing the need to design them for minimal insertion loss and optimal center frequency responses. A typical structure of a $LiNbO_3$-based SAW device is shown in Fig. 2.13. Key parameters, such as electrode width, spacing, aperture, and the number of fingers, govern frequency bandwidth and electrical impedance, highlighting the intricacies of tailoring IDTs to meet specific performance criteria and user requirements (Smith et al. 1969; White and Voltmer 1965).

Understanding the distinct characteristics and functionalities of the type of surface waves enables engineers and researchers to tailor SAW sensors to specific applications. The selection of the wave type is often dependent on the nature of the target application, the required sensitivity, and the desired interaction with the substrate. This intricate knowledge of wave types in SAW devices underscores the precision and customization achievable in the design of sensors for diverse and specialized sensing needs (Sigrist 1998).

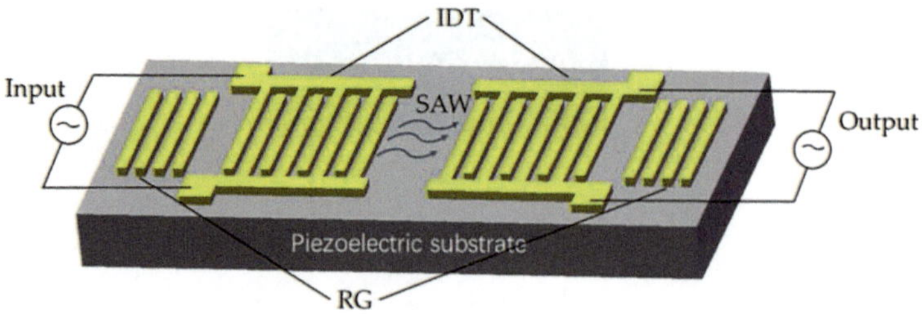

Fig. 2.12 Schematic of generation of SAW in piezoelectric wafer (Lei et al. 2022) (reprinted with permission)

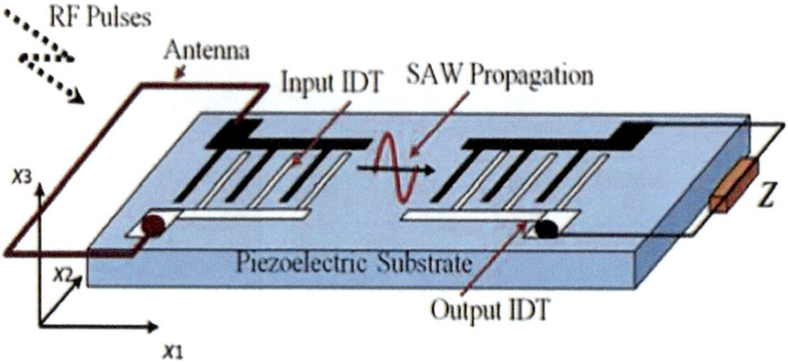

Fig. 2.13 Typical structure of a $LiNbO_3$ SAW device (Dissanayake et al. 2010) (reprinted with permission)

SAW sensors utilize piezoelectric components, such as crystals, known for their strong electromechanical coupling, providing the required robustness for SAW technology. Precision involves cutting crystals to specific orientations to optimize efficiency and generate distinct wave types. For example, a 36-degree YX cut in lithium tantalate produces shear horizontal waves, and a 128-degree YX cut in lithium niobate generates Rayleigh waves. The intentional orientation of crystal cuts is essential for a robust piezoelectric effect, with effectiveness tied to piezoelectric coefficients influencing the electromechanical coupling coefficient—a critical parameter for material efficacy.

Refer to Table 2.1 for details on wave types and materials used in SAW sensor fabrication (Mandal and Banerjee 2022).

SAW technologies have shifted focus from bulk substrates to thin film acoustic waves, garnering recent research attention. Thin-film approaches, especially using piezoceramics like ZnO, AlN, and PZT, offer benefits such as large piezoelectric constants and electromechanical coupling for sensing and actuation. Despite PZT's drawbacks, such as high energy loss and low wave velocity, thin-film piezoelectric

Table 2.1 Types of SAW waves and their corresponding materials (Mandal and Banerjee 2022) (reprinted with permission)

Wave type	Materials	Applicable frequency range	Attenuation	Misc. comments
Rayleigh wave	• 128° YX lithium niobate • 41° YX lithium niobate • ST-Quartz	• ~3 MHz to ~2 GHz • Most application 345 MHz	• Highly attenuative • With fluid load on the top of the substrate, attenuation is significant	• For Rayleigh wave generation, power consumption is lower • Can be of low cost
Lamb wave	Generated in any bounded plate-like structure. This wave is generated in layered media at lower frequencies. However, for high frequency applications, it is hard to find natural material where Lamb wave could be used for SAW sensors. Thus, artificial thin films—e.g., ZnO, AlN, PVDF thin film—could be used	• ~200 to 2 MHz in natural materials •200 MHz to 2 GHz in artificially made thin films	• Low attenuation, propagates long distance • In presence of liquid load on the substrate,leaky-Lamb wave propagates • Leaky waves attenuate faster	• Lamb waves consists of Antisym-metric and Symmetric wave modes • First antisym-metric (A_0) and first symmetric (S_0) wave modes are easy to detect with appropriate delay lines
SH-wave	• 64° YX lithium niobate • 36° YX lithium niobate • Quartz	• ~100 to ~450 MHz	• Nondispersive • Low attenuation; however, sometimes combined with bulk wave, so hard to detect	• Low cost • Wide application • Suitable to use with fluid loading on the substrate
Love wave	• Must have a guiding layer on top of the substrate • Possible substrates: similar to SH-wave • Possible guiding layers: SiO_2, ZnO, TiO_2, SU-8 photoresists, polymethyl methacrylate etc.	• ~−100 to ~450 MHz same as SH wave	• With increasing coating thickness insertion loss increases • Attenuation in guiding layer or the coating layer affects significantly	• Highly sensitive • Works in fluid loading environment

materials are gaining prominence in applications like lab-on-a-chip and acoustofluidic devices. The limitations of bulk substrates, coupled with the advantages of flexibility, selective film deposition, and cost efficiency, have led to a preference for thin-film-based SAW devices (Omori et al. 2001; Fflckernell 2000).

ZnO, a preferred thin-film material for piezoelectric devices, is distinguished by its higher piezoelectric coupling coefficient. ZnO films, known for biocompatibility, find popularity in biosensing applications due to low stress and improved adhesion, facilitating the deposition of thick films—a beneficial feature for surface acoustic wave devices. Aluminum nitride (AlN), another noteworthy piezoelectric thin-film material, excels in high-temperature resistance, making it valuable for SAW applications in elevated temperature environments. AlN's highest wave velocity among thin film piezoelectric deposits positions it as a favored material for high-frequency applications, primarily utilized in thin films and high-frequency-based applications, in contrast to ZnO (Xu et al. 2001; Liu et al. Feb. 2009; Schneider et al. 2017; Lee et al. 2004).

Other materials such as gallium arsenide (GaAs), polyvinylidene fluoride (PVDF), and gallium nitride (GaN) have also demonstrated excellence as piezoelectric materials in SAW devices. Recently, aluminum gallium nitride (AlGaN) has been explored as a potential piezo thin film. (Fflckernell 2000).

In conclusion, MEMS/NEMS-based Surface Acoustic Wave (SAW) devices epitomize innovation, integrating the precision of MEMS with robust SAW sensing. Yet, challenges in fabrication intricacies, material compatibility, and standardization persist. These obstacles serve as crucibles for breakthroughs. Future prospects are optimistic as researchers refine fabrication techniques, explore novel materials, and tackle scalability concerns. Envisioned applications span healthcare, environmental monitoring, and beyond, indicating a promising journey forward as MEMS/NEMS-based SAW devices evolve in tandem with technological progress.

Photoacoustic-Based Devices

In the dynamic and ever-advancing field of sensing technologies, a compelling synergy has emerged as MEMS/NEMS to converge with the fascinating domain of photoacoustic sensing. This collaboration brings forth a paradigm shift in precision sensing by leveraging the unique capabilities of MEMS and NEMS at both the micro and nanoscale.

At the heart of this technological convergence lies the utilization of photoacoustic effect within sensors. The photoacoustic effect manifests when a material absorbs light, resulting in the generation of acoustic waves. This phenomenon becomes the cornerstone of photoacoustic sensors, allowing them to detect and quantify various substances with unparalleled sensitivity. The integration of micro and nanotechnologies in this context introduces novel possibilities, offering a leap forward in the realm of precision sensing.

In 1880, Alexander Graham Bell made the groundbreaking discovery of photoacoustics (PA) by observing that sunlight could generate sound in a substance. Illuminating selenium with light, Bell noted audible sound, attributing it to temperature-induced expansion and contraction. Experiments with intense light beams onto solids or gases in sealed vessels revealed intense acoustical signals, detected by the human ear. PA spectroscopy's early limitations included weak signals from liquids due to low thermal expansion coefficients. Quantitative analysis became feasible in the 1930s with improved light sources and detectors. Theoretical foundations were laid in 1976 by Rosencwaig and Gersho, leading to a surge in PA spectroscopy applications. Despite extensive research, only a few commercial instruments, mainly for gas-phase analytics, are available (Razansky and Manohar 2016; Rosencwaig 1978).

As previously stated, the PA effect is based on sample heating caused by optical absorption. Periodic heating and cooling of the sample are required to generate pressure fluctuations that can be detected by pressure sensitive transducers in order to generate acoustic waves that can be detected by pressure sensitive transducers. The basic mechanism of a Photoacoustic Spectroscopy is shown in Fig. 2.14 and is described as follows:

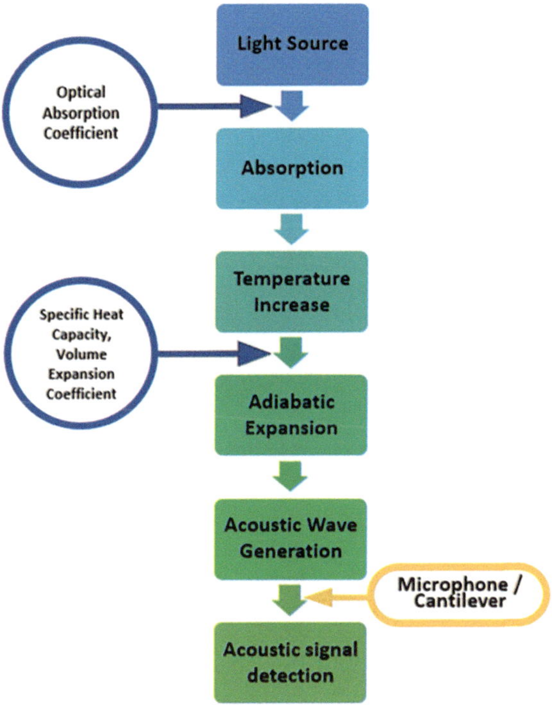

Fig. 2.14 Process-flow for photoacoustic spectroscopy

Excitation

Photoacoustic (PA) generation involves two main methods: modulated and pulsed excitation (Fig. 2.15). Modulated excitation adjusts the light source, improving signal detection for enhanced spatial resolution and penetration depth, minimizing acoustic scattering. Pulsed excitation relies on brief laser pulses for rapid sample heating, advantageous for high spatial resolution and detecting small features. Challenges include distinguishing signals from background noise and potential depth reduction due to acoustic scattering. Pragmatic use requires careful consideration of trade-offs in spatial resolution, penetration depth, and signal-to-noise ratio (Mandelis et al. 1991; Sigrist 1998).

The choice of excitation source is critical because it determines the specificity and sensitivity of the technique. Different materials have unique absorption spectra, so selecting the appropriate wavelength ensures that the excitation is selectively absorbed by the sample. The common excitation sources include pulsed lasers, light-emitting diodes (LEDs), and other tunable light sources. The development of versatile and powerful excitation sources has greatly contributed to the advancement and applicability of photoacoustic spectroscopy in various scientific and industrial fields (Palzer 2020).

Signal Generation

When light is absorbed, it transforms into heat, generating a proportional PA signal. Direct PA generation in gas samples involves modulated excitation with chopped

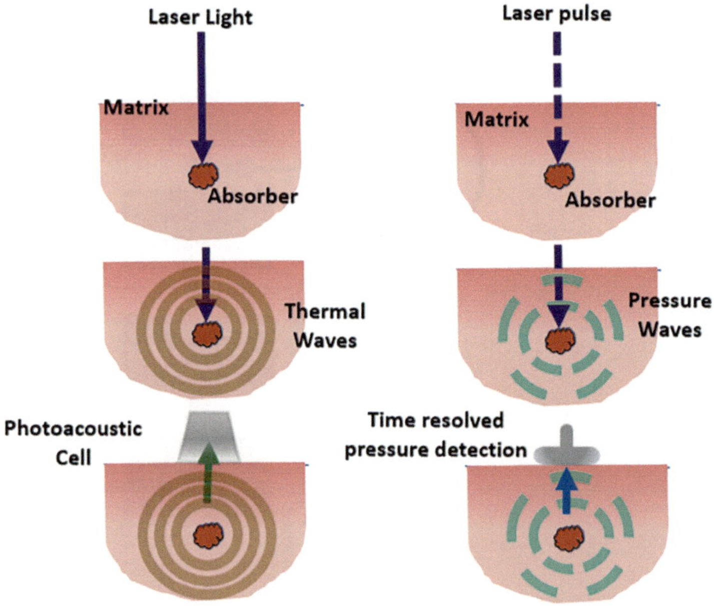

Fig. 2.15 Principle of PA depth profiling by modulated (left) and pulsed (right) laser excitation

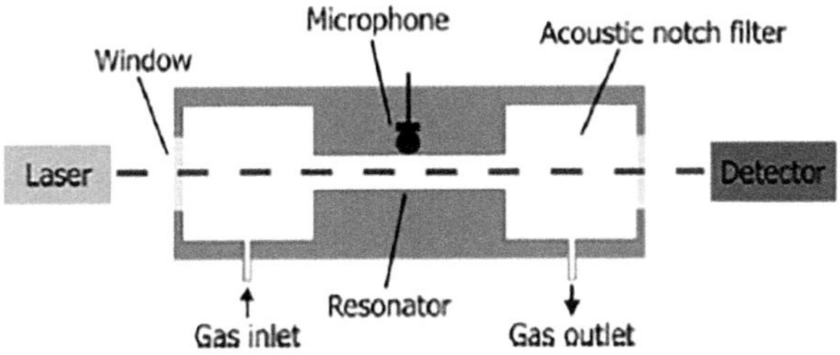

Fig. 2.16 Principal set-up of a photoacoustic cell for gas phase (Schmid 2006) (reprinted with permission)

lasers. For solid samples, challenges arise due to acoustic impedance mismatches, making modulated excitation impractical. Indirect PA generation involves modulated excitation causing sample warming, followed by heat transport to the sample's interface with the adjacent gas phase (Fig. 2.16) (Mcdonald 1981; Monolithic integrated SAW filter based on AlN for high-frequency applications 2013).

Detection

An acoustic detector, encompassing either a microphone or a cantilever, captures pressure waves, and the electronic processing system amplifies and processes the PA signal to unveil spectral details. The acoustic source's shape depends on the absorber type. In solid sample analysis, backward mode detection is common, using setups with a transparent prism or optical fiber for illumination and piezoelectric transducers for pressure pulse detection (Jackson and Amer 1980; Thakur et al. 2023).

Photoacoustic sensors have established a specialized role in a multitude of applications, underscoring their adaptability in tackling analytical challenges. Within environmental monitoring, these sensors exhibit proficiency in the detection and quantification of trace gases, serving as a robust tool for assessing air quality and controlling pollution. Their utility extends into medical diagnostics, where they facilitate non-invasive monitoring of blood components, presenting promising strides in the realms of personalized medicine and disease detection. Challenges persist, particularly in the optimization of sensitivity and mitigation of interference within intricate sample matrices. The incorporation of advanced materials and fabrication techniques, including MEMS and NEMS, introduces both prospects and complexities for enhancement. Looking forward, the future trajectory of photoacoustic sensors is compelling. As ongoing research unveils innovative materials and fabrication methodologies, these sensors are poised to redefine domains such as wearable health monitoring, industrial safety, and even space exploration, solidifying their role as pivotal components in the evolution of sensing technologies (Palzer 2020; Haisch 2012; Tam 1986).

Gas Sensing Using Acoustic Devices

The current air quality concerns pose substantial risks to well-being due to various pollutants, including H_2S, SO_x, NO_x, CO, VOCs, PM2.5, PM10, and greenhouse gases (CO_2, CH_4). The demand for reliable gas sensors has surged, especially in industrial settings, where they play a crucial role in optimizing manufacturing conditions, ensuring product quality, and averting the release of harmful gases. Gas sensors are integral to fire detection systems, alerting to smoke, heat, or combustion gases for swift hazard response. They also enhance energy efficiency in combustion processes, aligning with environmental impact reduction goals. Gas sensors provide accurate real-time data which facilitates compliance with gas emission regulations. Additionally, gas sensors serve as early warning systems for gas leaks, chemical spills, and deteriorating air quality (Yunusa et al. 2014).

The primary challenge is developing cost-effective, compact, and selective gas sensors. In conventional methods, cross-sensitivity and response time constraints are key issues. Additionally, electronic sensors may face constraints in terms of response time, a critical factor in scenarios where rapid detection of hazardous gases is imperative for preventing accidents or mitigating environmental impact. In response to the limitations of traditional electronic sensors, there has been a discernible shift towards exploring alternative sensing technologies. Among these emerging techniques, MEMS/NEMS has opened avenues to explore alternative sensing techniques, and acoustics has emerged as a promising avenue (Dhall et al. 2021).

Acoustic gas sensing presents a fundamentally distinct approach from electronic methods, leveraging the interaction between sound waves and gas molecules rather than relying on direct sensor-gas contact. This approach offers non-intrusiveness, reducing contamination risks and enhancing applicability in challenging scenarios. Acoustic sensors excel in detecting a variety of gases, demonstrating versatility in industrial and environmental settings. Their sensitivity to changes in sound speed enables precise identification and quantification of specific gases, and they can be manufactured in submillimeter sizes for analyte mass detection. These sensors provide highly sensitive, rapid, cost-effective per-sample testing with label-free operation and seamless integration into wireless communication systems. The growing interest in acoustic gas sensing reflects a practical acknowledgment of the need for innovative solutions, with researchers and industries recognizing its potential to enhance safety protocols, environmental monitoring, and overall operational efficiency (Ippolito et al. 2009). In this chapter, we have chosen to discuss specifically about gas sensors for detection of VOCs, Hydrogen, Ammonia, Carbon dioxide, Nitrogen dioxide. Table 2.1 lists some of the latest development in acoustic-based gas sensors with their limit of detections.

Volatile Organic Compounds (VOCs)

Volatile Organic Compounds (VOCs) are diverse carbon-based chemicals that can evaporate from industrial processes, household products, and vehicle emissions, posing health risks and contributing to pollution. VOC sensors are essential for monitoring indoor and outdoor air quality, detecting and quantifying VOC presence. Early detection helps mitigate health risks, ensure occupational safety, and contribute to environmental well-being. These sensors are indispensable in industries like manufacturing, construction, and environmental monitoring, safeguarding both human health and the environment (Rumchev et al. 2007; Miekisch et al. 2004).

In the realm of acoustic sensors for VOC detection, a Quartz Crystal Microbalance (QCM)-based sensor was meticulously engineered to detect methyl mercaptan (CH_3SH). This sensor, designed with heightened sensitivity and rapid response capabilities, incorporated Poly(ethylene imine) (PEI) as the polymeric layer to augment the surface area of the sensing film. An additional Al_2O_3 porous film, applied using the sol–gel method, further increased the surface area. The resulting sensor successfully detected 100 parts per billion (ppb) of CH_3SH gas (Kikuchi and Shiratori 2005).

Functionalizing the surface of acoustic sensors has proven effective in enhancing selectivity and sensitivity. An example is the development of zinc oxide nanorods incorporated into Surface Acoustic Wave (SAW) sensors for ethanol detection. In this study, ZnO nanorods were grown using a solution growth method on a ZnO/128° rotated Y-cut $LiNbO_3$ crystal. The device showed a frequency shift of 24 kHz when tested at an operating temperature of 270 °C with an ethanol concentration of 2300 ppm (Giffney et al. 2012). Another approach to improve ethanol sensing sensitivity involved using SnO_2 as the sensing layer, achieving a limit of detection of 15 ppm (Jahanshahi et al. 2018).

Expanding the scope to include SAW sensors coated with carbon nanotubes, significant progress has been made in room-temperature detection of VOCs. When exposed to ethanol, the SAW sensor coated with single-wall carbon nanotubes (SWCNTs) or multi-wall carbon nanotubes (MWCNTs) demonstrated substantial resonance frequency downshifts. This showcased a sensitivity up to 3 times greater than existing organic layer-coated SAW sensors. Importantly, the selectivity for tested VOCs was influenced by the choice of organic solvents used to disperse the carbon nanotubes (Penza et al. 2005).

In a pioneering effort, Zeng and his team introduced an electronic nose (e-nose) system with a novel temperature modulation-based FBAR multiparameter virtual sensor array (VSA) (Fig. 2.17). Utilizing a temperature-compensated film bulk acoustic wave resonator (TC-FBAR) coated with analyte-specific thin films, including self-assembled poly(sodium4-styrenesulfonate) (PSS)/poly(diallyl dimethyl ammonium chloride) (PDDA), the system exhibited over 97% accuracy in distinguishing and classifying six different VOC vapors. This work represents a significant advancement in compact and effective e-nose systems for commercial

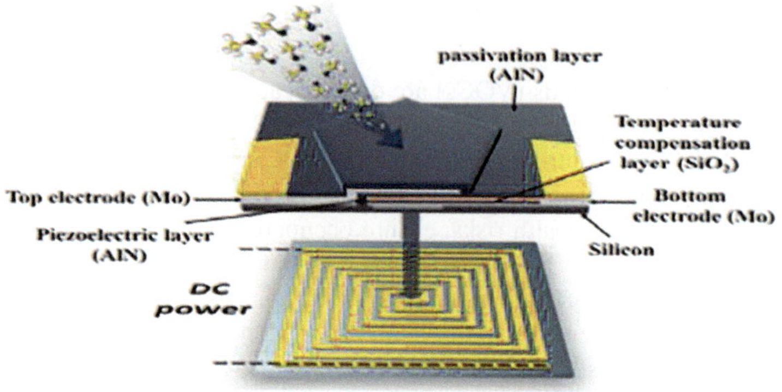

Fig. 2.17 A diagram illustrating the gas sensor using FBAR, where the sensor's surface is chemically treated with PSS and PDDA. The heater positioned between the FBAR and evaluation boards alters the FBAR temperature by utilizing the programmable DC power supply module (Zeng 2023) (reprinted with permission)

electronics, with the FBAR's ultrahigh-frequency sensitivity surpassing that of QCM and SAW devices by orders of magnitude (Zeng 2023).

Compared to QCM, FBAR, and SAW devices, photoacoustic sensors have been able to achieve sub ppb level detection of VOCs. In a recent study, Pangerl et al. developed a photoacoustic sensor that could quantify ethanol and acetone within a breath mixture containing CO_2 and water vapor. The sensing setup is claimed to have achieved a detection limit of 6.5 ppb for ethanol (Pangerl 2023).

Ammonia

Ammonia sensors are vital for protecting human health and the environment. Commonly used in industry, refrigeration, and agriculture, ammonia, a colorless gas with a pungent odor, poses health risks and environmental threats. These sensors act as vigilant guardians, promptly detecting and alerting us to its presence. Early warnings facilitate quick response measures, preventing illnesses and minimizing environmental impact in various settings. The demand for reliable ammonia sensors is crucial for ensuring safety and sustainable practices in industries, agriculture, and confined spaces.

In the domain of ammonia detection, a cutting-edge gas sensing device has been developed by incorporating electrospun nanofibrous membranes onto a quartz crystal microbalance (QCM). These nanofibers, ranging from 100 to 400 nm in diameter, are applied to the QCM surface through electrospinning blends of cross-linkable poly(acrylic acid) (PAA) and poly(vinyl alcohol) (PVA) (Ding et al. 2004).

Gas sensors designed around a quartz crystal microbalance (QCM) featuring a coating of ZnO nanorods offer an innovative approach for NH_3 detection at ambient

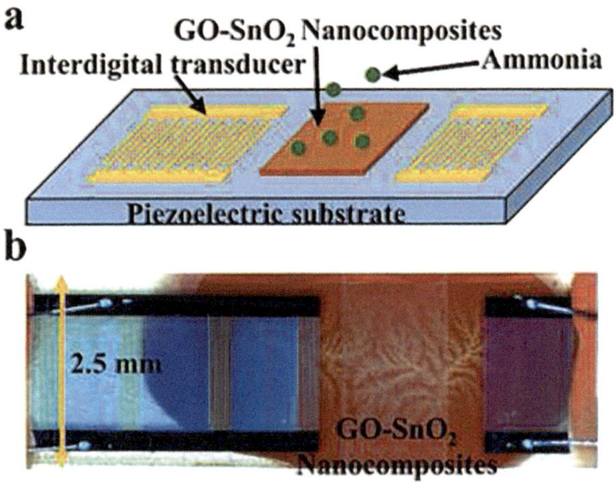

Fig. 2.18 **a** The SAW ammonia sensing device employing GO-SnO_2 nanocomposites, **b** The picture of the developed SAW sensing device (Sun 2023) (reprinted with permission)

temperature. Synthesized using a unique wet chemical method at a low temperature of 90 °C, vertically aligned ZnO nanorods were grown directly on the QCM for gas sensor applications. The QCM, coated with ZnO nanorods, displayed exceptional sensitivity to NH_3 gas, showing a frequency shift of approximately 9.1 Hz for 50 ppm NH_3 at room temperature. Response and recovery times varied with ammonia concentration, and the gas sensors exhibited commendable reproducibility and high stability (Quy et al. 2011).

Additionally, SAW sensors utilizing thin films of metal oxides and metal sulfides in a delay line configuration have demonstrated remarkable selectivity and sensitivity for ammonia detection. A SAW-based sensor enhanced with a GO-SnO_2 nanocomposite achieved a low detection limit of 40 ppb and a high sensitivity of 0.098 mV/ppb at room temperature (25 °C), with a rapid response time of less than 16.4 s (Fig. 2.18) (Sun 2023). In another development, mucosal-like ZnS nanostructures were synthesized on the surface of an ST-cut quartz SAW device through a chemical bath deposition method for ammonia sensing. The SAW device with ZnS mucosal nanostructures exhibited significant sensitivity, evidenced by a frequency shift of approximately 190 Hz upon exposure to 1 ppm ammonia gas (Long 2020).

Hydrogen

For the safe and effective use of this adaptable yet dangerous gas in a variety of applications, from fuel cells to industrial processes, hydrogen sensors are essential. Because hydrogen is flammable, these sensors serve as essential defenders, quickly identifying leaks and high quantities in real time. By means of prompt notifications, they enable prompt actions, reducing any risks and guaranteeing a safe atmosphere. Reliable hydrogen sensors are necessary for the safe and sustainable integration of

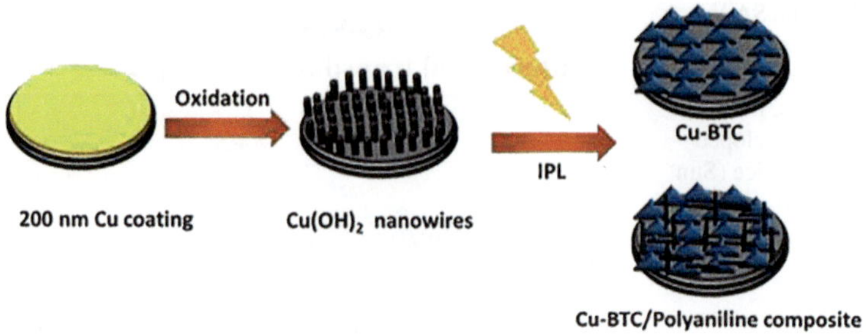

Fig. 2.19 Schematic of fabrication process flow of Cu-BTC/PANI nanocomposite incorporated QCM for Hydrogen sensing (Shokry et al. 2019) (reprinted with permission)

hydrogen technology in labs, factories, and hydrogen-powered automobiles (Hübert et al. 2011).

A high-performance hydrogen sensor was developed on a quartz crystal microbalance (QCM) using intense pulsed light and a Cu-BTC/polyaniline (PANI) nanocomposite film. The QCM, coated with a 200-nm-thin copper layer and subjected to oxidation and deposition of a solution containing the organic ligand (BTC) and PANI, facilitated the formation of the Cu-BTC/PANI nanocomposite film through intense pulsed light (Fig. 2.19). The sensor demonstrated a low limit of detection (15 ppm), rapid response times (2–5 s), and the ability to operate at room temperature even in high humidity (up to 60%) (Shokry et al. 2019).

A wireless MEMS-based hydrogen sensor was developed with a QCM, achieving a limit of detection around 10 ppm. The sensor chip comprises glass and silicon substrates, with an AT-cut quartz resonator embedded in a 70 μm depth microchannel (Fig. 2.20). The resonator, measuring 1.8×1.6 mm^2, is positioned between glass and silicon micropillars. A 5-nm chromium bonding layer and a 200-nm palladium hydrogen sensing material were deposited on one side of the resonator. The sensor chip underwent packaging via anodic bonding after resonator insertion, followed by a 2-min air-plasma treatment using a plasma cleaner (Zhou et al. 2021).A hydrogen sensor is proposed, featuring a SAW device coated with Pd/Ni nanowires (NWs) film. Gas adsorption in the film alters conductivity, influencing SAW velocity, with changes in phase serving as the sensing signal. A 200 MHz SAW device with a delay line pattern is fabricated on a Y35o X quartz substrate. Pd/Ni NW's enhanced specific surface area improves response time to less than 2 s. The sensor exhibits high sensitivity (1.65 mV/% in nitrogen gas concentration of 0.3–3.5 v/v %) and a low detection limit of 7 ppm at room temperature (25 °C) (Wang 2022).

A hydrogen (H_2) photoacoustic sensor has been developed to optimize resonant frequency sensitivity, leveraging distinctions between H_2 and air. Resonance frequency tracking is achieved through a field-programmable gate array (FPGA) device. The optimal photoacoustic (PA) cell includes a central cylindrical acoustic resonator (4 mm diameter, 60 mm length) and two buffer volumes (20 mm diameter, 30 mm length). Under specific conditions, the resonant frequency is 2452.36 Hz, with a Q-factor of approximately 120.8. The sensor shows versatile application potential,

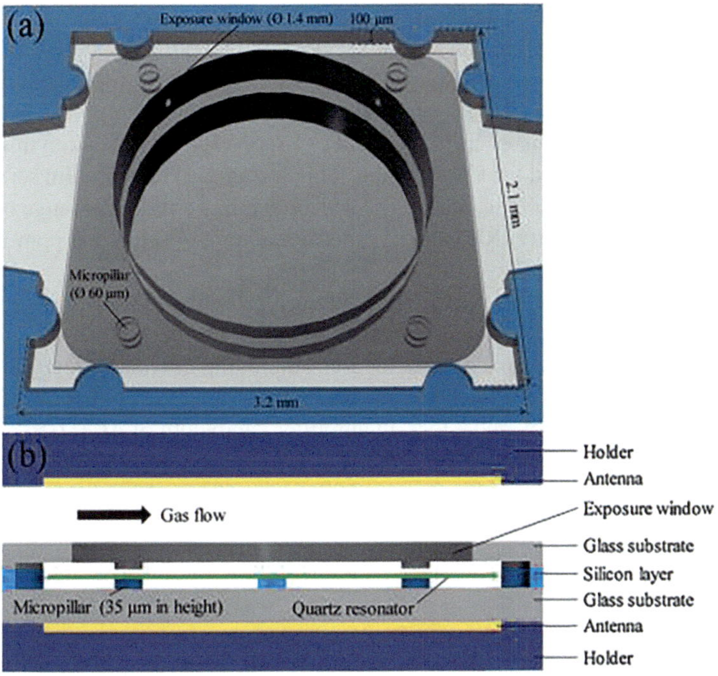

Fig. 2.20 Structure of MEMS-based hydrogen sensor chip **a** top view **b** bottom view (Zhou et al. 2021) (reprinted with permission)

detecting concentrations from 74 ppm to 100% at 1 atm pressure, and exhibits a response time of approximately 5 s (Wang et al. Oct. 2022).

Carbon Di-Oxide

CO_2 sensors are crucial for monitoring indoor air quality and environmental conditions, continuously measuring levels in real time to ensure health and safety in enclosed spaces. These sensors, essential for preventing discomfort and health issues associated with elevated CO_2 levels, also contribute to environmental research, aiding our understanding and mitigation of the broader impact of CO_2 emissions. In a sustainability-focused world, reliable CO_2 sensors are vital for fostering healthier indoor environments and addressing the implications of carbon emissions (Neethirajan et al. 2009).

A room-temperature CO_2 gas detection device has been developed using a QCM coated with a thin vanadium oxide film. The film, generated by depositing V_2O_5 powder onto AT-cut 6 MHz piezoelectric quartz crystals with Ag electrodes, exhibited increased sensitivity with longer annealing times, attributed to heightened surface roughness. The structure heated at 200 °C for 3 h demonstrated significant resonance frequency shift under a CO_2 pressure of 40 mbar, with a rapid response time (57 s), brief recovery time (43 s), and displayed good stability, linearity, reproducibility, and reversibility (Berouaken 2018).

A SAW based sensor utilizing Au/TiO_2/poly(3,4-ethylenedioxythiophene) (PEDOT) was developed for CO_2 sensing. This specialized sensing film served as a delay line of a 204 MHz (Fig. 2.21). The sensor exhibits a response range of 5% to 90% for relative humidity (RH) and a response range of 500 to 2000 ppm for CO_2 gas (Wang 2022).

An online CO_2 and CH_4 detection system utilized a photoacoustic setup with a sphere–tube coupled cell as the core unit (Fig. 2.22). The design incorporated time division multiplexing for mixed gas detection, employing distributed feedback lasers (DFB) at 1653 and 2004 nm as excitation sources. With an integration time of 300 ms, a filter slope of 12 dB/oct, and a lock-in amplifier bandwidth of $\Delta f = 0.833$ Hz, the system demonstrated no mutual interference among mixed gas photoacoustic signals. Minimum detection limits within 5 s were 509 ppb for CH_4 and 23 ppm for CO_2, with normalized noise equivalent absorption coefficients of 2.1×10^{-9} cm^{-1} $WHz^{-1/2}$ for CH_4 and 8.98×10^{-9} cm^{-1} $WHz^{-1/2}$ for CO_2 (Zhu et al. 2023).

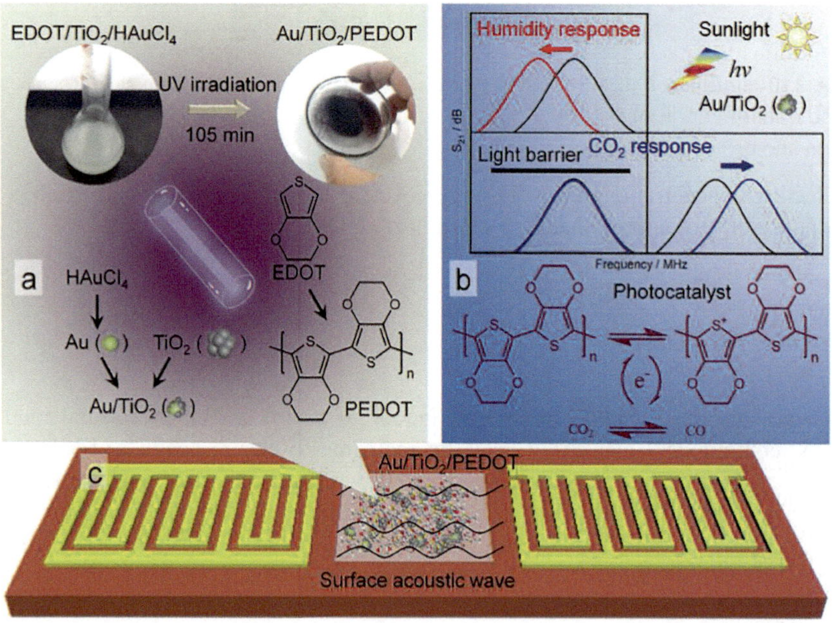

Fig. 2.21 Schematic of **a** fabrication process of Au/TiO_2/PEDOT sensing material, **b** sensing mechanism of CO_2 and humidity, and **c** Au/TiO_2/PEDOT-SAW sensor (Wang 2022) (reprinted with permission)

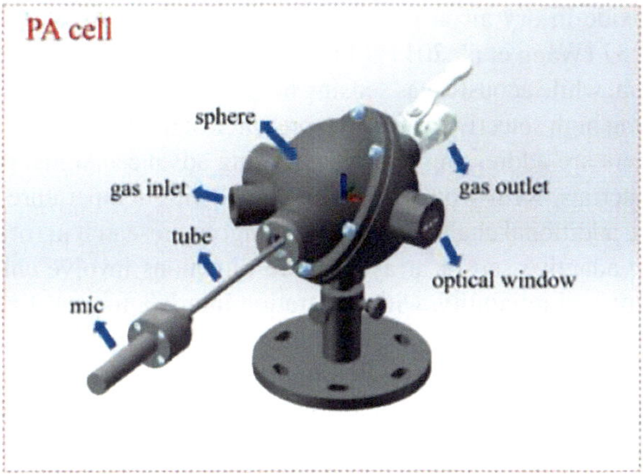

Fig. 2.22 Structure of the spherical photoacoustic cell for CO_2 detection (Zhu et al. 2023) (reprinted with permission)

Nitrogen Dioxide and Nitric Oxide

NOx sensors are vital for monitoring and addressing air quality challenges linked to emissions from vehicles, industries, and power plants, which contribute to environmental degradation and health risks. These sensors, crucial for real-time detection and quantification of NOx emissions, support regulatory efforts to reduce air pollution and protect public health in our era of heightened focus on sustainable and clean technologies (Srivastava 2021).

A gas sensor for detecting low concentrations of nitrogen dioxide (NO_2) was developed by modifying a quartz crystal microbalance (QCM) transducer with a synthetic graphene composite. The gas sensing characteristics of the QCM, coated with SnO_2-rGO and AgNPs-SnO_2-rGO composites, were examined under exposure to a low NO_2 concentration range (2.05–20.5 mg/m^3) at room temperature (Qi et al. 2016).

A nitric oxide (NO) gas sensor has been developed using a composite of indium tin oxide (ITO) and a quartz crystal microbalance. The sensor, created on a polished AT-cut quartz crystal, features an ITO thin film deposited through radio frequency magnetron sputtering. Experimental findings show a significant negative frequency shift of 110 Hz over 600 s when exposed to 60 ppm of NO (Hu et al. 2003).

Copper-ion-doped polyaniline/tungsten trioxide (Cu^{2+}/PANI/WO_3) films, featuring varying concentrations of Cu^{2+}, were applied as coatings on surface acoustic wave resonators to serve as nitric oxide sensors operating at room temperature. The sensing material underwent morphological characterization through scanning electron microscopy and functionalization via Fourier transform infrared spectroscopy. The optimal configuration was identified as a 0.2 weight percent Cu^{2+}/PANI/WO_3 film. The sensor exhibited a response of 165 Hz to the detection of 1 part per billion nitric oxide in dry air at room temperature, with a signal-to-noise ratio of approximately 37 (Wang et al. 2017) (Table 2.2).

In conclusion, while acoustic gas sensing holds promise, challenges exist, particularly in achieving high selectivity amid the presence of multiple gases, limiting accuracy. Researchers are addressing this by exploring advanced signal processing and specialized materials. Real-world conditions, including temperature and humidity variations, pose additional challenges, driving ongoing research in robust calibration techniques and adaptive sensor arrays. Future directions involve enhancing sensitivity, selectivity, and reliability, with integration into IoT for real-time monitoring. Novel materials and designs are being explored to overcome current limitations, making acoustic gas sensors a compelling choice for diverse applications, with the potential to revolutionize environmental monitoring for increased safety and protection.

Table 2.2 Some c-based gas sensors

Type of acoustic sensor	Analyte gas	LOD	References
Quartz crystal microbalance	Ethanol	0.018 ppm	Sukowati et al. (2023)
	Ammonia	2.8 ppm	Aflaha (2023)
	Hydrogen	15 ppm	Shokry et al. (2019)
	Carbon di oxide	80 ppm	Pang (2021)
	Nitrogen di oxide	50 ppm	Starbova et al. (2022)
	Sulphur di oxide	2.5 ppm	Thanh et al. (2022)
FBAR	Ethanol	100 ppm	Zhu et al. (2022)
	Ammonia	1 ppm	Liu et al. (2022)
	Carbon di oxide	50 ppm	Hoffmann et al. (2017)
	Nitrogen di oxide	300 ppm	Cheng et al. (2007)
	Sulphur di oxide	20 ppm	Barauskas et al. (2019)
SAW	Ethanol	15 ppm	Jahanshahi et al. (2018)
	Ammonia	40 ppb	Sun (2023)
	Hydrogen	1 ppm	Kim et al. (2022)
	Carbon di oxide	500 ppm	Wang (2022)
	Nitrogen di oxide	100 ppm	Sairam Pasupuleti et al. (2021)
	Sulphur di oxide	0.25 ppm	Yang et al. (2023a)
PAS	Ethanol	6.5 ppb	Pangerl (2023)
	Ammonia	3.2 ppb	Guo et al. (2019)
	Hydrogen	3.7 ppm	Ye (2023)
	Carbon di oxide	23 ppm	Zhu et al. (2023)
	Nitrogen di oxide	68 ppt	Li et al. (2022)
	Sulphur di oxide	1 ppm	El-Safoury (2021)

Application of Acoustic Devices for 5G/6G Communications

MEMS and NEMS have opened up new frontiers in technology, offering unprecedented miniaturization and precision. In the realm of 5G and the upcoming 6G communications, the integration of MEMS/NEMS-based acoustic devices marks a revolutionary leap. This subchapter explores the application of MEMS/NEMS in the development of acoustic devices, focusing on oscillators and filters for enhanced frequency stability and signal processing. As wireless communication progresses from 4 to 5G and anticipates the advent of 6G, the standards delineate an expanding array of frequency bands, depicted in Fig. 2.23 (Alraih 2022).

Filters for Signal Processing

5G mm-wave faces challenges due to high propagation losses and a lack of high-performance Radiofrequency Front-Ends (RFFE) hardware. Frequency Range 3 (7–20 GHz) emerges as an attractive alternative for 5G and 6G, but stringent bandwidth requirements call for innovative passive RFFE hardware, particularly band-pass filters. EM resonance-based filters have large footprints and are impractical for mobile handsets. Microstrip filters are integrated into commercial products, but they face issues like manufacturing variability, temperature sensitivity, and suboptimal band roll-off performance. State-of-the-art RFFE passive band-pass filters in sub-6 GHz band employ piezoelectric MEM resonators, which combine high-Q factors with compact footprints. MEM resonators exhibit mechanical resonance at specific frequencies, creating a standing acoustic wave in the device's body and ensuring minimal impedance across its terminals (Making_5G_Millimeter-Wave_Communications_a_Reality_Industry_Perspectives; Giribaldi et al. 2024).

The contemporary landscape of RF (Radio Frequency) filters predominantly revolves around two key implementations: Surface Acoustic Wave (SAW) filters and Bulk Acoustic Wave (BAW) filters. Figure 2.23 encapsulates a pivotal representation of micro-RF filter technologies, showcasing their evolution and significance. Among these, BAW filters stand out as a milestone due to their notable advantages in

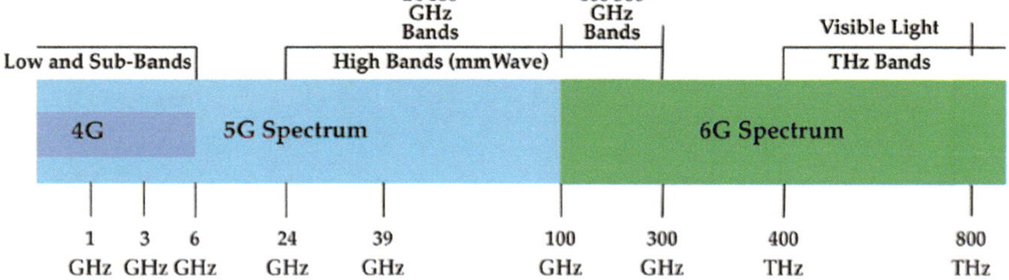

Fig. 2.23 Spectrum bands for 4G, 5G and 6G (Alraih 2022) (reprinted with permission)

operating frequency and power capacity when compared to SAW filters (Yang et al. 2023b).

BAW filters are widely recognized for their high-Q factor, which enables them to achieve steeper filter skirts and superior filtering performance. Their distinctive attributes of high acoustic velocity and thermal conductivity confer commendable power-handling capabilities. Furthermore, BAW filters exhibit low power consumption, high isolation, and compatibility with CMOS technology, solidifying their dominance as RF communication devices. The continuous advancement and deployment of BAW filters underscore their pivotal role in modern RF communication systems. Their unique amalgamation of high performance, power efficiency, and compatibility with semiconductor manufacturing processes positions BAW filters as essential components in the progression of wireless communication technologies (Liu et al. 2020).

A pivotal breakthrough occurred in 1999 when Agilent, an American company, successfully developed the first commercially viable film BAW duplexer marking a significant milestone in micro RF filter technologies. Following this success, several renowned international semiconductor companies, including Broadcom, Samsung, Qorvo, Taiyo, STMicroelectronics, EPCOS, and Fujitsu, have dedicated substantial efforts and resources to advance BAW filter technology (Liu et al. 2020) (Fig. 2.24).

The dominant segment in the Bulk Acoustic Wave (BAW) market is characterized by Thin-Film Bulk Acoustic Resonators (FBARs), which utilize Aluminum Nitride (AlN) or lightly-doped (LD) Scandium Aluminum Nitride (ScAlN) thin films as the piezoelectric layer. The use of these materials, particularly LD ScAlN, offers cost advantages compared to single-crystal alternatives and is compatible with post-Complementary Metal–Oxide–Semiconductor (CMOS) processes. Despite this cost-effectiveness, LD ScAlN thin films maintain a state-of-the-art resonator performance (Krishnaswamy et al. 1990; Schneider et al. 2017).

FBARs are notable for their combination of high quality factor (Q) and coupling coefficient (k). This distinctive feature allows for the creation of exceptionally

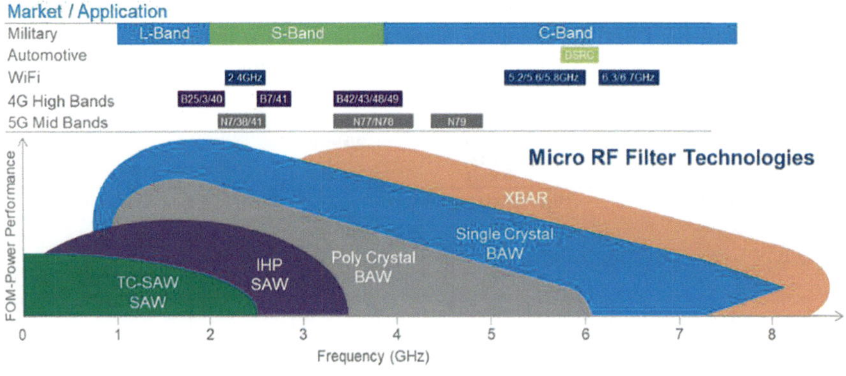

Fig. 2.24 Market application and band allocation of radio frequency (RF) filter technologies (Liu et al. 2020) (reprinted with permission)

discerning and minimally disruptive band-pass filters with low Insertion Loss (IL). The performance of FBAR overtones has been proven to reach 33 GHz, and microacoustic filters have effectively demonstrated their functionality beyond the 20 GHz mark (Mujahid et al. 2019).

Schaffer and colleagues devised an innovative Overmoded Bulk Acoustic Resonator (OBAR) MEMS resonator tailored for filtering applications in the 5G millimeter-wave spectrum. OBAR operates in the second overtone thickness mode, ensuring a uniform distribution of acoustic energy (refer to Fig. 2.25). It features a thicker piezo layer and metal electrodes compared to a fundamental mode resonator, making it suitable for manufacturing devices within the 30–60 GHz range. A Pt-AlN-Al device, measuring 14 μm × 14 μm (196 μm^2), demonstrated an electromechanical-coupling coefficient (k_2t) of 1.7% and a series resonance quality factor (Qs) of 110 at 33 GHz. To enhance OBAR performance, an optimized process for depositing very thin AlN on Pt layers has been developed (Schaffer et al. 2022).

Researchers have proposed a new technology platform for fabricating compact, wideband, and low-loss microacoustic filtering devices in the 6–20 GHz range using high-crystallinity 30% ScAlN piezoelectric thin films and a straightforward fabrication process. The platform is based on lateral field-excited (LFE) cross-sectional

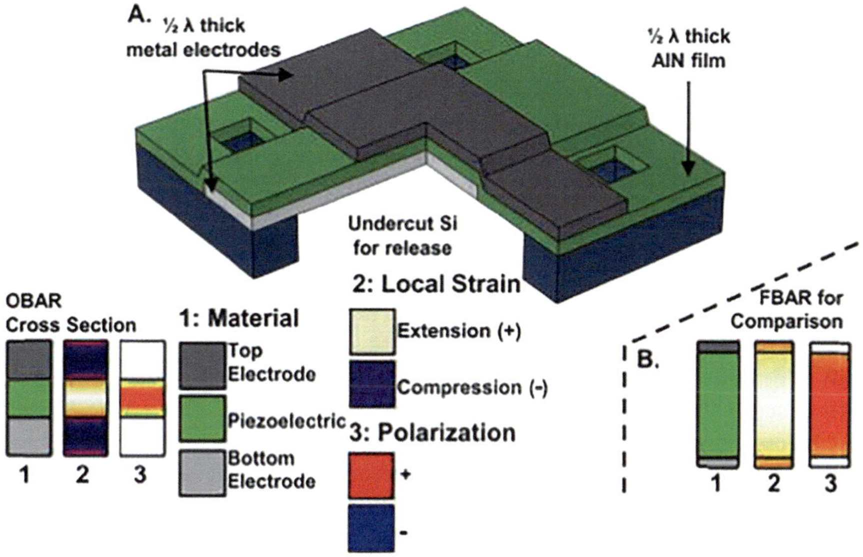

Fig. 2.25 **a** Isometric 3-D rendering of OBAR with cutaway cross-section views indicated by colors 1. material, 2. local strain, and 3. polarization. **b** FBAR cross-section views provided for comparison (Schaffer et al. 2022) (reprinted with permission)

Lame' mode resonators (CLMRs), which are fabricated using a three-mask post-CMOS compatible process. The platform has been demonstrated to produce microacoustic resonators with unprecedented performance metrics in the 6–11 and 14–20 GHz range. These resonators exhibit high Q factors (>1000) and high electromechanical coupling coefficients (>20%), surpassing the current state-of-the-art. The proposed technology platform has the potential to revolutionize the design and fabrication of RF front-ends for handheld devices. By enabling the synthesis of compact, wideband, and low-loss passive passband filters in the 6–20 GHz range, the platform can help to improve the performance and efficiency of next-generation 5G and 6G systems (Giribaldi et al. 2023).

Oscillators for Frequency Stability

MEMS-based acoustic oscillators excel in their ability to generate precise timing and frequency control, crucial for synchronizing and enabling efficient data transfer in 5G networks. Their miniature size facilitates integration into compact electronic systems, overcoming the spatial limitations of modern telecommunications devices. In addition, these oscillators exhibit superior frequency stability and reduced phase noise compared to traditional oscillator technologies, essential for fulfilling the demands of reliable and high-speed data transmission in 5G applications.

Piezoelectric materials are ideal for MEMS resonators because they allow for both electrical excitation and sensing of mechanical vibrations. Some common piezoelectric materials used in MEMS acoustic oscillators include quartz (SiO_2), aluminum nitride (AlN), lithium niobate ($LiNbO_3$), and gallium arsenide (GaAs) (Karim et al. 2014).

Kourani et al. developed a 12.9 GHz silicon germanium (SiGe) Pierce oscillator using a third antisymmetric overtone in a lithium niobate ($LiNbO_3$) resonator for 5G communications. The oscillator achieves a measured phase noise of −70 and −111 dBc/Hz at 1 and 100 kHz offsets from a 12.9 GHz carrier while consuming 20 mW of dc power. The oscillator achieves a figure-of-merit of 200 dBc/Hz at 100 kHz offset, surpassing the state-of-the-art electromagnetic (EM) and overtone MEMS oscillators. This work enabled low-power 5G transceivers featuring high speed, high sensitivity, and high selectivity in small form factors (Kourani et al. 2020).

Due to their high electromechanical coupling coefficient and the capability to cover a broad frequency range (1–10 GHz) with varying film thicknesses (ranging from 5 μm to 500 nm or less), MEMS filters based on thin films of Aluminum Nitride (AlN) show great potential for 5G communication applications (Pinto et al. 2022). Assylbekova and team successfully created the first experimental Lateral-Field-Excited (LFE) Aluminum Nitride (AlN) Cross-Sectional Lamé Mode Resonator (CLMR) that operates at 11 GHz (Fig. 2.26). The device, designed using Finite Element Analysis (FEA) and manufactured through a simple 2-mask process, exhibited a loaded quality factor (Ql) of 615, an electromechanical coupling coefficient (k_2t) of 1.3%, and a high FOM of 8. These characteristics make it a promising option

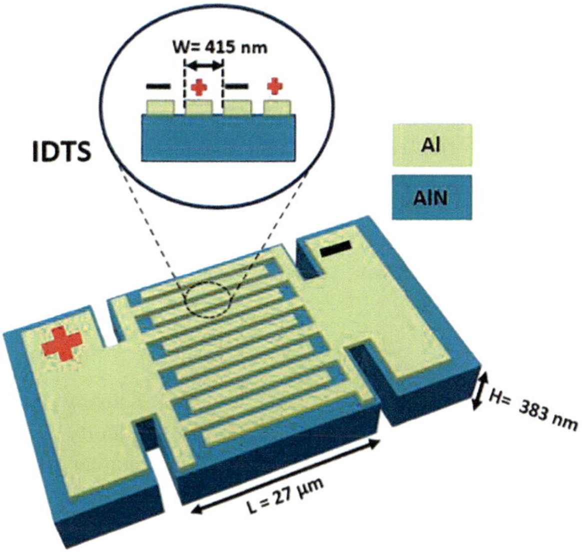

Fig. 2.26 Schematic of AlN-based Cross-Sectional Lamé Mode Resonator (CLMR) (adapted from Assylbekova et al. (2020))

for developing cost-effective, high-performance filters for X-band applications (Pillai and Li 2021; Assylbekova et al. 2020).

Challenges and Future Prospects

The exploration of MEMS/NEMS-based acoustic devices for 5G/6G communications reveals a landscape marked by challenges and promising prospects. The integration of these miniature systems, driven by advancements in MEMS/NEMS systems, has ushered in a new era of unprecedented miniaturization and precision. Challenges within the 5G mm-wave spectrum, such as high propagation losses and the need for advanced RFFE hardware, have spurred innovation in passive RFFE solutions, notably band-pass filters. MEM resonators, particularly those utilizing BAW filters with FBARs, play a crucial role in overcoming these challenges, showcasing a balance between performance and cost-effectiveness (Iannacci 2023).

However, the journey is not without obstacles. Challenges include the intricate manufacturing of MEMS/NEMS devices, sensitivity to environmental factors affecting long-term reliability, and cost considerations. These hurdles emphasize the necessity for continuous innovation and research to address fabrication complexities, enhance reliability, and reduce production costs, paving the way for widespread adoption. Looking to the future, technological platforms like lateral field-excited CLMRs can offer unprecedented performance metrics for compact, wideband,

and low-loss microacoustic filtering devices. In the realm of oscillators, MEMS-based acoustic oscillators show promise in providing precise timing and frequency control, crucial for meeting the demands of 5G networks. Despite formidable challenges, ongoing efforts and technological innovations position MEMS/NEMS-based acoustic devices as transformative components shaping the future of wireless communication technologies in the evolving landscape of 5G/6G communications (Wu et al. 2020).

Conclusion

In the investigation of "Acoustic Devices for Gas Sensing and RF Applications in MEMS/NEMS," the fusion of scientific ingenuity and technological progress unfolds a spectrum of opportunities. This chapter explores the intricate domain where acoustic devices play a dual role, tackling challenges in gas sensing and contributing to the advancement of Radio Frequency (RF) applications in Micro-Electro-Mechanical Systems (MEMS) and Nano-Electro-Mechanical Systems (NEMS).

As we traverse the diverse landscape of gas sensing, the discourse reveals the distinctive attributes of acoustic devices at the forefront of innovation. From their non-intrusive characteristics to the versatility in detecting various gases, acoustic devices emerge as precise and reliable tools in environmental monitoring and industrial safety. The examination of fabrication techniques and materials highlights the miniaturization capabilities of MEMS and NEMS, ushering in an era where gas-sensing devices seamlessly integrate into portable, wearable, and efficient solutions.

Concurrently, the chapter explores the realm of RF applications, where the seamless integration of acoustic devices with MEMS and NEMS technologies heralds a new era. The oscillators ensuring frequency stability and filters facilitating signal processing stand as evidence of the transformative potential of this integration. MEMS-based acoustic oscillators redefine the paradigm with unmatched stability and low phase noise, addressing the challenges of high-frequency domains inherent in 5G and anticipated in 6G communication technologies. The adaptability of MEMS technology in crafting intricate structures enables the design of filters tailored to the specific requirements of the millimeter-wave spectrum, contributing to enhanced selectivity and reduced signal distortion.

In concluding this chapter, the resonance of acoustic devices in gas sensing and RF applications within the MEMS/NEMS domain signifies not just a scientific breakthrough but a synergistic collaboration pushing the boundaries of possibility. The harmonization of innovation in these acoustic devices indicates a promising trajectory, where environmental monitoring, industrial safety, and wireless communication converge in a crescendo of advancements. As we wrap up this exploration, the echoes of these breakthroughs reverberate, hinting at a future where the symphony of acoustic devices continues to compose the melody of progress in MEMS/NEMS applications.

Competing Interests The authors have no conflicts of interest to declare that are relevant to the content of this chapter.

References

"Making_5G_Millimeter-Wave_Communications_a_Reality_Industry_Perspectives".

Monolithic integrated SAW filter based on AlN for high-frequency applications (2013) Semicond Sci Technol 28:7. https://doi.org/10.1088/0268-1242/28/6/065013

Aflaha R (2023) Improving ammonia sensing performance of quartz crystal microbalance (QCM) coated with nanofibers and polyaniline (PANi) overlay. Biosens Bioelectron X 13(2023). https://doi.org/10.1016/j.biosx.2022.100300

Alanazi N, Almutairi M, Alodhayb AN (2023) A review of quartz crystal microbalance for chemical and biological sensing applications. Sens Imaging 24(1). https://doi.org/10.1007/s11220-023-00413-w

Ali WR, Prasad M (2020) Piezoelectric MEMS based acoustic sensors: a review. Sens Actuators, A: Phys 301. Elsevier B.V. https://doi.org/10.1016/j.sna.2019.111756.

Alraih S (2022) Revolution or evolution? technical requirements and considerations towards 6G mobile communications. Sensors 22(3). MDPI. https://doi.org/10.3390/s22030762

Alton Everest F (2001) Master handbook of acoustics, 4 th edn. McGraw-Hill

Assylbekova M, Chen G, Michetti G, Pirro M, Colombo L, Rinaldi M (2020) 11 GHz lateral-field-excited aluminum nitride cross-sectional Lamé mode resonator. In: IFCS-ISAF 2020—joint conference of the IEEE international frequency control symposium and IEEE international symposium on applications of ferroelectrics, proceedings. Institute of Electrical and Electronics Engineers Inc.. https://doi.org/10.1109/IFCS-ISAF41089.2020.9234874

Athanasopoulos GA, Pelekis PC, Anagnostopoulos GA (2000) Effect of soil stiffness in the attenuation of Rayleigh-wave motions from field measurements. Soil Dyn Earthq Eng 19(4):277–288. https://doi.org/10.1016/S0267-7261(00)00009-9

Auld BA (1973) Acoustic fields and waves in solids. Wiley, United States

Barauskas D, Pelenis D, Vanagas G, Viržonis D, Baltrušaitis J (2019) Methylated poly(Ethylene)imine modified capacitive micromachined ultrasonic transducer for measurements of CO_2 and SO_2 in their mixtures. Sensors (Switzerland) 19(14). https://doi.org/10.3390/s19143236

Bean A (2009) Engineering acoustics. Global Media

Berouaken M (2018) Quartz crystal microbalance coated with vanadium oxide thin film for CO_2 gas sensor at room temperature. Arab J Sci Eng 43:5957–5963. https://doi.org/10.1007/s13369-018-3153-y

Cheng, H, Qin L, Wang QM (2007) P1D-2 high temperature langasite BAW gas sensor based on ZnO nanowire arrays. In: 2007 IEEE ultrasonics symposium proceedings, New York, NY, USA, 1361–1364. https://doi.org/10.1109/ULTSYM.2007.342

Dhall S, Mehta BR, Tyagi AK, Sood K (2021) A review on environmental gas sensors: materials and technologies. Sens Int 2:100116. https://doi.org/10.1016/j.sintl.2021.100116

Ding B, Kim J, Miyazaki Y, Shiratori S (2004) Electrospun nanofibrous membranes coated quartz crystal microbalance as gas sensor for NH_3 detection. Sens Actuators B Chem 101(3):373–380. https://doi.org/10.1016/j.snb.2004.04.008

Dissanayake D, Al-Sarawi S, Abbott D (2010) Surface acoustic wave based wireless MEMS actuators for biomedical applications. In: Acoustic waves Sciyo. https://doi.org/10.5772/9713

Dixon MC (2008) F R A B A brief background and history of QcM quartz crystal microbalance with dissipation monitoring: enabling real-time characterization of biological materials and their interactions. J Biomol Tech 18:151–158. https://doi.org/10.1002/cm.21752100

El-Safoury M (2021) On-board monitoring of so_2 ship emissions using resonant photoacoustic gas detection in the uv range. Sensors 21(13). https://doi.org/10.3390/s21134468

Fflckernell FS (2000) Thin-films for saw devices. Int J High Speed Electron Syst 10(3). https://doi.org/10.1142/S012915640000060X

Fogel R, Limson J, Seshia AA (2016) Acoustic biosensors. Essays Biochem 60(1):101–110. https://doi.org/10.1042/EBC20150011

Giffney TJ, Ng YH, Aw KC (2012) A surface acoustic wave ethanol sensor with zinc oxide nanorods. Smart Mater Res 2012:1–4. https://doi.org/10.1155/2012/210748

Giribaldi G, Colombo L, Rinaldi M (2023) 6–20 GHz 30% ScAlN lateral field-excited cross-sectional lame' mode resonators for future mobile RF front-ends. IEEE Trans Ultrason Ferroelectr Freq Control 1–1. https://doi.org/10.1109/tuffc.2023.3312913

Giribaldi G, Colombo L, Simeoni P, Rinaldi M (2024) Compact and wideband nanoacoustic pass-band for future 5G and 6G cellular radios. https://doi.org/10.21203/rs.3.rs-2569732/v1

Guo M, Chen K, Gong Z, Yu Q (2019) Trace ammonia detection based on near-infrared fiber-optic cantilever-enhanced photoacoustic spectroscopy. Photonic Sens 9(4):293–301. https://doi.org/10.1007/s13320-019-0545-x

Haisch C (2012) Photoacoustic spectroscopy for analytical measurements. Measur Sci Technol 23(1). Institute of Physics Publishing. https://doi.org/10.1088/0957-0233/23/1/012001

Hoffmann R, Schreiter M, Heitmann J (2017) The concept of thin film bulk acoustic resonators as selective CO_2 gas sensors. J Sens Sens Syst 6(1):87–96. https://doi.org/10.5194/JSSS-6-87-2017

Hu J, Zhu F, Zhang J, Gong H (2003) A room temperature indium tin oxide/quartz crystal microbalance gas sensor for nitric oxide. https://doi.org/10.1016/S0925-4005(03)00186-2

Huang X, Bai Q, Hu J, Hou D (2017) A practical model of quartz crystal microbalance in actual applications. https://doi.org/10.3390/s17081785

Hübert T, Boon-Brett L, Black G, Banach U (2011) Hydrogen sensors—a review. Sens Actuators B 157:329–352. https://doi.org/10.1016/j.snb.2011.04.070

Iannacci J (2023) A perspective vision of micro/nano systems and technologies as enablers of 6G, super-IoT, and tactile internet. Proc IEEE 111(1):5–18. https://doi.org/10.1109/JPROC.2022.3223791

Ippolito SJ, Trinchi A, Powell DA, Wlodarski W (2009) Acoustic wave gas and vapor sensors. In: Solid state gas sensing. Springer, US, pp 1–44. https://doi.org/10.1007/978-0-387-09665-0_8

Jackson W, Amer NM (1980) Piezoelectric photoacoustic detection: theory and experiment. J Appl Phys 51(6):3343–3353. https://doi.org/10.1063/1.328045

Jahanshahi P, Wei Q, Jie Z, Zalnezhad E (2018) Designing a non-invasive surface acoustic resonator for ultra-high sensitive ethanol detection for an on-the-spot health monitoring system. Biotechnol Bioprocess Eng 23:394–404. https://doi.org/10.1007/s12257-017-0432-5

Karim J, Zahirul Alam AHM, Nurashikin Nordin A (2014) MEMS-based oscillators: a review

Kikuchi M, Shiratori S (2005) Quartz crystal microbalance (QCM) sensor for CH_3SH gas by using polyelectrolyte-coated sol-gel film. Sens Actuators B 108:564–571. https://doi.org/10.1016/j.snb.2004.12.122

Kim S, Singh G, Lee K, Kim S, Singh G, Lee K (2022) 2200180 (1 of 13) development of highly sensitive and stable surface acoustic wave-based hydrogen sensor and its interface electronics. https://doi.org/10.1002/admt.202200180

Kinsler L (2000) Fundamentals of acoustics, Wiley

Kourani A, Yang Y, Gong S (2020) A Ku-band oscillator utilizing overtone lithium niobate RF-MEMS resonator for 5G. IEEE Microw Wirel Compon Lett 30(7):681–684. https://doi.org/10.1109/LMWC.2020.2996961

Krishnaswamy SV, Rosenbaum J, Borwlitz S, Vale C, Moore RA (1990) Film bulk acoustic wave resonator technology. Ultrason Symp

Lee SH, Yoon KH, Lee JK (2002) Influence of electrode configurations on the quality factor and piezoelectric coupling constant of solidly mounted bulk acoustic wave resonators. J Appl Phys 92(7):4062–4069. https://doi.org/10.1063/1.1505977

Lee JB, Jung JP, Lee MH, Park JS (2004) Effects of bottom electrodes on the orientation of AlN films and the frequency responses of resonators in AlN-based FBARs. Thin Solid Films 610–614. https://doi.org/10.1016/j.tsf.2003.07.023

Lei L, Dong B, Hu Y, Lei Y, Wang Z, Ruan S (2022) High-frequency surface acoustic wave resonator with diamond/AlN/IDT/AlN/diamond multilayer structure. Sensors 22:6479. https://doi.org/10.3390/s22176479

Li Z, Liu J, Si G, Ning Z, Fang Y, Pan Y (2022) Optimization design of sphere-tube-coupled photoacoustic cell and application of nitrogen dioxide detection. https://doi.org/10.1002/mop.33321

Liang J, Huang J, Zhang T, Zhang J, Li X, Ueda T (2013) An experimental study on fabricating an inverted mesa-type quartz crystal resonator using a cheap wet etching process. Sensors (Switzerland) 13(9):12140–12148. https://doi.org/10.3390/s130912140

Liu WK, Tay KW, Kuo SC, Wu MJ (2009) Fabrication of piezoelectric AlN thin film for FBARs. Sci China, Ser G 52(2):226–232. https://doi.org/10.1007/s11433-009-0021-5

Liu Y, Cai Y, Zhang Y, Tovstopyat A, Liu S, Sun C (2020) Materials, design, and characteristics of bulk acoustic wave resonator: a review. Micromachines 11(7). MDPI AG. https://doi.org/10.3390/mi11070630

Liu Z, Lotfi A, Shamami MH, Hesketh PJ, Ayazi F (2022) A high-order membrane resonator array for gravimetric ammonia sensing with sub-ppm limit of detection. https://doi.org/10.1109/MEMS51670.2022.9699741

Long G (2020) Surface acoustic wave ammonia sensor based on ZnS mucosal-like nanostructures. Microelectron Eng 222. https://doi.org/10.1016/j.mee.2019.111201

Lurton X (2002) An introduction to underwater acoustics: principles and applications. Springer, London

Mandal D, Banerjee S (2022) Surface AcousticWave (SAW) sensors: physics, materials, and applications. Sensors 22(3). https://doi.org/10.3390/s22030820

Mandelis A, Peralta SB, Thoen J (1991) Photoacoustic frequency-domain depth profiling of continuously inhomogeneous condensed phases: theory and simulations for the inverse problem. J Appl Phys 70(3):1761–1770. https://doi.org/10.1063/1.349517

Mcdonald FA (1981) Theory of photoacoustic signal generation for optimized photoacoustic cells articles you may be interested in. J Appl Phys 52:1462. https://doi.org/10.1063/1.329782

Miekisch W, Schubert JK, Noeldge-Schomburg GFE (2004) Diagnostic potential of breath analysis—focus on volatile organic compounds. Clin Chim Acta 347(1–2):25–39. https://doi.org/10.1016/J.CCCN.2004.04.023

Mujahid A, Afzal A, Dickert FL (2019) An overview of high frequency acoustic sensors—QCMs, SAWs and FBARs—chemical and biochemical applications. Sensors (Switzerland) 19(20). MDPI AG. https://doi.org/10.3390/s19204395

Neethirajan S, Jayas DS, Sadistap S (2009) Carbon dioxide (CO_2) sensors for the agri-food industry—a review. https://doi.org/10.1007/s11947-008-0154-y

O'Sullivan CK, Guilbault GG (1999) Commercial quartz crystal microbalances—theory and applications. Biosens Bioelectron 14:663–670. https://doi.org/10.1016/S0956-5663

Omori T, Hashimoto K, Yamaguchi M (2001) PZT thin films for SAW and BAW devices. In: International symposium on acoustic wave devices for future mobile communication systems

Palzer S (2020) Photoacoustic-based gas sensing: a review. Sensors (Switzerland) 20(9). MDPI AG. https://doi.org/10.3390/s20092745

Pang J (2021) Highly precision carbon dioxide acoustic wave sensor with minimized humidity interference. Sens Actuators B Chem 338:129824. https://doi.org/10.1016/j.snb.2021.129824

Pangerl J (2023) A sub-ppbv-level acetone and ethanol quantum cascade laser based photoacoustic sensor—characterization and multi-component spectra recording in synthetic breath. Photoacoustics 30. https://doi.org/10.1016/j.pacs.2023.100473

Penza M, Antolini F, Vittori-Antisari M (2005) Carbon nanotubes-based surface acoustic waves oscillating sensor for vapour detection. Thin Solid Films 472(1–2):246–252. https://doi.org/10.1016/J.TSF.2004.07.071

Pillai G, Li SS (2021) Piezoelectric MEMS resonators: a review. IEEE Sens J 21(11):12589–12605. Institute of Electrical and Electronics Engineers Inc. https://doi.org/10.1109/JSEN.2020.3039052

Pinto RMR, Gund V, Dias RA, Nagaraja KK, Vinayakumar KB (2022) CMOS-integrated aluminum nitride MEMS: a review. J Microelectromech Syst 31(4):500–523. Institute of Electrical and Electronics Engineers Inc. https://doi.org/10.1109/JMEMS.2022.3172766

Qi P, Wang Z, Wang R, Xu Y, Zhang T (2016) Studies on QCM-type NO_2 gas sensor based on graphene composites at room temperature. Chem Res Chin Univ 32(6):924–928. https://doi.org/10.1007/s40242-016-6129-z

Van Quy N, Anh Minh V, Van Luan N, Ngoc Hung V, Van Hieu N (2011) Gas sensing properties at room temperature of a quartz crystal microbalance coated with ZnO nanorods. Sens Actuators B 153(1):188–193. https://doi.org/10.1016/j.snb.2010.10.030

Razansky D, Manohar S (2016) Photoacoustics: a historical review. Adv Opt Photon 8(4):586–617. https://doi.org/10.1364/AOP.8.000586

Rosencwaig A (1978) Photoacoustic spectroscopy. Adv Electron Electron Phys 46(C):207–311. https://doi.org/10.1016/S0065-2539(08)60413-8

Rossing T (2007) A brief history of acoustics. In: Springer handbooks, pp 9–24. https://doi.org/10.1007/978-0-387-30425-0_2/COVER

Rumchev K, Brown H, Spickett J (2007) Volatile organic compounds: do they present a risk to our health? Rev Environ Health 22(1):39–55. https://doi.org/10.1515/REVEH.2007.22.1.39/MACHINEREADABLECITATION/RIS

Sairam Pasupuleti K et al (2021) High performance langasite based SAW NO 2 gas sensor using 2D g-C 3 N 4 @TiO 2 hybrid nanocomposite. https://doi.org/10.1016/j.jhazmat.2021.128174

Schaffer Z, Simeoni P, Piazza G (2022) 33 GHz overmoded bulk acoustic resonator. IEEE Microw Wirel Compon Lett 32(6):656–659. https://doi.org/10.1109/LMWC.2022.3166682

Schmid T (2006) Photoacoustic spectroscopy for process analysis. Anal Bioanal Chem 384(5):1071–1086. https://doi.org/10.1007/s00216-005-3281-6

Schneider M, DeMiguel-Ramos M, Flewitt AJ, Iborra E, Schmid U (2017) Scandium aluminium nitride-based film bulk acoustic resonators. In: MDPI AG, p 305. https://doi.org/10.3390/proceedings1040305

Shokry A, Khalil MMA, Ibrahim H, Soliman M, Ebrahim S (2019) Highly luminescent ternary nanocomposite of polyaniline, silver nanoparticles and graphene oxide quantum dots. Sci Rep 9(1). https://doi.org/10.1038/s41598-019-53584-6

Sigrist MW (1998) Laser generation of acoustic waves in liquids and gases. J Appl Phys 60(7):R83. https://doi.org/10.1063/1.337089

Smith WR, Gerard HM, Collins JH, Reeder TM, Shaw HJ (1969) Analysis of interdigital surface wave transducers by use of an equivalent circuit model. IEEE Trans Microw Theory Tech 17. https://doi.org/10.1109/TMTT.1969.1127075

Srivastava S (2021) Study of gas sensor detection for NOX Gas: a review. Mater Today Proceed 37:3709–3712. https://doi.org/10.1016/j.matpr.2020.10.161

Starbova K, Georgieva V, Starbov N, Stefanov P, Georgieva B, Lazarov J (2022) Quartz crystal microbalance sensor for NO2 detection based on electrospun amorphous titanium oxide fibers. J Phys Conf Ser 2240(1):012017. https://doi.org/10.1088/1742-6596/2240/1/012017

Strutt JW (2011) The theory of sound. Theory Sound 9781108032209:1–326. https://doi.org/10.1017/CBO9781139058087

Sukowati R, Rohman YM, Agung BH, Hapidin DA, Damayanti H, Khairurrijal K (2023). An investigation of the influence of nanofibers morphology on the performance of QCM-based ethanol vapor sensor utilizing polyvinylpyrrolidone nanofibers active layer. Sens Actuators B Chem 386. https://doi.org/10.1016/j.snb.2023.133708

Sun X (2023) Enhanced sensitivity of SAW based ammonia sensor employing GO-SnO_2 nanocomposites. Sens Actuators B Chem 375. https://doi.org/10.1016/j.snb.2022.132884

Tam AC (1986) Applications of photoacoustic sensing techniques. Rev Mod Phys 58:381–431. https://doi.org/10.1103/RevModPhys.58.381

Thakur SN, Rai VN, Singh JP (2023) Physics and techniques of photoacoustic spectroscopy of liquids. In: Photoacoustic and photothermal spectroscopy: principles and applications, pp 51–68. https://doi.org/10.1016/B978-0-323-91732-2.00006-9

Thanh VN, Tuan TQ, Van Cuong N, Truong CX, Van Quy N (2022) Application of an artificial neural network and QCM sensor coated with γ-Fe2O3 nanoparticles for estimation of SO2 gas sensing characteristics. J Sci Transp Technol 2(1):22–31. https://doi.org/10.58845/JSTT.UTT.2022.EN.2.1.22-31

Triyana K (2018) Chitosan-based quartz crystal microbalance for alcohol sensing. Electronics (Switzerland) 7(9). https://doi.org/10.3390/electronics7090181

Uchino K (2017) The development of piezoelectric materials and the new perspective. In: Advanced piezoelectric materials, pp 1–92. https://doi.org/10.1016/b978-0-08-102135-4.00001-1

Wang X (2022) Pd/Ni nanowire film coated SAW hydrogen sensor with fast response. Sens Actuators B Chem 351:130952. https://doi.org/10.1016/J.SNB.2021.130952

Wang C (2022) Surface acoustic wave sensor based on Au/TiO_2/PEDOT with dual response to carbon dioxide and humidity. Anal Chim Acta 1190:339264. https://doi.org/10.1016/j.aca.2021.339264

Wang X, Cui F, Lin J, Ding B, Yu J, Al-Deyab SS (2012) Functionalized nanoporous TiO_2 fibers on quartz crystal microbalance platform for formaldehyde sensor. Sens Actuators B Chem 171–172:658–665. https://doi.org/10.1016/j.snb.2012.05.050

Wang S-H, Shen C-Y, Lien Z-J, Wang J-H (2017) Nitric oxide sensing properties of a surface acoustic wave sensor with copper-ion-doped polyaniline/tungsten oxide nanocomposite film. Sens Actuators B 243:1075–1082. https://doi.org/10.1016/j.snb.2016.12.101

Wang J, Chen M, Chen Q, Wang H (2022) A resonant photoacoustic cell for hydrogen gas detection. Int J Hydrogen Energy 47(84):35940–35946. https://doi.org/10.1016/J.IJHYDENE.2022.08.166

White RM, Voltmer FW (1965) Direct piezoelectric coupling to surface elastic waves. Appl Phys Lett 7(12):314–316. https://doi.org/10.1063/1.1754276

Woo J (2002) A short history of the development of ultrasound in obstetrics and gynecology. Hist Ultrasound Obstet Gynecol 3:1–25

Wu G, Xu J, Ng EJ, Chen W (2020) MEMS resonators for frequency reference and timing applications. J Microelectromech Syst 29(5):1137–1166. https://doi.org/10.1109/JMEMS.2020.3020787

Xu X-H, Wu, Hai-Shun, Zhang C, Jin Z (2001) Morphological properties of AlN piezoelectric thin films deposited by DC reactive magnetron sputtering. Thin Solid Films 388:62–67. https://doi.org/10.1016/S0040-6090(00)01914-3

Yang J, Wang T, Zhu C, Yin X, Dong P, Wu X (2023a) AgNWs@SnO2/CuO nanocomposites for ultra-sensitive SO_2 sensing based on surface acoustic wave with frequency-resistance dual-signal display. Sens Actuators B Chem 375. https://doi.org/10.1016/j.snb.2022.132966

Yang Y, Dejous C, Hallil H (2023b) Trends and applications of surface and bulk acoustic wave devices: a review. Micromachines 14(1). MDPI. https://doi.org/10.3390/mi14010043

Ye W (2023) Photoacoustic dual-gas sensor for simultaneous detection of hydrogen and water vapor. https://doi.org/10.1016/j.ijhydene.2023.01.374

Yunusa Z, Hamidon MN, Kaiser A, Awang Z (2014) Gas sensors: a review. Sens Transducers 168:61–75

Zeng G (2023) Detection and discrimination of volatile organic compounds using a single film bulk acoustic wave resonator with temperature modulation as a multiparameter virtual sensor array. 29:41. https://doi.org/10.1021/acssensors.8b01678

Zhang G (2022) Bulk and surface acoustic waves: fundamentals, devices, and applications, 1st edn. Jenny Stanford Publishing. https://doi.org/10.1201/9781003256625

Zhou L, Kato F, Nakamura N, Oshikane Y, Nagakubo A, Ogi H (2021) MEMS hydrogen gas sensor with wireless quartz crystal resonator. Sens Actuators B Chem 334:129651. https://doi.org/10.1016/J.SNB.2021.129651

Zhu Z, Li Z, Liu J, Fang Y (2023) Methane and carbon dioxide mixed gas detection based on sphere-tube coupled photoacoustic cell. Opt Commun 527:128977. https://doi.org/10.1016/j.optcom.2022.128977

Zhu J, Feng M, Lian G (2022) a film bulk acoustic resonator-based sensor with AlN piezoelectric material for detecting ethanol and acetone. Shock Vib 2022. https://doi.org/10.1155/2022/8620943

Part II
Micro-Nano Systems for Healthcare

Chapter 3
MEMS in Twenty-First-Century Healthcare

Tanzila Noushin, Mohammad Salman Parvez, Prosanto Biswas, Md. Emran Hossain Bhuiyan, Md. Abdullah al Rumon, and Shawana Tabassum

Abstract In the twenty-first century, Micro-Electro-Mechanical Systems (MEMS) have emerged as transformative technologies with profound implications for healthcare. These miniature devices, operating at the micro scale, offer unprecedented capabilities in monitoring, diagnosing, and treating medical conditions. This chapter delves into the multifaceted integration of MEMS in healthcare, highlighting their roles in wearables, biosensors, lab-on-a-chip devices, organ-on-a-chip devices, and artificial implants, thereby facilitating personalized medicine and remote patient monitoring. By enabling real-time data collection, enhancing precision in interventions, and enabling novel therapeutic approaches, MEMS are reshaping the landscape of healthcare, promising more effective and patient-centric outcomes. This chapter discusses different types of MEMS devices and the opportunities presented by these cutting-edge technologies.

Keywords MEMS · Wearables · Biosensors · Lab-on-a-chip · Organ-on-a-chip · Implants · Precision medicine

S. Tabassum (✉)
Department of Electrical and Computer Engineering, The University of Texas at Tyler, Tyler, TX, USA
e-mail: stabassum@uttyler.edu

T. Noushin · M. S. Parvez · P. Biswas · Md. E. H. Bhuiyan
Department of Electrical and Computer Engineering, The University of Texas at Dallas, Richardson, TX, USA

Md. A. Rumon
Department of Electrical, Computer and Biomedical Engineering, University of Rhode Island, Kingston, RI, USA

V. Pachkawade and K. Guha (eds.), *Micro- and Nano-Systems in 21st-Century*, Springer Tracts in Electrical and Electronics Engineering,
https://doi.org/10.1007/978-981-96-4659-3_3

Introduction

In the twenty-first century, the realm of healthcare has experienced a remarkable technological metamorphosis, with Micro-Electro-Mechanical Systems (MEMS) devices playing a pivotal role in shaping the future of medical practice. MEMS devices represent a fusion of microelectronics and mechanical engineering, resulting in miniature systems that possess an incredible range of capabilities (Mishra et al. 2019; Yang and Zhang 2021; Bogue 2013). Their integration into the healthcare landscape has ushered in a new era of personalized medicine, precise diagnostics, and innovative therapeutic interventions. MEMS devices have made significant contributions to diagnostic medicine. They have empowered healthcare providers with the ability to conduct rapid and highly accurate tests, often at the point of care. These devices can detect a wide array of biological markers, from glucose levels in diabetics (Ropmay et al. 2020; Huang et al. 2009a, 2013) to specific disease biomarkers in cancer patients (Hsiao et al. 2022; Gogoi et al. 2022; Kurmendra and Kumar 2019). By enabling early diagnosis and personalized treatment, MEMS devices are instrumental in improving patient outcomes and reducing the burden of chronic diseases.

Patient monitoring has also witnessed a revolution, thanks to MEMS technology. Wearable MEMS sensors and devices allow continuous real-time tracking of vital signs, such as heart rate (Kaisti et al. 2019; Nguyen and Ichiki 2019), blood pressure (Sun et al. 2018; Sun and Chang 2023), and oxygen saturation (Davies et al. 2020; Yang et al. 2015). These innovations provide both patients and healthcare professionals with a wealth of data for better-informed decisions and early intervention. Continuous monitoring can detect subtle changes in a patient's condition, enabling timely responses and enhancing patient safety.

The promise of MEMS devices extends to drug delivery systems, where precise, targeted treatments are becoming increasingly feasible (Cobo et al. 2015; Lee et al. Mar. 2018). MEMS-based drug delivery systems can control the release of medications with exceptional accuracy, ensuring that therapeutic agents are delivered precisely where and when they are needed. This approach not only enhances treatment effectiveness but also minimizes side effects, leading to improved patient comfort and adherence to therapy. MEMS-based drug delivery systems are particularly promising for chronic diseases, where long-term medication management is required (Trani et al. Jun. 2019; Jiang et al. 2022). By offering a new level of control over dosing and timing, these devices are set to revolutionize the way we approach drug delivery, fostering a future of personalized and patient-centric medicine.

Furthermore, MEMS devices have made significant inroads into the burgeoning field of telemedicine and remote healthcare (Ma et al. 2011; Osiadala and Kolakowski 2020; Digital MEMS 2023). These tiny yet sophisticated devices allow patients to monitor their health conditions at home while sharing critical data with healthcare providers in real time. Whether it's tracking chronic conditions, monitoring post-surgery recovery, or providing immediate care during emergencies, MEMS devices are at the forefront of this revolution (Ciuti et al. 2015). The ability to obtain critical medical data without the need for physical hospital visits opens up new possibilities for accessible and efficient healthcare. Telemedicine, empowered by MEMS, has the potential to reach underserved populations and improve access to medical expertise, even in remote areas.

MEMS technology, with its versatility and miniaturization, has brought about a paradigm shift in healthcare research. Laboratories can now employ MEMS-based lab-on-a-chip systems (Podbiel et al. 2020), which offer a high degree of automation, precision, and the ability to conduct multiple experiments simultaneously. This has streamlined the process of drug discovery, genomics research, and disease modeling, accelerating the pace of medical breakthroughs. MEMS technology is making a significant impact in the development of organ-on-chip devices (Koyilot et al. 2022; Leary et al. 2013), a cutting-edge approach in biomedical research. These devices, inspired by the structure and function of human organs, are instrumental in drug testing, disease modeling, and personalized medicine. MEMS-based sensors, pumps, and actuators are at the heart of these platforms, recreating the mechanical microenvironments and fluidic conditions of human organs (Hajare et al. 2022; Algamili et al. 2021). This innovation allows scientists to closely mimic human physiological responses on a miniature scale. MEMS devices ensure precise control over fluid flow, mechanical strain, and the exchange of nutrients and waste, enabling detailed studies of organ function and response to drugs. As a result, organ-on-chip devices are revolutionizing drug development, toxicology testing, and the understanding of complex diseases, offering the potential to reduce animal testing and deliver more effective therapies to patients. The synergy between MEMS technology and organ-on-chip systems is pushing the boundaries of biomedical research and propelling us closer to transformative breakthroughs in healthcare.

As the twenty-first-century healthcare landscape evolves, MEMS devices will continue to play a pivotal role in enhancing patient care, optimizing therapeutic interventions, and contributing to groundbreaking medical research. The integration of MEMS technology into healthcare practice heralds a future characterized by precision, personalization, and transformative advancements in medical science and patient well-being. This chapter discusses different types of MEMS devices including wearables, biosensors, lab-on-a-chip devices, organ-on-a-chip devices, and implantable devices. Each section provides a detailed understanding of the corresponding MEMS device and its role in improving healthcare. Hence, this chapter will be of great interest to engineers, physicians, and patients.

MEMS Sensors in Wearables

Micro-Electro-Mechanical Systems (MEMS) sensors, being a convergence of mechanical elements, sensors, actuators, and electronics on a common silicon substrate, have brought about a significant shift in the wearable technology domain. These tiny devices, capable of converting a measured mechanical signal into an electrical signal, find numerous applications in wearable devices, primarily due to their miniature size, low power consumption, and precision in capturing minute changes

(Lammel 2015; Khoshnoud and Silva 2012). The beginnings of today's MEMS technology can be traced back to 1954 when Smith from Bell Telephone Laboratories published a paper in Physical Review (Bogue 2007). In this paper, Smith introduced for the first time the concept of piezoresistance, highlighting how stress influences the electrical properties of silicon and germanium. One of the key attributes of MEMS is its ability to interact seamlessly with electronic parts on semiconductor chips. This has transformed the sensor industry, as MEMS technology merges electronic and mechanical operations. Typically, they are energy-efficient and highly sensitive because of their compactness (Ahmed et al. 2021; Xu et al. 2020). MEMS sensors, due to their lightweight, high resolution, and reliable performance, easily integrate with other systems. These features have led researchers to adopt them widely in wearable health devices for monitoring respiration, motion, ECG, EEG, wound conditions, and disease progression. There are multiple types of MEMS sensors and each one has different nature of sensing. Table 3.1 shows the generalized nature and variables of the MEMS sensor to measure wearable sensing.

Table 3.1 Sensing and application analogies

Nature	Sensing	Measurement	Wearable applications
Potential	Voltage	Vibrations, sound, force	ECG, EMG, Movement, Vibration Sensors. (Koivisto et al. 2015; Mathias et al. 2016)
Flow	Current	Pressure, temperature, humidity	Spirometer, Ventilator, Leak detection (Göreke et al. 2016; Laghrouche et al. 2016; Vutla et al. 1276)
Generalized resistance	Resistance	Pressure, force, temperature	Pressure Sensors, Temperature Sensors (Song et al. 2020; Meti et al. 2016; Guo et al. 2019)
Generalized Capacitance	Capacitance	Proximity, pressure, force	Respiration, ECG, Humidity, Heart Rate, Touchscreens, Pressure (Nguyen and Ichiki 2019; Xu et al. Nov. 2018; Li et al. Oct. 2017; Zhang et al. 2016)
Generalized Inductance	Inductance	Magnetic field	EMG, ECG, EEG, Pulse, Accelerometer. (Nguyen and Ichiki 2019; Bhavana et al. 2016; Hajare and Kadam 2021; Niu et al. 2018)

In the next section, we will discuss the different types of MEMS sensors, sensing mechanisms, and current research and commercial wearable applications.

Piezoelectric Sensors

In the realm of advanced materials research, piezoelectric materials exhibit high energy density characteristics, which are amplified upon miniaturization. This has subsequently fueled a growing interest in the integration of piezoelectric films in MEMS applications. Piezoelectric materials offer a direct transduction mechanism for the conversion between mechanical and electrical signals. Specifically, the linear and reversible piezoelectric effect is characterized by the generation of an electric charge (voltage) upon the exertion of mechanical stress (direct effect), and, conversely, the generation of strain (stress) when an electric field is applied (converse effect). Piezoelectric thin films are created through multiple methodologies, including sputtering, chemical solution processes, screen printing, and atomic layer deposition. Typically, these methodologies can be sliced down into three pivotal steps: (i) production of the desired atomic or molecular species, (ii) their transportation and subsequent condensation (or deposition) onto a substrate, and (iii) post-deposition crystallization and annealing of the film (Ali and Prasad 2020; Li et al. 2015; Chen et al. 2021).

Quartz can be found in nature, the majority utilized in practical applications is synthetic, specifically the AT-cut single crystal, which is right-handed and in the α-phase. Below its Curie temperature of 573 °C, quartz possesses a trigonal structure (Saigusa 2017). However, when the temperature exceeds this point, it transforms into β-quartz with a hexagonal configuration. The AT-cut is primarily chosen for its negligible temperature coefficient of frequency. However, alternative cuts, such as the Y-cut or the dual-mode SC-cuts, exhibit significant sensitivity to temperature changes in resonance frequency, making them apt choices for accurate temperature measurements. Langasite stands out as a material with temperature coefficients comparable to quartz. Yet, it presents a quality factor that's five times greater and a piezoelectric coupling coefficient triple that of quartz ("Characterization and Comparison of Lithium Niobate and Quartz Surface Acoustic Wave Devices—ProQuest" 2023). Lithium niobate and tantalate are renowned ferroelectric crystals identified in 1949. Since 1965, these have been effectively cultivated into single crystals from their melt using the Czochralski technique (Turutin et al. 2023). Much like quartz, these materials are grown in large volumes, with specific properties determined by their cutting orientation. $LiNbO_3$ often adopts the Y-cut, whereas $LiTaO_3$ is more commonly associated with the X-cut. Applications of MEMS piezoelectric sensors in wearable monitoring are described below (Fig. 3.1).

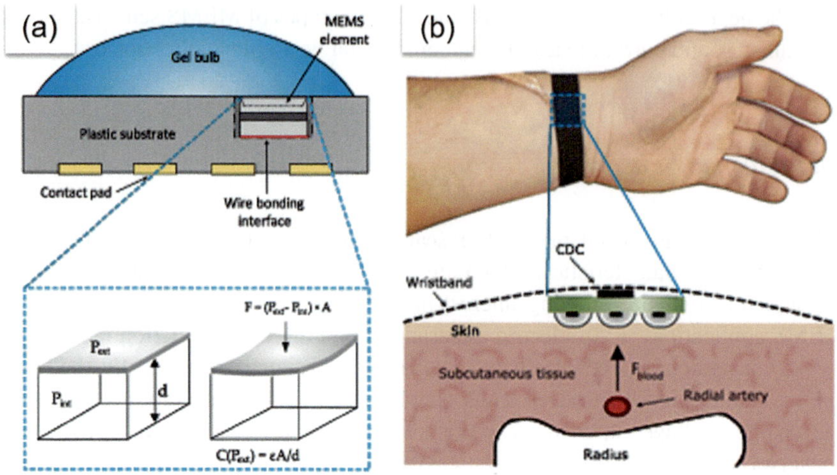

Fig. 3.1 Sensor system overview. **a** The capacitance of the element changes as a function of outer pressure, deforming the diaphragm. **b** An illustration of the obtained signals from the array and details of the pulse profile during one cardiac cycle (Kaisti et al. 2019)

Activity and Fitness Monitoring

MEMS sensors, adeptly integrated into fitness trackers and smartwatches, operate as accelerometers with the capacity to detect mechanical vibrations associated with motion (Qu et al. 2021). When a user engages in physical activity or exercise, MEMS piezoelectric sensors sense the accelerations and decelerations tied to each movement. Leveraging the piezoelectric effect, wherein certain materials generate an electrical charge in response to mechanical stress or deformation, these sensors convert these mechanical vibrations into electrical signals (Fig. 3.2) (Qu et al. 2021; Wang et al. 2021). These signals are then meticulously processed to provide users with an array of valuable information, including step counting, distance estimation, exercise intensity categorization, and calorie expenditure estimation. Additionally, during sleep, these sensors continue to contribute by detecting movements and translating them into data for sleep duration and quality assessment.

Heart Rate Monitoring

MEMS piezoelectric sensors embedded in wearable devices, often in wrist-worn form factors, enable unobtrusive and real-time monitoring of the wearer's heart rate (Unnikrishnan 2022). The sensors are strategically positioned to detect subtle mechanical vibrations associated with each heartbeat. As the heart contracts and blood is pumped throughout the body, these sensors pick up these minute mechanical movements and convert them into electrical signals. The resulting data is then processed and analyzed by the wearable device's onboard algorithms to provide users

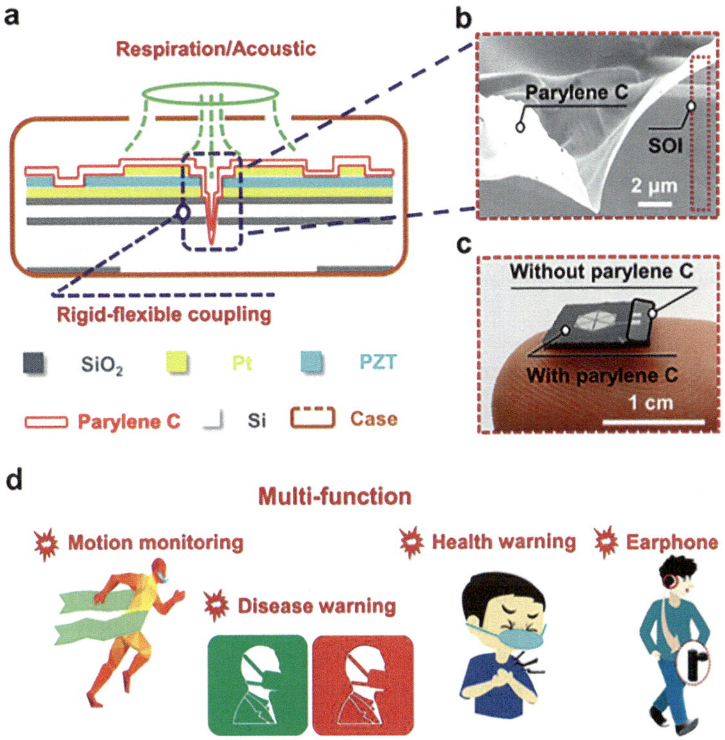

Fig. 3.2 The strategy and design of the multifunctional piezoelectric MEMS device. **a** The strategy and design. **b** The flexible parylene C thin membrane in the surface gap. **c** A photograph of the fabricated device is placed on a finger. **d** The multifunctional device for motion monitoring, disease warning, health warning, and earphones. Reprinted with permission from Wang et al. (2021), Copyright (2021), with Elsevier, License Number 6024820493067

with accurate and continuous heart rate measurements (Fig. 3.1) (Kaisti et al. 2019; Ji and Zhang 2022).

This continuous heart rate monitoring has far-reaching implications for health and wellness. For athletes and fitness enthusiasts, it allows for precise tracking of heart rate during exercise, helping individuals optimize their workouts, set target heart rate zones, and prevent overexertion. In medical contexts, continuous heart rate monitoring aids in the early detection of arrhythmias, irregular heart rhythms, and other cardiovascular anomalies, facilitating timely medical interventions.

Sleep Monitoring

MEMS offers precise and unobtrusive data collection during sleep cycles. These sensors, often integrated into wrist-worn wearables or bedding, leverage the piezoelectric effect to detect even subtle mechanical vibrations and movements associated with sleep. By meticulously capturing changes in body position, limb movements, and respiratory patterns, MEMS piezoelectric sensors enable the classification of sleep stages, including light sleep, deep sleep, rapid eye movement (REM) sleep, and wakefulness (He et al. 2022). They contribute to the calculation of sleep efficiency, offering insights into the effective use of sleep time and aiding in the identification of sleep disruptions and respiratory irregularities (Mallegni et al. 2022). These sensors also play a critical role in the early detection of sleep apnea events by monitoring chest and abdominal movements, offering users valuable insights into their sleep quality and facilitating timely interventions for sleep-related issues.

Stress and Anxiety Management

MEMS piezoelectric sensors are increasingly employed in wearable devices designed for stress and anxiety management, providing a scientifically grounded approach to monitoring physiological responses to stressors. These sensors, typically integrated into wearables worn on the wrist or other convenient locations, capitalize on their ability to capture subtle mechanical vibrations associated with stress-induced muscle tension (Khan et al. 2023). By measuring variations in muscle activity and tension, MEMS piezoelectric sensors offer insights into stress levels and anxiety-related physical manifestations. The collected data contributes to real-time stress tracking and can trigger personalized stress management interventions, such as guided breathing exercises or relaxation prompts (Byrom et al. 2018). This scientific approach empowers users to gain a deeper understanding of their stress responses, facilitating proactive stress management and fostering improved emotional well-being.

Fall Detection and Prevention

MEMS sensors are adept at detecting rapid and impactful body movements, characteristic of falls, due to their ability to convert mechanical vibrations into electrical signals. Scientifically validated algorithms analyze data from these sensors, discerning fall events from regular movements. Upon detection, these systems trigger immediate alerts, which may include notifications to caregivers or emergency services (Dziadak and Czumak 2022).

Gait Analysis

These sensors are adept at capturing subtle mechanical vibrations and accelerations associated with various gait parameters such as stride length, step frequency, and gait symmetry. Scientifically validated algorithms process the data, enabling the precise characterization of an individual's gait dynamics and the identification of abnormalities or irregularities that may indicate musculoskeletal or neurological issues (Awasthi and Joshi 2015; Tahir et al. 2020). Wearable gait analysis, powered by MEMS piezoelectric sensors, holds immense potential for clinical applications, rehabilitation monitoring, and personalized intervention strategies, offering valuable insights into mobility and musculoskeletal health for research and clinical purposes.

Piezoresistive Sensors

MEMS piezoresistive sensors are a class of micro-scale sensors that utilize the piezoresistive effect to measure mechanical stress or strain in various applications. The working principle of MEMS piezoresistive sensors is grounded in the unique property of certain materials to change their electrical resistance when subjected to mechanical deformation. Silicon is a commonly used material in MEMS piezoresistive sensors due to its excellent mechanical properties and compatibility with microfabrication processes (Li et al. 2017).

The key component of a MEMS piezoresistive sensor is a micro-sized mechanical structure, often in the form of a diaphragm or cantilever, made of silicon. Within this structure, small piezoresistive elements, typically diffused or implanted into the silicon substrate, are strategically placed. When mechanical stress or strain is applied to the sensor, the silicon structure deforms, causing a change in the resistance of the piezoresistive elements. This change in resistance is directly proportional to the magnitude of the mechanical deformation, allowing the sensor to convert mechanical signals into electrical signals. By measuring the resulting change in electrical resistance, MEMS piezoresistive sensors can precisely quantify parameters such as pressure, force, or acceleration (Meti et al. 2016; Li et al. 2017; Xu et al. 2019). These sensors are highly advantageous due to their small size, low power consumption, and suitability for integration into compact electronic devices and wearables, making them invaluable in various industries, including automotive, healthcare, and industrial monitoring. Applications of MEMS piezoresistive sensors in wearable monitoring are described below.

Heart Rate Monitoring

The integration of MEMS piezoresistive pressure sensors into wrist-worn devices has revolutionized continuous heart rate monitoring. MEMS piezoresistive pressure sensors are often integrated into heart rate monitors, particularly those worn on the wrist (Fig. 3.3) (Sun et al. 2022). These sensors can measure the subtle fluctuations in blood volume caused by the beating of the heart. By analyzing these variations, the wearable can calculate the user's heart rate in real-time, allowing for continuous monitoring during exercise or daily activities (Zhang et al. 2016; Cui et al. 2022). Some wearables can also provide heart rate variability (HRV) data, which can offer insights into stress levels and overall health.

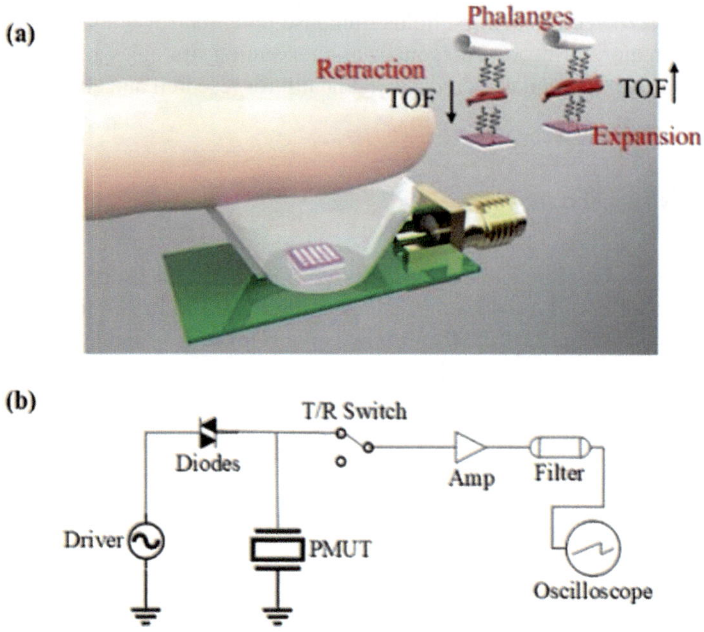

Fig. 3.3 Working principle and setup **a** Schematic of the working principle **b** Circuit diagram. Reprinted with permission from Sun et al. (2022), Copyright (2022), with IEEE, License Number 6024830120184

Fall Detection

In the realm of elderly care and safety, MEMS piezoresistive sensors are instrumental in fall detection. Advanced algorithms for fall recognition and differentiation from daily activities are discussed. These sensors can detect rapid changes in acceleration and orientation associated with a fall (He et al. 2023). When a fall is detected, the wearable can send alerts to caregivers or emergency services, providing assistance to the wearer in a timely manner. Advanced algorithms can distinguish between falls and normal movements to reduce false alarms.

Sleep Monitoring

The capacity of MEMS piezoresistive sensors to capture subtle body movements is harnessed in sleep monitoring wearables. Accelerometers can track movements during sleep, helping to determine sleep stages (e.g., light, deep, REM) and overall sleep quality (He et al. 2022). Coupled with heart rate monitoring, these devices can provide valuable insights into sleep patterns and help users identify issues like sleep apnea or insomnia. Sleep stage classification and sleep quality assessment are examined in detail.

Environmental Monitoring

Barometric Pressure Sensors

Barometric pressure sensors, a subset of MEMS piezoresistive sensors, operate on the fundamental principle that atmospheric pressure varies with changes in altitude and weather conditions. These sensors consist of a diaphragm exposed to the external atmosphere and a cavity beneath the diaphragm (Wada and Takahashi 2020; Kishore et al. 2021). As atmospheric pressure changes, it causes the diaphragm to deflect, and the resulting strain is measured by piezoresistive elements integrated into the diaphragm. The change in resistance due to this strain is then converted into a pressure reading.

Humidity and Temperature Sensors

In humidity sensing, a piezoresistive film is designed to absorb or adsorb moisture from the environment, causing dimensional changes that alter its electrical resistance. Similarly, in temperature sensing, the piezoresistive elements exhibit resistance changes in response to temperature fluctuations (Xu et al. 2019; Cai et al. 2018). This combination of piezoresistive-based MEMS technology with humidity and temperature sensors not only enhances their sensitivity but also allows for miniaturization and integration into compact wearables. These sensors find applications in various fields, including healthcare, industrial monitoring, and outdoor activities, where precise and simultaneous monitoring of humidity and temperature is imperative for user comfort, safety, and health management.

Health Monitoring and Beyond

Muscle Activity Sensors

Piezoresistive-based MEMS muscle activity sensors, in particular, have gained prominence in the fields of sports science, rehabilitation, and physical therapy. These sensors function on the principle that muscle contractions result in mechanical deformations and strain within muscle tissues (Fatema et al. 2021; O'Hare et al. 2022). By strategically integrating piezoresistive elements into wearable sensors, minute changes in resistance can be detected when muscles contract or relax. This allows for real-time monitoring of muscle engagement, muscle fatigue, and even fine-grained analysis of muscle activity patterns during physical activities or rehabilitation exercises. These sensors offer a non-invasive and unobtrusive means of assessing muscle health and performance, enabling athletes to optimize their training regimens, individuals to monitor rehabilitation progress, and healthcare professionals to tailor treatment plans for conditions related to muscle function (O'Hare et al. 2022). The integration of piezoresistive-based MEMS muscle activity sensors in wearables holds great promise for enhancing athletic performance, accelerating recovery from injuries, and improving the overall quality of life for individuals with musculoskeletal conditions.

Blood Pressure Monitoring

The convergence of MEMS technology and blood pressure monitoring has led to the development of highly accurate, non-invasive, and wearable blood pressure monitoring solutions. Piezoresistive-based MEMS sensors, integrated into wearable cuffs or patches, offer continuous blood pressure measurements without the need for invasive procedures or cumbersome equipment. These sensors function on the principle that blood pressure-induced mechanical deformation of the cuff membrane results in changes in electrical resistance within the piezoresistive elements (Panwar et al. 2017; Nguyen et al. 2020). This resistance variation is precisely correlated with systolic and diastolic blood pressure values. As such, these sensors provide real-time monitoring, offering individuals with hypertension or cardiovascular conditions the ability to track their blood pressure trends, allowing for early detection of anomalies and improved management of their health (Yi et al. 2022). Moreover, the non-invasive nature of these sensors promotes patient compliance, ensuring consistent monitoring, and reducing the risk of complications associated with hypertension. The integration of piezoresistive-based MEMS technology into blood pressure monitoring represents a significant leap forward in personalized healthcare, enhancing the precision and convenience of monitoring, and fostering better patient outcomes for individuals with cardiovascular concerns.

Respiration Monitoring

Respiration, an essential physiological process, encompasses the inhalation of oxygen from the atmosphere and the exhalation of carbon dioxide as a metabolic waste product. Monitoring respiration is crucial, as it offers insights into an individual's metabolic and respiratory health, ensuring timely interventions in potential respiratory ailments or life-threatening conditions. With advancements in sensor technology, piezoresistive MEMS sensors have emerged as invaluable tools in respiratory monitoring. These sensors, leveraging their piezoresistive properties, detect and measure the mechanical changes associated with chest and abdominal movements during breathing (Nguyen et al. 2020; Moradian and Abdolvand 2016). By converting these minute mechanical deformations into electrical signals, piezoresistive MEMS sensors facilitate accurate, continuous, and non-invasive monitoring of respiration, ensuring real-time assessment of an individual's respiratory pattern and health.

Capacitive Sensors

MEMS capacitive sensors are instrumental in wearable applications, with a working mechanism grounded in precise electrical capacitance measurements. These sensors consist of two conductive plates separated by a dielectric material. In wearables, they are often integrated into various devices, such as fitness trackers or smartwatches. When subjected to mechanical stimuli, such as changes in pressure or motion, the separation distance between the plates is altered, consequently affecting the capacitance. Scientifically, this change in capacitance is proportional to the applied force or displacement, enabling the sensors to detect and quantify physical parameters like touch, pressure, or motion. In wearable health and fitness applications, capacitive sensors play a pivotal role in touchscreens, enabling intuitive user interactions, and in inertial measurement units (IMUs), facilitating precise motion tracking for activities such as step counting and gesture recognition. Applications of MEMS capacitive sensors in wearable monitoring are outlined below.

Respiration Monitoring

MEMS capacitive sensors, often integrated into wearable devices such as smart clothing or adhesive patches, operate based on capacitance changes due to chest or abdominal movements during respiration. Scientifically, as a person breathes, their chest and abdomen expand and contract, altering the distance between the sensor's capacitive plates. This mechanical variation induces changes in capacitance, a fundamental electrical property, which can be precisely measured (Park et al. 2017; Mishra et al. 2021). MEMS capacitive sensors, with their high sensitivity and miniaturization, capture these capacitance fluctuations, enabling accurate and real-time monitoring of breathing rate, depth, and patterns.

ECG Monitoring

These sensors are seamlessly integrated into wearable devices, often in chest-worn configurations, to detect the minute electrical fluctuations generated by cardiac activity. Scientifically, as the heart contracts and releases electrical impulses, MEMS capacitive sensors detect these variations in electrical potential through capacitive coupling (Mathias et al. 2016). This process results in changes in capacitance, a fundamental electrical property, which is precisely measured and translated into ECG waveforms (Wang et al. 2023). This scientific mechanism allows for continuous, real-time ECG monitoring, offering invaluable insights into cardiac health, the identification of arrhythmias, and early intervention opportunities.

Wound Monitoring

These sensors are integrated into wearable wound care devices, such as smart bandages or adhesive patches, and function by detecting changes in capacitance due to variations in wound characteristics. As a wound heals or becomes infected, it undergoes alterations in electrical properties, such as conductivity and dielectric constant (Duell et al. 2018; Lin et al. 2022). MEMS capacitive sensors exploit these changes by measuring the corresponding variations in capacitance. This scientific mechanism enables continuous, non-invasive monitoring of wound status, including healing progress, infection detection, and moisture levels. By providing real-time data, MEMS capacitive sensors enhance wound management, allowing for timely interventions and promoting optimal healing outcomes, particularly in chronic wound care and postoperative monitoring.

MEMS Biosensors

Electrochemical MEMS Biosensors

Electrochemical MEMS sensors depict significant advantages, primarily attributed to their compact dimensions, seamless integration with electrical and electronic circuits, minimal weight and size, low energy consumption, and impressive resonance frequencies. Additionally, MEMS-based electrochemical sensors offer remarkable precision, cost-effectiveness, large-scale manufacturing with enhanced sensitivity, and improved overall performance, rendering them well-suited for self-powered and wearable electronic applications (Chircov and Grumezescu 2022). MEMS technology has helped push the limit of detection of the electrochemical sensors down to sub-micromolar concentrations (Coffel and Nuxoll 2018).

The electrochemical sensors employ the principles of electrochemistry to identify the target analytes or biomolecules. Typical electrochemical sensors are configured with three electrode systems; the working electrode is responsible for measuring the potential developed at the interface between the electrode and the electrolyte, in relation to a reference electrode (commonly an Ag/AgCl electrode), where completing the circuit is achieved with the counter electrode. In most sensors, the electrochemical signal emerges as a consequence of the diffusion of the target analyte, either from the surrounding environment toward the working electrode or toward an analyte-sensitive material positioned between the working and counter electrodes. This electrochemical signal is manifested as a shift in electrical signal or potential at the working electrode (potentiometric response) (Yunus et al. 2013), an accumulation of charge (coulometric response) (Han et al. 2021), an increase or decrease in current (amperometric response) (Hussain and Keçili 2020), or a modification in the conductance of the electrolyte between the working and counter electrode (Hussain and Keçili 2020). These changes in electrical signals indicate alterations in the concentration of the detected molecules. As many integral components of electrochemical sensors are similar to electrical circuit elements, these are exceptionally well-suited for optimization through MEMS technology. Metal/Au/Ag or other electrodes and insulating materials along with catalysts can be integrated into thin layers and precisely patterned at the micron scale. This allows for the creation of meticulously ordered arrays of electrodes that can be seamlessly integrated with other circuit components (Coffel and Nuxoll 2018).

Electrochemical biosensors tend to depict fewer interference issues compared to optical and acoustic wave techniques, which often contend with interference issues related to light and sound. A diverse range of materials serves as the electrodes in these sensors, including graphene, carbon, and gold electrodes. Different functionalization materials such as graphene-based materials, single/multi-walled carbon nanotubes, glassy carbon, and nanoporous carbon are exceptionally well-suited for use in electrochemical sensors for having great sensing properties (Noushin and Tabassum 2021). These sensors can be employed for the detection of various viral and chronic diseases, enzymes, glucose, inflammatory biomarkers (Noushin and Tabassum 2021), ions (Pereira et al. 2022), pH (Noushin et al. 2022), and gases (Dutta et al. 2023). As these sensors exhibit a high degree of selectivity, remarkable sensitivity, and rapid response times, having significant potential for various biomedical diagnostic applications, researchers are focusing on different types of electrochemical MEMS biosensors design and fabrication, as outlined below.

Glucose Sensing

The signal transduction mechanism for the MiniMed 670G system, used for automatic feedback in an insulin pump, as well as most other continuous glucose sensors operates based on electrochemical principles (Coffel and Nuxoll 2018). The process commences with the enzymatic oxidation of glucose at the working electrode's surface, resulting in the production of hydrogen peroxide as a byproduct. Immobilized glucose oxidase is used as an oxidizing enzyme in most glucose-sensing electrodes; however, glucose-1-dehydrogenase and hexokinase have also been used. Enzymes contribute to the selectivity of the system, ensuring that only the target analyte influences the output signal (Yoo and Lee May 2010).

The amount of hydrogen peroxide generated corresponds to the local glucose concentration. Subsequently, hydrogen peroxide undergoes oxidation, leading to the generation of an electrical signal that can be detected through various methods such as amperometric and potentiometric techniques. Typically, the immobilized enzyme and electrode are covered with a membrane, which serves to regulate the transport of glucose to the electrode. This design maintains a consistent and dominant transport resistance, effectively overshadowing any variable physiological transport resistance. This ensures that the rate at which glucose reaches the electrode for oxidation remains directly proportional to the blood glucose concentration. Integration of MEMS has played a pivotal role in lowering the detection limit of electrochemical glucose sensors (Coffel and Nuxoll 2018).

Another work reported a MEMS glucose sensor fabricated on a bare platinum strip that detects the oxidation of hydrogen peroxide (Kim et al. 1999). Detection of hydrogen peroxide has been enhanced in another work by dividing this strip into an array of Pt working electrodes to increase the total charge accumulation (Sassa et al. 2010). The spacing, size, and number of Pt strips have been meticulously optimized to maximize the total charge in this coulometric detection technique. In addition to platinum, various other metal catalysts have been micro-patterned to facilitate hydrogen peroxide decomposition, including iridium, palladium, gold, silver, and manganese dioxide (Chen and Chatterjee 2013). One glucose sensor currently undergoing clinical trials utilizes an array of eight electrodes made from carbon paste, glucose oxidase, and microfabricated manganese dioxide within a compact 1×6 mm^2 area that demonstrated an oxidation potential of 0.35 V vs. Ag/AgCl for hydrogen peroxide, potentially mitigating interference from other analytes in the body (Schmelzeisen-Redeker et al. 2013; Zschornack et al. 2013).

In the quest to eliminate interfering oxidation currents, researchers have explored novel materials beyond noble metals and their oxides, such as transition metal dichalcogenides (TMD), including MoS_2, $MoSe_2$, WS_2, and WSe_2. These materials can be deposited in thin films through magnetron sputtering, making them ideal for MEMS fabrication (Chia et al. 2015; Koçak et al. 2016). These materials have emerged as cost-effective metal electrodes for detecting hydrogen peroxide (Toh et al. 2017). WS_2, in particular, exhibited the highest current densities and boasted a remarkably low detection limit of 2.0 nM to H_2O_2. Furthermore, interfering analytes displayed minimal to negligible oxidation currents on this WS_2 electrode in the presence of hydrogen peroxide oxidation at 0.5 V vs. Ag/AgCl. Nonenzymatic mechanisms for hydrogen peroxide detection can be incorporated into MEMS devices using carbon-based materials, conductive polymers, and nanostructured metals (Coffel and Nuxoll 2018).

PH Sensing

pH serves as a vital physiological chemical indicator, typically detected through electrochemical means. Given the varying ideal pH values in different parts of the body, many of which are not easily accessible via wired connections, these sensors often necessitate wireless signal transmission integration using MEMS technology with all the requisite driving, conditioning, and telemetry circuitry.

One notable development involves the fabrication of an IrOx electrode on a flexible polyimide circuit for pH detection within the range of 2.0–11.0, boasting impressively rapid response times as quick as 0.9 s (Cao et al. 2012). This sensor was situated alongside an impedance sensor, resulting in a dual-sensing, implantable device tailored for monitoring esophageal reflux events. The impedance component tracked occurrences of reflux where the sensor's exposure to moisture altered the ion concentration between gold electrodes, leading to changes in impedance. As a result, the pH sensor discerned acidity shifts based on the potential difference between the IrOx electrode and the Ag/AgCl reference electrode. Another presently available commercial sensor, the Bravo pH monitoring device by Medtronic, employs antimony but lacks the impedance sensing component. It has shown limitations in detecting reflux events in the esophagus, particularly when consecutive events possess similar or neutral pH characteristics (Kwiatek and Pandolfino 2007).

Gas Sensing

The MEMS electrochemical gas sensor operates by detecting chemical reactions on the surface of a sensing material through absorption and de-absorption processes. These reactions lead to changes in the conductivity of the sensing material, which are transduced into measurable electrical signals. Two types of electrochemical gas sensors are commonly used: Metal Oxide Semiconductor (MOS) sensors and Chemiresistive gas sensors (Dutta et al. 2023).

Metal Oxide Semiconductor (MOS), invented in the 1960s, employs metal oxide materials, like zinc oxide (ZnO), to sense gases. They consist of sensing layers, substrates, contact electrodes, and a heating element. MOS sensors generally operate at high temperatures (100–450 °C). The sensing mechanism relies on a shift in the equilibrium of surface chemisorbed oxygen reactions in the presence of the target gas. MOS sensors can be of n-type or p-type, exhibiting different behaviors when exposed to oxidizing or reducing gases. Their performance is measured by sensitivity and recovery time. SnO_2 is a commonly used sensing material due to its stability in reducing gases. Researchers have enhanced MOS sensor performance by using nanostructures and doping with other materials. These sensors are stable, cost-effective, and widely used in gas-sensing applications but consume a lot of power due to their high operating temperatures.

Chemiresistive Gas Sensors, introduced by Seiyama, operate based on changes in the electrical resistance of a sensing material when it interacts with target gases. Chemiresistive sensors have a simple design without the need for a micro-heater. Typically, a chemiresistive gas sensor features a fundamental structure comprising two electrodes with sensitive materials positioned in between (Fig. 3.4a, b) (Tang et al. 2019). Nanostructured materials are often used for their high sensitivity, and doping methods can further enhance their performance, and increase response and recovery times. It’s worth noting that flexible chemiresistive gas sensors can be implemented on polymer substrates. When the chemiresistive sensor comes into contact with the target gas, interactions occur between the gas molecules and the sensing material, resulting in alterations in electrical resistance. Chemiresistive sensors are employed in environmental monitoring due to their simple structure, low power consumption, and sensitivity to toxic gases. Their main limitation is sensitivity to humidity, which affects their performance (Asri et al. 2021).

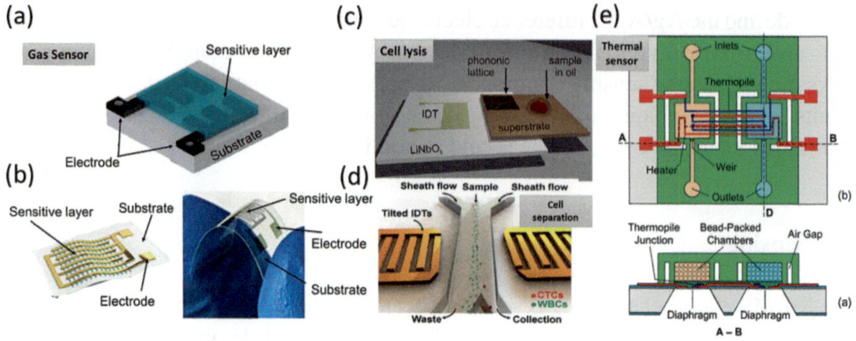

Fig. 3.4 **a** Schematic design of a chemiresistive gas sensor (Tang et al. 2019) **b** Flexible chemiresistive gas sensor. Reprinted with permission from Tang et al. (2019), Copyright (2019), with American Chemical Society **c** A high-throughput SAW device for cancer cell separation (Reboud et al. 2012), Copyright (2012), National Academy of Sciences. **d** An acoustic-based nanoparticle separation device (Li et al. 2015), Copyright (2015), National Academy of Sciences. **e** Schematic of the MEMS thermal biosensor's top view and cross-sectional view (as defined by line A-B indicated in the top view). Reproduced with permission from Wang et al. (2008), Copyright (2008), with IEEE, License Number: 6024840870655

Inflammation Sensing

For inflammatory biomarkers detection in infectious/viral diseases, antibodies of target biomarkers are immobilized followed by different functionalization coatings, capturing the antigen, and converting the chemical reaction on the sensor electrode's surface into electrical properties, resulting in distinct current–voltage characteristics (Noushin and Tabassum 2021; Pereira et al. 2021). A recent work reported a MEMS biosensor for inflammatory biomarker, interlukin-6 (IL-6) detection by combining two types of biorecognition elements on the silicon nanowire field-effect transistors (SiNW-FETs) (Hu et al. 2023). Upon evaluating the response to various concentrations of IL-6, they observed that the aptamer-functionalized SiNW-FET displayed a broader detection range in comparison to the antibody-functionalized SiNW-FET. Additionally, the aptamer-functionalized SiNW-FET exhibited a low detection limit of 2.1 pg/mL, which is two orders of magnitude lower than that of the antibody-functionalized SiNW-FET. Furthermore, the detection range of the aptamer-functionalized SiNW-FET encompassed IL-6 concentrations relevant for predicting severe outcomes in COVID-19 cases.

Surface Acoustics Wave (SAW)-Based MEMS Sensors

The utilization of MEMS in SAW (Surface Acoustic Wave)-based sensors is crucial for seamless integration with microelectronics and flexible antennas. This integration leads to the creation of programmable microsensors that find utility in a multitude of engineering and biomedical applications. SAW-based MEMS sensors facilitate simple technological fusion with wireless power transmission systems and the sensor data can be wirelessly monitored without requiring power at the sensor location as passive sensors (Zheng et al. 2019). Furthermore, SAW-based MEMS biosensors are advantageous in many applications, where the sensors are expected to meet specific criteria, including high precision, a wide dynamic range, and a broad frequency spectrum. Overall, the typical advantages of these devices can be complemented by robustness, exceptional sensitivity, surface adaptability, and long-lasting durability (Varadan 2000).

SAW-based MEMS biosensors rely on the propagation of surface acoustic waves where the sensing material is situated on a MEMS platform. These sensors can resonate with specific surface acoustic waves, making them exceptionally valuable for biomedical applications due to their sensitivity and rapid response time. SAW sensors operate by allowing acoustic waves to propagate along the substrate's surface. When these sensors come into contact with specific molecules, the resonant frequency of the surface acoustic wave changes, enabling the detection of these substances. SAW-based sensors typically incorporate interdigitated transducers and an analyte is placed between the two electrodes. When a voltage is applied to the piezoelectric material through the electrodes, the material undergoes mechanical deformation, generating an acoustic wave. When biomolecules interact with the piezoelectric material, it alters the generated acoustic wave's properties, causing a shift in the material's resonant frequency. Unlike other methods such as enzyme-linked immunosorbent assays and radio-immunoassays, SAW sensors do not require labeled molecules. They stand out as the sole technique capable of detecting biomolecules and manipulating fluids. These sensors are instrumental in detecting numerous infectious diseases that demand early identification. In point-of-care (POC) testing, SAW-based devices are advantageous because they simplify the diagnostic process.

Vapor and Small Molecule Detection

Surface Acoustic Wave (SAW) technology has versatile applications in detecting vapors and small molecules, primarily by tracking minute shifts in frequency resulting from the slight addition of analytes. These frequency shifts can be finely calibrated, allowing detection down to the picogram level (Mujahid and Dickert 2017). One effective approach involves covalently attaching biomolecules to the SAW surface, offering benefits like enhanced stability and reproducibility of bioreceptors (Ai et al. 2013). For instance, by applying three distinct odorant binding proteins (OBPs) onto an array of five SAW resonators, changes in resonant frequency can be leveraged to gauge the concentrations of octanol and carvone vapors (Pietrantonio et al. 2015). SAW sensors customized with olfactory receptors to boost sensitivity are also capable of detecting diacetyl. In another work, a 10 nm gold coating on a lithium niobate ($LiNbO_3$)-based SAW sensor, specifically modified with the ODR-10 olfactory receptor from Caenorhabditis elegans, was utilized to capture diacetyl. This resulted in a high sensitivity to diacetyl concentrations, with frequency shifts as high as 2 kHz/ng and a lowered detection limit of 10×10^{-10} mM (Wu et al. 2011). Potential research avenues in SAW-based gas sensors include enhancing substrate stability, selecting circuit electrode materials capable of withstanding harsh sensing environments, and employing advanced sensing materials or multiple SAW resonator arrays for vapor and small molecules detection and quantification (Huang et al. 2021).

Cell Lysis

In certain biosensing applications, it's necessary to break open cells to release internal materials like RNA, DNA, or proteins for pathogen detection and point-of-care diagnostics (Fig. 3.4c) (Reboud et al. 2012). Biomedical demands for miniaturized systems have led to advanced methods of cell lysis in small-scale volumes. Acoustic waves, particularly surface acoustic waves (SAW), provide a robust solution. While ultrasound is traditionally used for cell lysis, it requires bulky, high-power equipment and relatively large sample volumes (~ml). Efforts have been made to reduce sample size and power consumption using bulk acoustic waves (BAW), but SAWs offer even greater efficiency and require smaller sample volumes due to their ability to trap acoustic energy at the surface.

A recent work utilized SAWs generated on a $LiNbO_3$ piezoelectric substrate coupled with a phononic superstrate acting as a frequency-tunable filter. They placed a 5–20 μl droplet containing blood cells on the superstrate and actuated the SAW at 9.5 MHz with 3.1 W power. They reported lysis rates exceeding 98% for sample volumes ranging from 10 to 20 μl, with 5 μl samples achieving lysis rates between 50 and 98%, depending on dilution. The study also examined the lysis of various cell types at different SAW frequencies and power levels, achieving 98% lysis at 0.8 W power (Fig. 3.4c) (Reboud et al. 2012).

Researchers have explored the use of SAWs for exosome lysis and RNA detection in a pancreatic cancer study. They employed twenty pairs of titanium/aluminum interdigitated transducers (IDTs) to generate Rayleigh SAWs on a $LiNbO_3$ substrate. At an operating frequency of 28.3 MHz and with a 1 W power for 30 s, they achieved a 38% lysis rate, sufficient for obtaining exosomal RNA (Taller et al. 2015). However, SAW-based lysis has limitations, as it generates heat that can denature proteins. Therefore, precise adjustments of device frequency and operating power are essential for effective cell lysis while minimizing heat-induced damage.

Cell Adhesion

Cell adhesion is a crucial aspect of understanding cell behavior, with significant applications in fields like bone repair and tissue assembly. It involves measuring the binding affinity of proteins found on a cell's surface. SAW-based methods are effective for characterizing cell adhesion by detecting various aspects of the binding process. SAW techniques can capture binding kinetics, including coupling and decoupling rates, equilibrium dissociation rates, and conformational changes resulting from binding. When biomolecules bind to the SAW sensor's surface, they induce mass loading, leading to a phase change. Additionally, changes in viscosity caused by the binding process result in an amplitude change (Huang et al. 2021).

An example involves the development of a Love-wave biosensor to monitor cell adhesion. In this system, a Polymethyl Methacrylate (PMMA) polymer guiding layer is placed on top of a quartz-based Love-wave biosensor, followed by a thin gold layer to immobilize a biorecognition layer. Anti-HLA antibodies are then immobilized to Protein G on the gold surface. As ligand receptors like HLA-42 bind to the surface, it increases the Love wave sensor's amplitude (dB), providing real-time sensitivity to the mass loading of small molecules during the binding process (Saitakis et al. 2008).

Separation of Cells and Particles

Cell and particle sorting are crucial processes in various biomedical applications, and substantial research efforts have gone into improving their efficiency. Traditional invasive methods like filtration and pinch flow margination are relatively simple but lack the flexibility needed for microscale operations (Huang et al. 2021). Noninvasive methods employing magnetic, electric, or acoustic fields have become more prevalent, with acoustic field-based techniques being particularly attractive due to their scalability and cost-effectiveness.

Acoustic wave modes like SAW and shear mode are commonly used for particle and cell separation (Fig. 3.4d) (Li et al. 2015). SAW-based devices create a standing wave field via pairs of interdigitated transducers (IDTs), generating pressure nodes and antinodes within fluids (Yosioka and Kawasima 1955). The primary acoustic radiation force, caused by pressure waves, either attracts or repels particles based on their size and density, with positive and negative contrast factors directing them toward pressure nodes and antinodes, respectively. These SAW devices have been used to separate platelets from whole blood with high purity, isolate circulating tumor cells (CTCs) from clinical samples, and separate bacteria from blood or sputum samples, potentially aiding in pathogen identification or sepsis diagnosis (Li et al. 2015, 2017; Ai et al. 2013; Nam et al. 2011). SAW-based platforms have also shown promise in separating different types of nanoparticles. Furthermore, researchers have explored various techniques, such as virtual deterministic lateral displacement (vDLD), dielectrophoresis, and microstreaming, to enhance particle separation with SAW devices. Additionally, SAW systems have been used to regulate the speed of catalytic micromotors, offering precise control over their motion (Huang et al. 2021).

Fluid Mixing

In microscale lab-on-a-chip biosensing applications, efficient fluid mixing is crucial to reduce incubation times in biomarker detection. Microscale flows are typically laminar, resulting in slow mixing through diffusion. However, SAW has emerged as a powerful tool to significantly enhance mixing rates. SAW-generated acoustic streaming offers an effective solution for improving microscale fluid mixing, and its performance can be fine-tuned by adjusting various parameters, including IDT design, SAW amplitude, and SAW power (Huang et al. 2021).

Studies have demonstrated the effectiveness of SAWs in improving fluid mixing, reporting a 9.6 MHz frequency SAW device with 35 V (peak to peak) to achieve fluid mixing in a microchannel, primarily through acoustic fluid interactions (Tseng et al. 2006). Another work employed a SAW mixer chip with tapered Interdigitated Transducers (TIDTs) to generate a narrow SAW beam, inducing substantial acoustic streaming for efficient mixing (Frommelt et al. 2008). Shilton et al. utilized a 128° rotated Y-cut X-propagating $LiNbO_3$ medium to generate SAWs and focused them into a droplet using concentric single-phase unidirectional transducers (SPUDTs), achieving rapid fluid mixing through micro acoustic streaming (Frommelt et al. 2008). Liu et al. integrated SAW with metal-enhanced fluorescence for the rapid detection of cancer biomarkers, where SAWs were employed to enhance the microscale mixing of antigens with antibodies. Their study demonstrated that SAW-driven mixing significantly reduced incubation times (Liu et al. 2018).

Immunosensors

Immunosensors rely on the specific binding of antibodies to antigens and are commonly used in detecting a wide range of substances, including hormones, drugs, bacteria, and environmental pollutants. SAW-based immunosensors detect mass changes on their surface associated with antigen–antibody binding (Huang et al. 2021).

The early attempts to use SAW devices for antigen detection were promising but faced challenges related to sensitivity and the detection of small molecules. Subsequent research addressed these limitations. For example, the utilization of shear horizontal (SH)-SAWs on a $LiTaO_3$ substrate allowed the quantification of Escherichia coli L-asparaginase (Welsch et al. 1996). Metal nanoparticles, such as gold nanoparticles (AuNP), were introduced to enhance detection sensitivity. The addition of AuNPs significantly reduced the limit of detection for various antigens, such as carcinoembryonic antigens and exosomes. AuNP amplification enabled the detection of exosomes at a concentration of 1100 particles/mL, demonstrating high sensitivity and specificity, especially in cancer diagnostics (Wang et al. 2020).

However, SAW biosensors have some drawbacks, such as nonspecific binding that can degrade test signals and difficulties in directly identifying small molecular weight molecules due to their limited mass-induced changes on the sensor surface. Despite these limitations, SAW biosensors have shown considerable potential for various immune-sensing applications.

DNA Sensing

Piezoelectric devices have been employed for detecting DNA hybridization events. A significant development in this area was the creation of a SAW sensor based on $LiTaO_3$, designed to detect 15-mer oligonucleotide DNA in liquid solutions through complementary DNA hybridization. This sensor featured a dual delay line device with probe DNA immobilized on an Au layer deposited on the SAW delay lines. The hybridization process led to an oscillation frequency shift between the reference and sensing channels, achieving a maximum sensitivity of 155 ng/ml/Hz (Hur et al. 2005). Another study by Zerrouki et al. also utilized a dual delay-line SAW sensor based on $LiTaO_3$ to monitor the frequency responses during the hybridization and grafting of synthesized oligonucleotides with their natural DNA counterparts (Zerrouki et al. 2010). To improve DNA detection limits, Zhang et al. developed a Love wave DNA sensor functionalized with deoxynucleoside transferase and incorporated synthesized silver nanoparticles to amplify the phase shift. This method enhanced the detection limit by three orders of magnitude, reaching 0.8 pM (Zhang et al. 2017).

Furthermore, SAW sensors have demonstrated the ability to discriminate DNA mismatches. For instance, Mazouz et al. used a 104 MHz leaky-SAW sensor to detect single-base mismatches and single-nucleotide polymorphisms in synthetic oligonucleotides (Mazouz et al. 2013). Gronewold et al. utilized a Love-wave-based SAW sensor to identify individual point mutations in the cancer gene fragment human BRCA1 (Gronewold et al. 2006). Liu et al. proposed a graphene oxide-modified SAW sensor for detecting the CYP2D6*10 mutation, demonstrating high sensitivity and specificity in clinical samples (Liu et al. 2015). These advances highlight the potential of SAW-based DNA sensors for various applications in genomics and diagnostics.

Bulk Acoustics Wave (BAW)-Based MEMS Sensors

BAW-based MEMS sensors can be easily designed to have extended battery life and compact device size/weight. Lower power consumption, high-frequency sensor design, high Q mechanical gain, and smaller drive-mode displacements confer a distinct advantage in BAW-based MEMS sensors. Furthermore, these sensors offer optimal durability, remarkable resistance to vibrations, minimal noise, and linearity (Shi et al. 2022). Bulk Acoustic Wave (BAW)-based MEMS resonators are commonly used for various sensing applications. These resonators consist of a thin piezoelectric film, often made of materials like aluminum nitride (AlN) or zinc oxide (ZnO), placed between two metallic electrodes (Liu et al. 2020). When an alternating current (a.c.) signal is applied across these electrodes, it generates a standing wave within the piezoelectric layer, resonating at a frequency determined by the film's thickness. Two main types of BAW resonators are Film Bulk Acoustic Resonators (FBARs), which use air for acoustic isolation, and Solidly Mounted Resonators (SMRs), which employ an acoustic reflector made of alternating layers with varying acoustic impedance (Liu et al. 2020). Typically, thin film BAW resonators operate in the Thickness Longitudinal Mode (TLM), where the mechanical deformation aligns with the direction of wave propagation. However, TLM resonators are not suitable for liquid environments due to severe damping, making them unsuitable for chemical and biological sensing in liquid conditions (Mujahid and Dickert 2017). As an alternative, the Thickness Shear Mode (TSM) resonance is employed for in-liquid biosensing. TSM resonators are effective because they do not compress in low-viscosity liquids and have lower attenuation (Rughoobur et al. 2018). To attain TSM resonances, piezoelectric films with a c-axis inclined at angles ranging from 20° to 40° relative to the surface normal are employed (Rughoobur et al. 2018).

Multi-Analyte Detection

BAW devices have found applications in gas sensing, liquid sensing, and biosensing. These devices encompass BAW sensors utilizing thickness-shear modes. The ability of these sensors to detect specific analytes relies on the coatings applied to them. When detecting a group of analytes or when only selective coatings are available (i.e., coatings responsive to multiple analytes), the use of multi-sensor arrays becomes advantageous. These arrays enable the qualitative and quantitative characterization of samples by evaluating the resulting signal patterns to detect multiple analytes. In cases where only one sensor is employed, virtual sensor arrays come into play. Virtual sensor arrays combine a single sensor with advanced signal evaluation methods or preliminary sample separation techniques, yielding outcomes similar to those achieved with multi-sensor arrays. Both types of arrays, multi-sensor and virtual, offer promising prospects for system integration and cost-effectiveness (Länge 2019).

In-Liquid Wave Resonators

BAW-TSM resonators are successful for in-liquid sensing, however, they have limitations such as lower mass sensitivity compared to TLM resonators and still exhibit damping in more viscous liquids. Additionally, fabricating inclined c-axis films presents challenges in terms of scalability. Recent work presented a BAW resonator involving the use of an interface layer between the BAW resonator and the liquid to enable TLM devices for in-liquid sensing (Rughoobur et al. 2018). Instead of complete isolation, this interface layer partially separates the TLM and TSM from the liquid by engineering its thickness. Additionally, the interface layer can be functionalized to allow mass attachment of molecules from the liquid, which can then be detected by the resonant acoustic wave. Carbon nanotubes (CNTs) are employed as an example of such an interface layer in this study, grown directly on the active area of the resonator. This approach allows the resonant device to remain decoupled from the process of mass attachment while retaining mass sensitivity in the liquid environment (Rughoobur et al. 2018).

Thermal MEMS-Based Biosensors

Thermal MEMS-based Biosensors offer improved thermal isolation, reduced thermal mass, enhanced sensitivity, broader linear range, reduced power consumption, smaller sample volume, and faster measurement time. When produced in arrays through batch fabrication, these devices can further facilitate high-throughput operations, allowing for the parallel measurement of multiple samples. MEMS technology enables cost-effective batch production and integration of miniaturized features in thermal biosensors that have led to extensive bioanalytical applications of these sensors (Wang et al. 2008).

MEMS thermal sensors primarily rely on temperature detection, with two common types being thermistors and thermopiles. Thermistors operate by changing their electric resistance in response to temperature variations, allowing for the measurement of absolute temperatures. However, they tend to have limited sensitivity. On the other hand, thermopiles consist of a series of thermocouple junctions connected in sequence, enabling the measurement of temperature differences between two distinct regions where these junctions are positioned. Thermopiles excel in common-mode noise cancellation and zero offset, making them highly sensitive. Various materials, including semiconductors, metals, and semiconductor compounds, have been used to create thin film thermopiles on diverse substrates. To enhance thermal isolation and minimize thermal mass, suspended diaphragms or bridges have been employed, resulting in thermal detection sensitivities at the nanojoule or nanowatt level and time responses approaching microseconds. These thermal components have been applied in diverse fields such as infrared detection, material characterization, and gas sensing (Fig. 3.4e) (Wang et al. 2008).

Metabolic Monitoring

MEMS thermal sensor applications typically involve solid or gaseous samples and do not facilitate the measurement of liquid samples, which is a common requirement in biosensors. Research in the realm of MEMS thermal sensors for liquid samples has been relatively scarce but has encompassed the detection of substances like glucose, uric acid, and insulin, as well as investigations into living systems such as microorganisms, human or animal cells, and tissues (Bataillard et al. 1993; Verhaegen et al. 2000). Moreover, there have been inquiries into the development of multianalyte thermal biosensors incorporating MEMS thermopiles. Recent work presented a MEMS differential thermal biosensor that integrated microfluidics with thermal sensing to achieve sensitive measurements of small volumes of analyte solutions (Wang et al. 2008). The integration involves a polymer microfluidic structure and a silicon-based thermal sensor chip. The microfluidic structure is designed with chambers and channels to handle and measure significantly reduced biological sample volumes, as small as 0.8 μl. The key innovation is the incorporation of a bimetallic thermopile onto the microfluidic structure, enabling low-noise temperature measurements of the sample. Additionally, a differential measurement approach is employed, offering excellent common-mode noise cancellation and thereby significantly improving sensitivity. The device's performance is demonstrated in the context of enzymatic glucose detection, a critical aspect of managing diabetes and other diseases (Wang et al. 2008).

Food Temperature Detection

Commercial MEMS thermal sensors incorporate a silicon lens, a MEMS thermopile sensor, a dedicated analog circuit, and a logic circuit to convert the data into digital temperature readings ("D6T | MEMS Thermal Sensors | OMRON Device and Module Solutions—Americas" 2023). The silicon lens plays a crucial role by focusing the radiant heat, particularly far-infrared rays, emitted by the object onto the thermopile sensor within the module. When the focused radiant heat (far-infrared rays) reaches the thermopile sensor, it generates an electromotive force. The sensor measures both the electromotive force and the internal temperature. Using these values, the module calculates the measured value, which corresponds to the temperature of the object. This calculation involves interpolation based on a built-in lookup table. The resulting measured value is then transmitted from an upstream system via the I2C bus for further use or processing. This MEMS thermal sensor has the capability to detect the temperature of food in a refrigerator or microwave oven. This functionality enables energy conservation and enhanced performance by providing feedback for temperature control ("D6T | MEMS Thermal Sensors | OMRON Device and Module Solutions—Americas" 2023).

Optical MEMS Biosensors

MEMS-based optical biosensors offer mechanical structures with enhanced reliability and low cost. Optical MEMS biosensors involve the manipulation and detection of optical signals at the microscale, achieved by integrating mechanical, electrical, and optical components and systems.

In biomedical applications, optical micro-electromechanical systems, under the category of micro-opto-electromechanical systems (MOEMS) integrate micro-optics, mechanical components, and electronics to manipulate and sense optical signals at the micron scale (Motamedi 2005). Optical biosensors employ various methods to detect molecules, such as analyzing changes in optical properties or modifications in sensing layer characteristics caused by optical property shifts (Damborský et al. 2016).

One approach involves interferometry, where light is directed through the target analytes, leading to interference when some of the light reflects back to the source

(Takahashi et al. 2014). This interference is detected by a photosensor, enabling the measurement of analyte properties based on the interference patterns.

Another interferometric method employs a soft sensing layer that reacts with analytes (e.g., antigens binding to immobilized antibodies), resulting in deformation of the sensing layer's surface. One side of the sensor features a photodetector, while the other side houses a single-wavelength light source. As the sensing layer deforms, the photodetector registers changes in wavelength, facilitating molecule detection.

The small size, immunity to electromagnetic interference, cost-effectiveness, and lightweight nature make them easily integrable into the human body through appropriate configurations. As a result, these devices have discovered diverse applications in biomedical sensing within the realm of photonic MEMS (Upadhyaya et al. 2021).

Dental Application

Optical sensors are revolutionizing dental applications by enabling precise monitoring of dental health. These sensors are vital for assessing dental filler materials, tracking the structural behavior of braces and aligners, and ensuring comprehensive dental health management. Recent advancements in this field have leveraged Fiber Bragg Grating (FBG) technology, providing valuable insights into dental structural dynamics (Cheng et al. 2016). Photonic MEMS technologies have emerged as promising tools for monitoring orthodontic tooth movement and biting force, enhancing our understanding of dental processes (Upadhyaya et al. 2021).

Strain and Angular Motion Monitoring

Optical MEMS biosensors play a significant role in monitoring strain within human spinal discs, offering non-invasive pressure assessment for intervertebral discs (Newell et al. 2017). Angular joint motion monitoring benefits from goniometer devices equipped with FBG sensors (Umesh et al. 2018), providing sensitivity adjustments and improved flexibility in measuring angular rotation. Gait analysis, particularly in horses, utilizes these sensors embedded in limbs to diagnose injuries and assess performance. In human gait analysis, insole FBG-based optical fiber pressure sensors, integrated with the Internet of Things, offer precise foot pressure assessment, particularly for patients with diabetes, enabling data transmission to physiotherapists for remote monitoring. These applications highlight optical MEMS sensors' versatility and potential in biomedical monitoring (Upadhyaya et al. 2021).

Pressure Monitoring (Respiratory, Intraocular, Blood, Bladder)

Pressure monitoring in various biomedical applications is crucial for diagnosing and managing health conditions. In the context of respiratory pressure monitoring, optical MEMS sensors have been employed to monitor respiratory movements and body motion during sleep, providing valuable insights into breathing patterns and sleep habits (Nishiyama et al. 2010). Intraocular pressure (IOP) monitoring is essential to detect and prevent glaucoma. Researchers have developed micro-sized optical implants and flexible photonic crystal membranes to continuously monitor IOP (Karrock and Gerken 2015), offering early detection and potential treatment for glaucoma. Optical aberrations have also been studied in relation to IOP measurement (Han et al. 2018), emphasizing the importance of minimizing optical aberrations for accurate readings.

Blood pressure measurement is crucial for cardiovascular health. Optical non-invasive sensors have been developed and integrated into wearable devices for accurate measurements of blood pressure, offering high sensitivity and reliability (Upadhyaya et al. 2021).

Bladder pressure monitoring is essential for evaluating lower urinary tract dysfunction. Current methods involve fluid-based systems with limitations, such as delays and measurement errors (Upadhyaya et al. 2021). Photonic MEMS technology offers small, precise, and non-invasive pressure sensors that can continuously monitor bladder pressure, providing valuable diagnostic information (Upadhyaya, et al. 2021; Poeggel et al. 2015).

Viscometric MEMS Biosensors

The viscometric MEMS sensor measures the viscosity of a sample. It can determine both the concentration and the nature of a sample, often by examining the sensor's mechanical properties. A microcantilever was integrated into MEMS for viscometric sensing and it experienced deformation when the sample was applied to it (Fernandez et al. 2007). When a fluid sample is introduced between these cantilevers, it acts as the dielectric material of the capacitor. Changes in the dielectric properties of the sample occur as a result of variations in sample viscosity. In another approach, the resonance frequency of the cantilever is measured where an external frequency matching the resonance frequency is applied, inducing vibration in the cantilever (Huang et al. 2009b). The introduction of an analyte to the cantilever changes the vibration frequency due to variations in sample viscosity. This vibration can be precisely measured using the optical lever method, which excels at measuring minute object displacements. Monitoring different parameters such as amplitude, phase shift, and vibrating frequency using viscometric MEMS biosensors aids in sample analysis or detecting sample properties.

MEMS in Lab-On-A-Chip Applications

Lab-On-A-Chip (LOC) Devices

Lab-on-a-chip (LOC) devices have opened another dimension in the world of miniaturized technology by incorporating the capability of performing a wide range of biochemical analyses at the pl/nl scale in a single device. Like the traditional laboratory, LOC devices can perform genetic analysis, drug development, and specialized diagnosis at high speed, with less reagent consumption and reduced size. Micro total analysis (μ-TAS) is a robust and comprehensive system for the development of not only LOC devices but also Point-of-Care (POC) diagnostic/testing devices and organ-on-a-chip (Jakaratanopas et al. 2021).

Micro Total Analysis (μ-TAS)

A micrototal analysis system (μ-TAS) is a compact apparatus that combines a sensing mechanism with miniaturized devices to conduct microscale biological or chemical analyses within a single integrated package (Sharma and Sharma 2022). Previously, emphasis was placed solely on the sensing mechanism for biological and chemical analyses. However, the introduction of miniaturized technologies, commonly referred to as MEMS, has allowed for the incorporation of devices such as micropumps, mixers, separators, and reactors into the sensors. This advancement enables not only the sensing through chemical reactions but also the preprocessing of samples before and the filtration of various components after the chemical or electrochemical reaction. In essence, a μ-TAS comprises four essential elements: (a) a sample pretreatment unit (Berg and Lammerink 1998), (b) a microfluidic unit (microfluidic channels), (c) a detector system (Sharma and Sharma 2022), and (d) a signal processing unit (Berg and Lammerink 1998).

Elements of Micro Total Analysis (M-TAS)/Lab-On-A-Chip Devices

Before delving into the micro-total-analysis system's components, it's essential to explain microfluidics. Microfluidics is a science that focuses on manipulating fluids within the range of picolitres to microliters, using channels sized in the tens to hundreds of micrometers. The utilization of microfluidic devices in fields such as diagnostics, biology, and agriculture, has enabled the testing and analysis of tiny sample volumes. This is possible owing to the reduced feature size characteristic of microfluidic devices. The physics of microfluidics is different from the fluids of macro scale. For microfluidics, the surface-to-volume ratio is the key determinant that makes its physics different from macro-fluidics. Effects of viscosity, stiction force, interfacial surface tension, diffusion, evaporation, electrokinetics, and electrical double layer play an important role in regulating the micro-scale fluid dynamics. The following properties are significant in microfluidic devices (Korvink et al. 2006).

i. Physical properties: Compressibility and thermal expansion
ii. Surface tension
iii. Vapor pressure
iv. Electrical properties
v. Optical properties
vi. Transport phenomena: Diffusion, viscosity, and thermal conductivity

The following subsection provides more details on μ-TAS.

Sample Pretreatment Unit

The key components of the sample pretreatment unit include:
(a) Micromixers, (b) Microvalves, (c) Micropumps, and (d) Microfilters.

a. Micromixers

Micro mixers play an important role in detecting cells and biological targets by increasing the efficiency and sensitivity of microfluidic devices. Due to the microfluidic channel's small size, the flow inside it is predominantly laminar, making it necessary to introduce turbulence to optimize heat and mass transfer (Ward and Fan 2015). Generally, there are two types of micromixers:

i. Active Micromixers
ii. Passive Micromixers

i. Active Micromixers: They require an external energy source to generate turbulence or disturbance in the fluid flow and induce chaotic advection to maximize the contact area thus increasing the mixing efficiency. The active micromixers can be further categorized based on the external energy sources (Ward and Fan 2015; Cai et al. 2017):

(a) Pressure field driven micromixer: based on alternate perturbation
(b) Electrical field driven micromixer: based on electro-hydro dynamic instability
(c) Sound field driven micromixer: based on acoustic resonance disturbance
(d) Magnetic field driven micromixer: based on magneto hydro dynamics and magnetic stirring
(e) Thermal field driven micromixer: based on electrothermal effect

ii. Passive Micromixers: Passive Micromixers, also known as static micromixers, do not require any external power, rather they are designed by altering channel dimensions, geometry, or by modifying surface properties. They can be sub-classified as the following based on the dimension of the structure (Ward and Fan Sep. 2015; Cai et al. 2017):

(a) Two-dimensional passive micromixers: It is a planar structure fabricated with the lithography technique. Some two-dimensional micromixers include:

(1) Obstacle-based micromixer: Different grooves or barriers of different shapes act as obstacles in the channel.
(2) Unbalanced collision-based micromixers: The asymmetric structure of the channel and different flow rates of the fluids are the bases of this type of micromixer.
(3) Spiral micromixers: They work on the principle of Dean vortices that form inside the curved channels due to centrifugal force. This causes the fluid to circulate and mix.
(4) Convergence-divergence-based micromixer: The disturbance is caused due to expansion vortices in the laminar flow channel which eventually increases the mixing efficiency.

(b) Three-dimensional passive micromixers: It is a complex structure, that requires complicated fabrication. Some of the three-dimensional micromixers are (Ward and Fan 2015):

(1) Lamination-based micromixers: They are composed of multilayer structures and fast mixing capabilities.
(2) Chamber-based micromixers: Specially structured chambers for recirculation, counterflow, and convergence or divergence of the fluid.
(3) Three-dimensional spiral micromixers: The combination of two spiral microchannels with an erect channel or oppositely placed double helical channels is used to form this type of micromixer.
(4) Overbridge-based micromixers: Three-dimensional structures are connected by a bridge-shaped channel to form this type of micromixer (Ward and Fan 2015)

Another classification of the passive micromixers is:

i. Diffusion and stream splitting/combination: The concept is to put different fluids in contact to mix them by molecular diffusion. e.g., T-shaped and Y-shaped micromixers.
ii. Slanted wells ridges or grooves: The mixing occurs due to slanted walls and grooves in the system.
iii. Multiphase mixing: It's a process of mixing different phases of fluid using an immiscible carrier to enhance the mixing efficiency.

b. Microvalves

The function of the microvalve is to control and restrict the fluid flow in the microfluidic channel. It can be categorized as (Oh and Ahn Mar. 2006):

1 Active Microvalves

(a) Mechanical: This type of microvalve uses magnetic, electric, piezoelectric, or thermal actuation principles to actuate the tiny mechanical moving parts.

(b) Non-mechanical: This type of microvalve does not contain tiny mechanical moving parts but uses bistable, electrochemical, or phase change mechanisms for operation.
(c) External: This type of microvalve uses external actuation forces by indirect contact with polymer membranes. Examples include rheological, modular, and pneumatic.

2 Passive Microvalves

(a) Mechanical: Check valve (Flap, membrane, and ball)
(b) Non-Mechanical: Capillary (Abrupt, liquid triggered, burst, and hydrophobic valves)

c. Micropumps

In μ-TAS, micropumps are employed for drug delivery, manipulation of small volumes of fluid for monitoring single cells, and cell culturing. Micropumps can be classified as follows based on the operation and moving parts (Korvink et al. 2006):

i. Mechanical micropumps
ii. Non-mechanical micropumps

(i) Mechanical micropumps: The pumping action is solely based on physical actuation. It contains moving parts. This can be further classified as:

(1) Piezoelectric (PZT) micropumps: This type of pump works on the reciprocating mechanism to generate suction and discharge flow. The essential components of these pumps include the pumping membrane or diaphragm, along with check valves, responsible for producing the reciprocating motion. Further, the diaphragm is actuated by a piezoelectric mechanism.
(2) Electrostatic micropumps: These micro pumps use Coulomb's law of attraction by using parallel plate configuration to exert a force or induce a displacement.
(3) Thermal actuation micropumps: It works on the mechanism of thermal deformation of the material which helps to drive fluid flow due to expansion and contraction. Thermal actuation micropumps consist of a diaphragm, a heater, and a chamber.
(4) Electromagnetic (EM) actuation micropumps: They are based on the electromagnetic mechanism or magneto-strictive method. The process involves the creation of a Lorentz force to induce a reciprocating movement in the membrane. This is achieved by placing two magnets with opposite poles in a chamber to establish a consistent magnetic field and passing a low-frequency square wave electric current through small wires.

(ii) Non-mechanical micropumps: They convert non-mechanical energy into kinetic momentum to drive fluid in the microchannel. They can be categorized as:

(1) Magnetohydrodynamic (MHD) micropumps: These micropumps work by the interaction between the working fluid (which is conductive in nature) and electrical and magnetic fields. Out of the four walls of the channels, two walls are surrounded by electrodes, and two walls are surrounded by permanent magnets to generate Lorentz force.
(2) Electrohydrodynamic (EHD) micropumps: The fluid flow is manipulated by introducing an external electric field that generates electrostatic forces to interact with charges in the dielectric fluids.
(3) Electroosmotic (EO) micropumps: The application of an electric field across the conductive liquid causes charge accumulation near the channel walls. This attracts oppositely charged particles in the liquid to create a flow along the channel popularly known as electroosmotic flow which is the basic working principle of the EO pumps. These are of two types:

    a. DC electroosmotic pump
    b. AC electroosmotic pump

(4) Bubble-type micropumps: The basic mechanism lies in the periodic expansion and collapse of the bubbles to create the pumping effect. The bubble generation in the microchannel is controlled by voltage input.
(5) Electrowetting micropumps: These pumps operate by reducing the interfacial energy between the solid–liquid interface by the application of a voltage, which promotes fluid transport by utilizing surface tension.
(6) Electrochemical micropumps: The gas bubbles generated by the electrolysis of water are the driving force for electrochemical micropumps. These pumps consist of electrodes, channels, inlets, outlets, and an electrolysis chamber.

d. Microfilters

In a microfluidic system, a microfilter is employed to enhance the performance of the valves and protect them from potential blockages caused by contaminants like dust particles and living cells. This microfilter can be equipped with a polymer membrane or a microfabricated silicon filter.

Microfluidic Unit

The microfluidic unit is the channel or chamber where all kinds of actions such as preconcentration, mixing, separation, and reactions occur. Also, micropumps, micromixers, and microvalves are part of the microfluidic unit. This sub-section explains different types of microfluidic channels and chambers. They can be classified as:

a. Glass-based microfluidic channel: Borosilicate glass is commonly employed in microfluidic chips due to its ability to withstand elevated temperatures and its transparency to light.

a. Silicon-based microfluidic channel: Silicon-based microfluidic channels are manufactured by microfabrication technique, primarily by etching.
b. 3D printed microfluidic channel: Microfluidic channels are produced via 3D printing by taking a design specified in CAD software and then translating it into the desired structure and size using materials like Acrylonitrile butadiene styrene (ABS), resins, and other materials.
c. Polymer-based microfluidic channel: Polydimethylsiloxane (PDMS), a polymer of silicon, is the most popular material for microfluidic channels due to its biocompatibility, opacity, and molding capability. Other materials in this category include polymethyl methacrylate (PMMA) and polystyrene.

Detector System

The detection and sensing is the most important element of the μ-TAS. The detection methods can be categorized into the following (Sharma and Sharma 2022):

(i) Electrochemical methods, (ii) Mass spectroscopy methods, and (iii) Optical methods.

a. Electrochemical Methods: The electrochemical methods have advantages such as low power consumption, excellent sensitivity, cost-effectiveness, and high biocompatibility. These methods can be further categorized as:

(i) Amperometry, (ii) Potentiometry, and (iii) Impedimetric.

i. Amperometry: In amperometry, the current generated by redox reactions due to the interactions of the analyte at the surface of the working electrode is measured. In this case, a constant potential is applied between the reference electrode and the working electrode. The result is plotted in a cyclic voltammogram (CV) and current–time graph.
ii. Potentiometry: In this method, the voltage or potential difference between the reference electrode and the ion-selective electrode is measured. The ion-selective electrode contains a membrane that allows a specific ion to pass through. As a result, an equilibrium is reached leading to a charge separation which builds up potential across the electrodes.
iii. Impedimetric: In this case, an AC signal is applied to the working electrode, and the resulting current is measured. It is performed over a wide range of frequencies and finally, the results are plotted in a Nyquist or a Bode plot.

b. Mass Spectroscopy (MS) Methods: Mass spectroscopy is capable of performing highly specific detection by observing the direction of ions in an electrical or magnetic field which can demonstrate the structure of chemical compounds based on their mass-to-charge ratio. In microfluidic devices, two types of MS configurations are used (Wu and Gu 2011):

i. Electrospray ionization (ESI)-MS
ii. Matrix-assisted laser desorption ionization (MALDI)-MS

c. **Optical Methods**: The properties of light are used to collect information about biological samples for detection. The light properties include absorption, reflection, refraction, emission, and scattering. Some commonly used optofluidic sensing methods and their applications are discussed below (Sharma and Sharma 2022; Wu and Gu 2011):

i. **Fluorescence Spectrometry**: This technique is based on the absorption and emission of light wavelengths and is used to measure temperature, flow velocity, etc.

   i. **Polarization Interferometry**: It uses the interference of light waves and the principles of polarization to gather information on the index of refraction, thickness, and density of the medium.
   ii. **Surface Enhanced Plasmon Resonance**: It measures the changes in the refractive index of the surrounding medium, as well as chemical reactions and dynamics.
   iii. **Fiber Optics**: They measure the rotation and acceleration of bioparticles.
   iv. **Planar Waveguides**: Measure refractive index of target particles. They provide label-free detection capability and good sensitivity.
   v. **Photonic Crystals**: They are composed of micro/nanostructuring of substrates including optical fibers to measure protein–protein interactions, antibody-antigen interactions, and their concentrations.

Signal Processing Unit

Following the interaction between the target analyte and the biological receptor (e.g., enzyme, DNA probe, antibody), a detectable and quantifiable signal (e.g., optical, electrochemical, piezoelectric) is produced, which then undergoes additional processing, modification, and amplification to make it suitable for analysis (Tetyana et al. 2021). This task is accomplished by a signal processing unit which mainly consists of the following elements:

(a) Amplifier, (b) Filter, (c) Processor/Microcontroller, and (d) Display.

a. Amplifier: The generated signal is very low which requires amplification to increase the strength of the signal to improve the precision, accuracy, and sensitivity.

b. Filters: If the output has a high-frequency noise signal, a filter is used to remove unwanted interference.

c. Processor/Microcontroller: The processor implements specific programs to analyze and interpret the data.

d. Display: A display shows the output of the interpreted and analyzed data through an LED screen or some other interface.

Point of Care (POC) Application

The COVID-19 pandemic has served as a harrowing reminder that conventional, in-person lab testing is no longer a feasible option in moments of crisis. The advent of COVID testing kits has become a worldwide point of reference, pointing toward the indispensability of Point-of-Care (POC) testing and diagnostics. This global health crisis has illuminated the limitations of relying solely on centralized lab facilities. The swift development and widespread adoption of COVID-19 testing kits have demonstrated their effectiveness and efficiency. Not only do they provide rapid and accessible testing, but they also play a pivotal role in decentralizing healthcare, ensuring that even remote or underserved areas have timely access to crucial diagnostics (Sachdeva et al. 2021).

The inclusion of micro-electromechanical systems (MEMS) with Point-of-Care (POC) devices, especially incorporating microfluidic systems, has significantly simplified the process of analyzing a diverse range of biological samples, rendering traditional laboratory testing equipment replaceable. A POC testing device should possess the following features (St John and Price 2014):

i. Simple and quick
ii. User friendly
iii. The reagents and consumables should demonstrate durability in both storage and utilization.
iv. The outcomes should align with the established laboratory methods.
v. The device should be safe to use.

MEMS in Organ-On-A-Chip Applications

Organ-on-a-chip (OOAC) technology has revolutionized biomedical research by faithfully replicating human organ conditions in vitro. At its core lies the remarkable contribution of MEMS. This section explains the pivotal role of MEMS in OOAC design and modeling which involves four fundamental components (Wu et al. 2020).

MEMS-microfluidics synergy: Microfluidic systems are the lifeblood of OOAC design, and MEMS technology is seamlessly integrated with them to provide precise control over the microenvironment. MEMS-driven microfluidics regulate nutrients, enhance waste removal, and replicate dynamic organ conditions such as blood flow elevating their precision and functionality for researchers.

MEMS-enhanced concentration gradients: Creating and sustaining stable biochemical concentration gradients is vital for faithfully replicating in vivo conditions. MEMS-based microfluidics provide exceptional control over gradient generation and stability, which is crucial for processes such as angiogenesis and cell migration, thereby improving the accuracy of disease modeling and drug testing.

Dynamic mechanical stress through MEMS: Organ-on-a-chip systems aim to recreate the mechanical stresses that human organs endure. MEMS technology plays a pivotal role in achieving this goal by incorporating elastic membranes that can apply periodic mechanical stress to cultured cells (null Kshitiz et al. 2012). This dynamic mechanical stimulation, driven by MEMS, is essential for influencing cell differentiation and accurately replicating the physiological conditions of organs.

Precision cell patterning with MEMS: The intricate arrangements of cells found in human organs are faithfully recreated using MEMS-based techniques for cell patterning. MEMS technology, whether through surface modifications, templates, or advanced 3D printing, provides the precision required for accurate cell placement (Zhou et al. 2012; Tibbe et al. 2018). MEMS-driven cell patterning is fundamental for constructing functional tissue models within OOAC systems, pushing the boundaries of what is achievable in biomedical research.

MEMS-driven OOACs are poised to transform drug testing, disease modeling, and various areas of biomedical research, ultimately advancing our understanding of human biology and diseases while ethically reducing reliance on animal testing (Al-Lamki et al. 2017). As we transition to the next section, we will explore practical applications and emerging trends that further underscore the role of MEMS in shaping the future of OOAC technology.

Mimicking Physiological Conditions

Heart-On-A-Chip

Cardiotoxicity, a leading cause of drug recalls, presents substantial challenges for pharmaceutical companies (Wadman 2007). While mathematical models and animal studies have been traditional approaches, limitations persist. Human pluripotent stem cells hold promise, yet differentiation methods often yield immature cardiomyocytes (Kattman et al. 2011). A breakthrough approach leverages electrical and mechanical stimulation, facilitated by MEMS components, to enhance tissue maturation, thereby improving cell organization and electrophysiology (Nunes et al. 2013; Mathur et al. 2015). The integration of MEMS technology into microscale organ-on-a-chip devices has revolutionized cardiac research. These devices offer cost-effective, high-throughput measurements of cardiac contraction dynamics, paving the way for more effective drug testing and reducing research costs.

Liver-On-A-Chip

MEMS technology is driving a paradigm shift in liver-on-a-chip platforms, addressing the significant issue of liver toxicity in drug development. Liver toxicity, responsible for 32% of post-approval drug withdrawals, poses a formidable challenge for the pharmaceutical industry (Watkins et al. 2014). While hepatocytes can be sourced from human biopsies, maintaining their functionality in vitro is a complex task. MEMS-based innovations enable crucial heterotypic and homotypic interactions with stromal cells to preserve hepatocyte function (Shlomai et al. 2014; Khetani and Bhatia 2008). Leading this charge is HepatoPac, a MEMS-driven micropatterning technology, boasting a remarkable 65% success rate in drug toxicity screening over 14 days (Khetani et al. 2013). Integration of MEMS technology enhances accuracy through the inclusion of pro-inflammatory cytokines and advanced 3D models, heralding a transformative era in drug development.

Body-On-A-Chip

Innovative body-on-a-chip devices, powered by MEMS technology, represent a groundbreaking approach to simulating the intricate functions of the human body and predicting drug pharmacokinetics (PK). These systems, when coupled with PK models, aspire to replicate the complexity of inter-organ interactions, enabling assessments of drug absorption, distribution, metabolism, and elimination (ADME), as well as the study of hormonal signaling across multiple organs (Wikswo et al. 2013). However, a pivotal challenge revolves around achieving precise scaling of each organ component to faithfully mimic physiologically relevant interactions. This scaling process must account for critical factors such as tissue barrier surface area, organ mass, blood residence time, and variations in tissue metabolism and functionality (Wu et al. 2020). In concert with computational modeling, MEMS-powered body-on-a-chip devices hold immense promise for illuminating these intricate interactions, offering valuable insights into organs where knowledge gaps persist.

Cross-Organ Signaling and Cellular Migration

Innovative microfluidic cell culture devices, driven by MEMS technology, are advancing our understanding of cross-organ signaling and cellular migration in body-on-a-chip systems (Zhang et al. 2018; Miller and Shuler 2016). These devices enable the study of inter-organ communication and the relationship between organ volume and blood residence time. They simulate drug passage through barriers like skin, the gastrointestinal tract, and lungs, maintaining cell viability for extended periods (Miller and Shuler 2016). MEMS-based micropumps, like those developed by TissUse, facilitate separate circulation systems for multi-organ interactions and drug ADME (absorption, distribution, metabolism, and elimination) profiling (Maschmeyer et al. 2015). Additionally, computational analysis and innovative biofabrication techniques, including 3D printing, enhance the capabilities of these micro-physiological systems.

While many multi-organ-on-a-chip systems emphasize biochemical signaling, a few are now addressing living cell trafficking between organs. Platforms like InVADE provide continuous vascular connections between organs, enabling the study of cell migration (Lai et al. 2017). Moreover, the Allergy Test-on-a-Chip uses chemical gradients to simulate immune cell migration for allergy research (Yarmush et al.

2017). Although organ scaling remains challenging, current platforms effectively model inter-organ biochemical crosstalk, drug ADME, and cellular trafficking. They also hold promise for addressing sex bias in research, with a crucial need for validation using human cell types in clinical applications (Nawroth et al. 2018).

Early Drug Discovery: Basic Research and Disease Modeling

Organ-on-a-chip technology is a groundbreaking approach that promises to revolutionize our understanding of human pathophysiology and disease mechanisms. It addresses a critical issue in drug development: the high failure rate of new drugs in clinical trials. Unlike traditional animal models, organ-on-a-chip systems offer a controlled and highly traceable environment that closely mimics human physiology.

These innovative platforms have proven invaluable for investigating complex processes underlying cancer progression and treatment. Organ-on-a-chip models provide insights into cancer cell behavior, extracellular signaling, biophysical factors in the tumor microenvironment (TME), and tumor heterogeneity (Marturano-Kruik et al. 2018; Xiao et al. 2019). Researchers have developed various 3D cancer-on-a-chip models that faithfully replicate different aspects of solid and liquid TMEs, including stromal components, immune suppressor cells, and chemokines (Hassell et al. 2017). These models are instrumental in uncovering the mechanisms driving resistance to chemotherapy and immunotherapy.

Tailored models for specific cancer types, such as pancreatic ductal adenocarcinoma (PDAC) and glioblastoma (Nguyen et al. 2019), shed light on how tumors interact with the vascular network and immune system, offering insights into treatment challenges. Moreover, 3D models simulating leukemia bone marrow niches have unveiled the intricacies of chemotherapeutic resistance, involving stromal niche cells, signaling pathways, and immune cell subsets (Ma et al. 2020). Organ-on-a-chip systems integrated with the circulatory system have been instrumental in unraveling disease development, including cancer initiation and metastasis. For example, a multi-organ chip system successfully replicated lung cancer metastasis to various organs, revealing the destructive impact of tumors on targeted tissues (Xu et al. 2016).

Beyond the realm of biological signaling, organ-on-a-chip technology facilitates exploration into mechanobiological factors influencing disease progression and therapeutic resistance. Studies employing lung chips have demonstrated how mechanical forces generated during breathing can induce dormancy and drug resistance in lung cancer cells. Similarly, investigations in bone perivascular niche models have underscored the role of extracellular physical factors in tumor dormancy and chemoresistance.

In conclusion, organ-on-a-chip technology stands as a promising alternative to traditional animal models, offering invaluable insights into critical disease mechanisms, therapeutic biomarkers, and targets to improve disease outcomes.

Revolutionizing Drug Development with Microfluidic OOAC Technology

In the pursuit of more efficient drug discovery, Organ-on-a-Chip (OOAC) technology, enriched with the pivotal contribution of MEMS, has emerged as a revolutionary approach. The current landscape of drug development heavily relies on in vitro 2D or 3D cell cultures and in vivo animal experiments (Cook et al. 2014; Dugger et al. 2018). However, these conventional methods often produce results that inadequately predict drug efficacy and toxicity in humans. Animal studies, long regarded as the gold standard for preclinical drug validation, frequently fall short due to inherent species differences between animals and humans (Jang et al. 2019). This disconnect leads to a high rate of drug candidates failing during clinical trials, resulting in significant financial and logistical challenges for pharmaceutical companies.

Organ-on-a-chip technology, empowered by MEMS, offers a transformative solution by replicating crucial aspects of human organ physiology on a minute scale. This innovative approach encompasses cellular architecture, 3D extracellular matrix, biochemical factors, and biophysical cues that define human tissues and organs (Ma et al. 2021). Organ-on-a-chip devices are typically crafted from materials such as polydimethylsiloxane (PDMS) and feature microchannels that enable precise cell patterning and manipulation of various fluidic and chemical parameters. This level of control, facilitated by MEMS technology, empowers the creation of highly accurate culture conditions that closely mimic the in vivo microstructural and functional characteristics of human organs (Whitesides 2006).

While organ-on-a-chip technology, fortified by MEMS, holds immense promise, its full potential is still unfolding. Challenges persist in transitioning from fundamental research to seamlessly integrating this platform into the drug development pipeline. Despite these obstacles, MEMS-enhanced organ-on-a-chip can significantly enhance drug discovery by providing more precise and efficient preclinical testing. Ultimately, this advancement stands to benefit both the pharmaceutical industry and patients worldwide.

Harnessing Organ-On-A-Chip for PK-PD Insights

Pharmacokinetics (PK) and Pharmacodynamics (PD) investigations are critical in drug development, shedding light on drug behavior and its effects. The pivotal role of MEMS-enhanced organ-on-a-chip technology in delivering real-time PK-PD insights across diverse human organ sites is undeniable. Multi-organ chips, faithfully replicating vital organ functions, stand out as ideal platforms for these investigations. They can mimic the liver, kidney, and heart, playing crucial roles in drug metabolism, and accommodate various drug delivery methods (Abaci and Shuler 2015). For instance, a multi-organ chip offered valuable insights into the anticancer drug 5-fluorouracil, revealing notable liver resistance compared to other organ compartments (Sung and Shuler 2009; Sung et al. 2010). Similar chips accurately predicted PK parameters for nicotine and cisplatin, aligning with clinical data and thereby optimizing clinical trial drug regimens. The study of drug transport across these chips necessitates a fluidic circulatory system connecting various organ compartments, effectively addressing physiological barriers. Vascularized multi-organ chips bring precise quantification to drug distribution studies (Wevers et al. 2018). In order to comprehend drug transport intricacies, it becomes imperative to tackle tissue barriers, such as the blood–brain barrier (BBB) (Koo et al. 2018). A MEMS-enhanced microfluidic BBB model delved into organophosphate toxicity, providing vital insights into barrier damage and cellular viability. The integration of these interfaces into multi-organ chips, while challenging, holds great promise for advancing drug development. Collaboration among experts remains an essential driving force to fully harness the potential of MEMS-powered OOAC technology in PK-PD studies.

Assessing Drug Safety and Efficacy in Preclinical Trials and Translation

In the late stages of preclinical drug development and clinical trials, the critical assessment of drug safety and efficacy takes center stage. At the forefront of this process, MEMS-empowered human organ-on-a-chip technology emerges as a pivotal tool, offering an efficient and precise means to evaluate drug toxicity before advancing to clinical trials (Cirit and Stokes 2018). One notable application involves a biomimetic liver chip (Ma et al. 2016), intricately designed to mimic liver lobules, facilitating the early detection of adverse reactions resulting from drug interactions during liver metabolism, particularly in combination therapies. Multi-organ chips further enhance this process by investigating both intended and unintended drug effects, automating biomarker analysis to monitor toxicity in various organs. Recent research findings have illuminated instances where certain anticancer drugs initially impact the liver (Zhang et al. 2017), subsequently leading to cardiotoxicity during hepatocyte metabolism. Crucially, organ-on-a-chip models incorporate cells sourced from both healthy individuals and patients, providing a realistic environment to assess patient-specific responses to drugs. This approach effectively replicates clinical symptoms in conditions like chronic obstructive pulmonary disease (Benam et al. 2016) and offers a means to evaluate personalized reactions to cancer immunotherapies.

Nonetheless, developing complex multi-organ chips presents challenges, encompassing organ scaling and seamless integration. Striking a delicate balance between complexity and feasibility remains imperative. Future research endeavors should prioritize the incorporation of autologous patient-derived immune and tumor cells, a pivotal step toward predicting patient outcomes, particularly in the context of immunotherapy trials.

MEMS in Implants

Artificial Implants

Artificial implants are medical devices or prosthetic components that are surgically placed inside the human body to replace, support, enhance, monitor, and stimulate the function of a particular organ, tissue, or body part. For example, joint replacements like hip or knee implants are used to restore mobility and alleviate pain in individuals with arthritis or joint problems. Another example is implants equipped with sensors that can be used to monitor various physiological parameters within the body for medical diagnosis and treatment (Silemek et al. 2023). The requirements for artificial implants can vary significantly depending on the type of implant and its intended use. The most common requirements are biocompatibility, small footprint, durability and longevity, less infection risk, cost-effectiveness, etc. Artificial implants research continues to evolve with advancements in material science, bioengineering, and medical technology, and provides solutions to various medical conditions to improve the quality of life for countless individuals.

Importance of MEMS in Artificial Implants

MEMS plays a significant role in the field of artificial implants, particularly in the realm of medical devices and healthcare. Advances in microfabrication and bio/chemical engineering techniques are now enabling a large variety of miniaturized implantable systems for sensing, health monitoring, or deficiency treatments (Bhansali and Vasudev 2012). This progress is driving physicians and patients to express an increasing need for miniaturized implantable devices as they are offering less invasive implantation procedures, greater comfort for the patient, biocompatibility, wireless communication, easy integration of sensors to implants, less infection risk, improved performance, and often provide innovative measurements and treatment (Wei and Liu 2008). MEMS technology has revolutionized the field of artificial implants by enabling the development of smaller, smarter, and more sophisticated implantable devices. Figure 3.5 illustrates the recent remarkable expansion of the application field of these devices (Wei and Liu 2008).

Common MEMS Devices in Artificial Implants

Various types of MEMS devices find widespread application in artificial implants, each fulfilling specific functions. These devices include pressure sensors for monitoring blood pressure and intracranial pressure (Ghanbari et al. 2017), accelerometer utilized in activity tracking and fall detection (Chen et al. 2011), gyroscopes

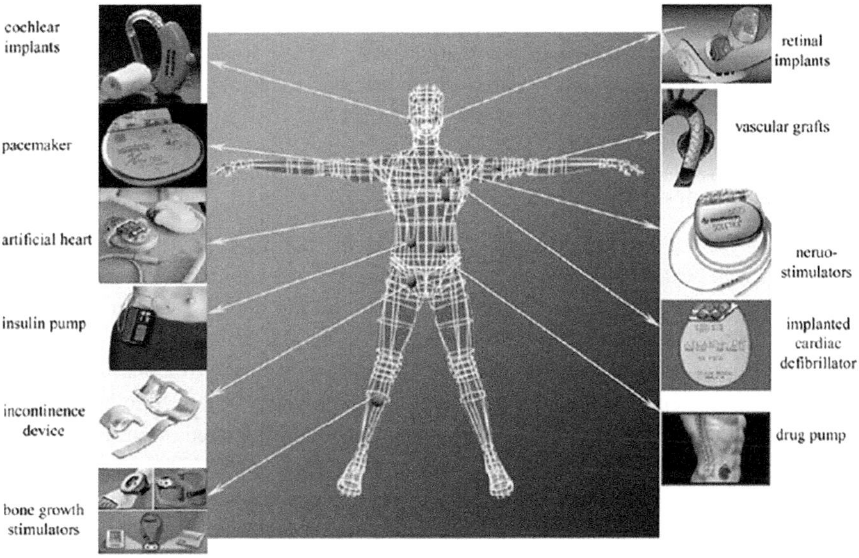

Fig. 3.5 Schematic illustration for applications of various implanted medical devices. Reproduced with permission from Wei and Liu (2008), Copyright (2008), with Springer Nature, License Number: 6024841294091

for measuring angular velocity and rotational motion (Valentin et al. 2013), and microphones integrated into cochlear implants to capture and process sound signals for individuals with hearing impairment (Mitchell-Innes et al. 2017). Additionally, resonators play a pivotal role in implantable medical devices like pacemakers and implantable cardioverter-defibrillators (ICDs), ensuring precise timing of electrical pulses delivered to the heart (Singhwal et al. 2022). Microvalves are employed in implantable drug delivery systems, enabling meticulous control over medication release. Energy harvesters are also employed to generate power from the implant's surrounding environment. These diverse MEMS devices operate on different actuation mechanisms, including electrostatic, piezoelectric, electrothermal, and polymer-based methods. Different types of MEMS devices according to the common actuation mechanism are explained next.

Electrostatic MEMS Sensors in Implant Applications

Electrostatic MEMS devices use electrostatic forces to actuate and move microstructures. This is achieved by applying voltage differences between conductive electrodes and the microstructure, which creates attractive or repulsive forces based on the electric field strength. Usually, this type of device consists of microscale components, such as microcantilevers, microbridges, microswitches, microvalves, and microactuators, and fabrication involves microfabrication techniques, including photolithography, etching, deposition, and bonding (Chuang et al. 2010). Electrostatic MEMS devices are particularly interesting due to low power consumption, fast response times, simple fabrication, high reliability, easiness of integration with CMOS platforms, and can be relatively easily upscaled to mass production (Chuang et al. 2010).

Energy Harvester

Energy harvesting or energy scavenging is the method of collecting energy from wasted energy sources such as heat, vibrations, and mechanical deformations, and it is converted into useful electrical energy. Supplying wireless power is a challenging technical problem of great importance for implantable biomedical devices. Nowadays, energy harvesting has become a promising solution to replace the use of conventional batteries. The batteries have a limited lifespan and need to be replaced or recharged regularly. This can be a problem for some implanted biomedical applications like cochlear implants and pacemakers (Deterre, et al. 2011). An electrostatic MEMS energy harvester consists of a movable microstructure and a fixed substrate separated by a small air gap or dielectric, both equipped with electrodes made of conductive materials. When the movable structure experiences mechanical vibrations or motion, it moves relative to the fixed substrate. This movement generates an electric field between the electrodes and a changing electric field induces an electrical charge on the electrodes. This induced charge creates an electrical potential difference (voltage) across the electrodes, which can be harnessed as electrical energy or stored or directly used to power low-power electronic devices or sensors (Deterre et al. 2011). To maximize energy harvesting efficiency, it's essential to design the harvester's resonance frequency to match the frequency of the external mechanical vibrations or motion source. The cantilever-based energy harvester (EH) structure is preferred because for a given input force in the cantilever structure, the average strain generated is higher and therefore the power output is much greater. Cantilever also results in the lowest resonant frequency for a given size (Bindu et al 2014). Moreover, cantilevers are popular mainly because the structure of cantilevers is easy to fabricate and implement. Reliability is another important design parameter to make the energy harvester operate reliably for an extended period of time as surgical removal or replacement is risky and costly.

Pressure Sensors

Pressure in various organs of the body (e.g., brain, eye, heart, bladder) is highly regulated and can indicate patient health or disease progression. In diseases where the ability to regulate these pressures is lost, impairment of function or death may result. Monitoring the absolute value of and variation of internal pressures allows diagnosis and tracking of the progress of medical intervention. Electrostatic pressure sensors consist of fixed and moveable electrodes separated by dielectric or air gap. In an artificial implant, the moveable electrode is typically exposed to the environment where the pressure needs to be measured (Yu et al. 2014). Any change in pressure causes the flexible electrode to deform and change the capacitance. By measuring the change in capacitance with the help of circuits, pressure can be measured. These electrostatic pressure sensors are commonly used in various medical applications including monitoring intracranial pressure, measuring blood pressure within artificial hearts, and assessing pressure in bladder and intraocular implants, among others (Yu et al. 2014). The sensing element must be able to transduce pressure over the range of clinically normal and abnormal values with sufficient resolution to inform treatment. A telemetry system is required to transmit the data to an external reader and may also serve to power the implanted system. In considering materials and packaging methods for implantable pressure sensors, several points must be considered: (1) acceptable tissue integration, (2) hermetic encapsulation for protection of active circuits, and (3) long-term mechanical stability in the in vivo environment.

Microelectrode Arrays

Electrostatic MEMS can be used to create microelectrode arrays for neural implants (Chen et al. 2011). These arrays can be inserted into the brain or peripheral nervous systems to record neural activity or provide electrical stimulation (Zoll et al. 2019). They consist of an array of tiny electrodes that can be used to monitor or stimulate the electrical activity of neurons. In electrostatic microelectrode arrays, the interaction between the neural tissue and the electrodes is primarily based on capacitive coupling. When an electrode is placed in proximity to neural tissue, a small capacitance is formed between the electrode and the tissue. When a neuron fires, it generates an electrical potential that results is charge distribution, which causes a shift in the electrical potential at the electrode-tissue interface, leading to a change in capacitance. The amplified signal is then collected, processed, and analyzed to extract relevant neural information, such as the firing rate or pattern of nearby neurons. Such implants are used in research, as well as for restoring lost sensory or motor functions.

Piezoelectric-Based MEMS Sensors for Implant Applications

The piezoelectric effect causes the piezoelectric material to generate an electric charge or voltage when mechanical stress or deformation is applied. Piezoelectric material is typically incorporated into the MEMS structure in a way that allows it to deform or generate strain when subjected to mechanical forces or pressure. If the generated electrical signal is very small, it can be amplified with the help of electric circuits. There are two modes of voltage generation in piezoelectric materials, d31 and d33. In d31 the voltage is generated in the direction perpendicular to the applied stress. In d33 mode, both the applied stress and the generated voltage are in the same direction (Kim et al. 2012). Conventional piezoelectric devices have the d31 mode in which the electrodes are put in the top and bottom of the piezoelectric layer. The voltage generated in a d31 mode piezoelectric energy harvester is given by Kim et al. (2012).

$$\mathrm{V} = 31 \times \ \mathrm{g31tp}$$

where 31 is the applied stress, g31 is the Piezoelectric voltage coefficient, and tp is the distance between the two electrodes which is equal to the thickness of the piezoelectric layer.

Piezoelectric materials exhibit high sensitivity to mechanical stimuli, making them capable of detecting even small forces, pressures, and vibrations. Piezoelectric sensors require very low power to operate. Different types of sensors for artificial implant applications can be made using the piezoelectric effect, such as pressure, strain, force sensors, acoustic transducers, and energy harvesters.

Acoustic MEMS Transducer

The piezoelectric acoustic transducer consists of a thin diaphragm incorporated with piezoelectric materials and electrodes. When sound waves strike the diaphragm, it vibrates in response to the changes in air pressure caused by the sound and generates an electrical charge across the electrodes. The generated electrical signal can be processed by an amplifier, analog-to-digital converter, and signal conditioner to make the signal representable as an audio signal, which can be used for recording, voice recognition, communication, or other acoustic sensing applications (Qu et al. 2021). For example, acoustic transducers can be integrated into cochlear implants to capture external sound and convert it into electrical signals that stimulate the auditory nerve, allowing recipients to perceive sound (İlik et al. 2018). Another example is in certain cardia implants, such as ventricular assist devices (VADs) or artificial hearts, acoustic MEMS transducers can monitor heart sounds (phonocardiography) to assess the performance of the device and detect potential issues like valve malfunction or blood flow anomalies (Qu et al. 2021).

Energy Harvester

The transformation of ambient mechanical energy into electrical energy using piezoelectric materials is known as piezoelectric energy harvesting. Piezoelectric energy harvesting systems show high flexibility, simple configuration, and higher conversion ability along with the proficiency of being integrated into a system. Different kinds of piezoelectric materials have been studied in the past decades. Among them, piezoceramics such as lead zirconate titanate (PZT) and $BaTiO_3$ found more interest due to their high piezoelectric coefficients and higher energy harvesting capability compared to other piezoelectric materials (You et al. 2019). The piezoelectric material used in this EH is zinc oxide, which is a biocompatible material and therefore suitable for biomedical implants. A heart is a highly efficient pump, and therefore a good source of reliable mechanical energy. Integrating an energy harvesting device on or in the heart will allow it to harvest a small amount of mechanical energy into usable electrical energy, which can be used to power an implantable device such as a pacemaker (Dagdeviren et al. 2014). Piezoelectric ribbons and films have been developed that attach to the outside of the heart to harvest energy. In a cochlear implant, the eardrum vibrates as the sound waves enter the ear. If the frequency of the vibration of the eardrum matches with the resonant frequency of the energy harvester, it generates maximum voltage output which can be used to power the battery of the cochlear implant. Many research groups have demonstrated different types of energy harvesters based on piezoelectric materials including EH for a leadless pacemaker (Jackson et al. 2017), hip replacement (Lange et al. 2021), and artificial heart (Niu et al. 2009).

Accelerometer

Piezoelectric accelerometers are devices that measure vibration through the piezoelectric effect exhibited by materials such as zinc oxide (ZnO), aluminum nitride (AIN), and PZT. Most MEMS piezoelectric uniaxial accelerometers are comprised of five elements: frame, seismic mass, a set of cantilevers, a piezoelectric layer, and a set of electrodes (Encapsulated sub-millimeter piezoresistive accelerometers for biomedical applications—ProQuest 2023). When the frame is excited by a base acceleration, the inertia of the seismic mass deforms the cantilever or membrane, upon which the piezoelectric material is placed, and therefore, deformed as well. The attached mass transforms applied acceleration into force in accordance with Newton's second law ($F = ma$). It also features a flexure that functions as a spring, deforming in proportion to the applied force, following Hooke's law ($F = kx$). Additionally, the surrounding medium serves as a damping element, reducing the amplitude over time, causing the structure to resonate at a damped frequency. Many accelerometers based on piezoelectric effect have been demonstrated for different artificial implant applications including accelerometer for implantable heart monitoring devices (Tjulkins et al. 2016), prosthetic arms (Kyberd and Poulton 2017), and deep brain stimulation (LeMoyne et al. 2015).

Electrothermal-Based MEMS Sensors for Implant Applications

Electrothermal MEMS sensors are typically composed of microscale or nanoscale structures, such as beams, diaphragms, or cantilevers, and a resistive element is integrated into the microstructure. With the application of voltage to the resistive element, it generates heat and causes thermal expansion of the microstructure, which can be used for sensing purposes (Potekhina and Wang 2019). Some limitations of electrothermal MEMS, such as the need for precise temperature control, susceptibility to thermal drift, and potential heat dissipation issues in high-power applications (Potekhina and Wang 2019).

Micro Actuator

Electrothermal microactuators can be used in various artificial implant applications to provide precise control and actuation within the body. These microactuators are typically miniature devices that can generate mechanical motion or deformation in response to electrical input and can be integrated into implantable devices. For example, in implantable microfluidic systems, electrothermal microactuators can manipulate microvalves or pumps to control the flow of fluids. This is valuable for applications like intravenous drug infusion, where accurate and controlled drug delivery is crucial (Li et al. 2010). Electrothermal microactuators can function as artificial muscles within implants, providing controlled mechanical motion. These actuators can be used in devices like artificial hearts or assistive devices for individuals with mobility impairments. In cochlear implants, microactuators can control the positioning of electrode arrays within the cochlea (Lee et al. 2005). Precise actuation can optimize the electrode-tissue interface, improving the quality of sound perception for individuals with hearing loss. In neurostimulation implants, such as deep brain stimulators (DBS) or spinal cord stimulation (SCS), electrothermal microactuators can be used to control the movement or positioning of electrodes (Anand et al. 2012). This enables precise targeting of neural tissues for therapeutic stimulation, offering improved treatment outcomes for conditions like Parkinson's disease or chronic pain.

Drug Delivery

Drug delivery in implantable systems is a critical aspect of medical technology, offering a way to administer medications or therapeutic agents directly to target areas within the body over an extended period. Implantable drug delivery systems typically contain a reservoir or container where the medication or therapeutic agent is stored, and a control mechanism is integrated into the system to regulate drug release (Prescott et al. Apr. 2006). The electrothermal activation process involves metal membranes that are subjected to an applied current. The generation of local resistive heating due to applied current causes failure of the membrane and, therefore, the release of the drug reservoir content. Typically, an array of reservoirs housed within a microchip is linked to a wireless communication system to create an implantable device for controlled pulsatile drug delivery [a]. In one research, implantable devices equipped with two microchips, collectively storing 20 doses of lyophilized hPTH91-34) were implanted in osteoporotic postmenopausal women for several months and wirelessly programmed to release daily doses (Sheybani et al. 2013). In another research, each microchip included 10 reservoirs (of 600 nL) containing 40 μg doses of hPTH(1–34), and a membrane composed of titanium and platinum was utilized for electrothermal activation. These 10 membranes were connected to metal contacts, which were used as the path for current pulses to melt the individual membranes, thereby exposing the contents of the reservoirs to the surrounding tissue fluid and promoting drug release (Maloney et al. 2005). Electrothermal MEMS drug delivery can be integrated into various implantable medical devices, such as insulin pumps, pain management systems, and neurostimulators. In cancer treatment, it can precisely deliver chemotherapy drugs to the tumor site, minimizing damage to healthy tissues.

Biodegradable MEMS Sensors for Implant Application

Biodegradable sensors can be made from biomaterial types such as metals (Zheng et al. 2014), polymers (Lanzani 2014), silicon (Si) (Chen et al. 2018), and their composites (Santis and Cacciotti 2020). To be suitable for use as implantable sensors, they have to meet certain requirements such as (i) cytocompatibility, (ii) biocompatibility, and (iii) functional and structural stability. A standard sensor is usually composed of transducers, a power source (for active transducers), a protective layer (encapsulation), several connectors to link internal components of the sensors, and a data acquisition system. Ideally, all the sensors' components of a biodegradable sensor should be made out of degradable materials. However, there are currently only a few examples of completely biodegradable sensors with the power supply and the sensor communication being the most challenging elements to be made of biodegradable material (Boutry et al. 2018). Natural degradation of biodegradable sensors eliminates the need for surgical removal after their intended use. These types of sensors are suitable for short-term implantable devices, drug delivery systems, and other biomedical applications where temporary monitoring is required. These sensors may not be suitable for long-term implantation because they degrade over time and may not possess the same mechanical properties as traditional MEMS materials, which can limit the range of applications. The small size and power of implants, and the limited selection of materials are the main constraints determining the capabilities of the biodegradable device. In contrast, polymer-based MEMS sensors use non-biodegradable polymers, such as polydimethylsiloxane (PDMS), SU-8, or various epoxy resins.

Temperature Sensors

Temperature monitoring is of great interest for detecting real-time changes in the local temperature of tissues and organs. The most widely used biodegradable temperature sensors are resistance-based sensors due to their fast response, high accuracy, and ease of fabrication. A biodegradable resistance-based temperature sensor has been fabricated by sputtering Zn on fully water-soluble galactomannan films (Yi et al. 2018). The measurement range and sensitivity of resistance-based temperature sensors can be adjusted by modifying the thickness and dimensions of the sending element or by the use of another biodegradable material.

Pressure Sensors

Monitoring various body pressures, such as intracranial intraocular, or diaphragmatic contraction pressure, is of great importance in diagnosing problems and following responses to therapy. The use of untethered implantable systems that can be used outside of the hospital settings, and that do not require a secondary procedure for their removal, is highly desirable for monitoring ICP in traumatic brain injury (TBI) patients (Boutry et al. 2015). A biodegradable pressure sensor is made of Si-nano membranes (NMs), and has a Poly (lactic-co-glycolic acid) (PLGA) membrane sealed against a nonoporous Si substrate, or an Mg foil relief etched onto its surface (Kang et al. 2016). The cavity permits membrane deflection in response to pressure. A capacitive type of pressure sensor has also been fabricated using polycarbonate (PCL) as a substrate on which gold (capacitors, inductors, and resistors) was printed (Annese and Hu 2022). The device is composed of two layers of PCL, one of which has the op-amp clock circuit, power supply, and one face of a pressure-sensitive capacitor, while the other layer has the transmitter and the second face of the pressure-sensitive capacitor. Pressure variations change the capacitance, shifting the oscillation frequency of the electronic circuit.

References

AVGK, Gogoi G, Behera B, Rila S, Rangarajan A, Pandya HJ (2022) RapidET: a MEMS-based platform for label-free and rapid demarcation of tumors from normal breast biopsy tissues. Microsyst Nanoeng, 8(1). https://doi.org/10.1038/s41378-021-00337-z.

Abaci HE, Shuler ML (2015) Human-on-a-chip design strategies and principles for physiologically based pharmacokinetics/pharmacodynamics modeling. Integr Biol (Camb) 7(4):383–391. https://doi.org/10.1039/c4ib00292j

Ahmed O, Wang X, Tran M-V, Ismadi M-Z (2021) Advancements in fiber-reinforced polymer composite materials damage detection methods: Towards achieving energy-efficient SHM systems. Compos B Eng 223:109136. https://doi.org/10.1016/j.compositesb.2021.109136

Ai Y, Sanders CK, Marrone BL (2013) Separation of Escherichia coli Bacteria from Peripheral Blood Mononuclear Cells Using Standing Surface Acoustic Waves. Anal Chem 85(19):9126–9134. https://doi.org/10.1021/ac4017715

Algamili AS et al (2021) A Review of Actuation and Sensing Mechanisms in MEMS-Based Sensor Devices. Nanoscale Res Lett 16(1):16. https://doi.org/10.1186/s11671-021-03481-7

Ali WR, Prasad M (2020) Piezoelectric MEMS based acoustic sensors: A review. Sens Actuators, A 301:111756. https://doi.org/10.1016/j.sna.2019.111756

Al-Lamki RS, Bradley JR, Pober JS (2017) Human organ culture: updating the approach to bridge the gap from In vitro to In vivo in inflammation, cancer, and stemcell biology. Front Med (Lausanne), 4, p. 148, https://doi.org/10.3389/fmed.2017.00148

Anand S, Sutanto J, Baker MS, Okandan M, Muthuswamy J (2012) Electrothermal microactuators with peg drive improve performance for brain implant applications. J Microelectromech Syst 21(5):1172–1186. https://doi.org/10.1109/JMEMS.2012.2203789

Annese VF, Hu C (2022) Integrating microfluidics and electronics in Point-of-Care diagnostics: current and future challenges. Micromachines, 13(11), https://doi.org/10.3390/mi13111923

Asri MIA , Hasan Md N, Fuaad MRA, Yunos Y Md, Ali MSM (2021) MEMS gas sensors: a review. IEEE Sens J, 21(17), pp 18381–18397. https://doi.org/10.1109/JSEN.2021.3091854

Awasthi S, Joshi A (2015) MEMS accelerometer based system for motion analysis. In: 2015 2nd international conference on electronics and communication systems (ICECS), pp 762–767. https://doi.org/10.1109/ECS.2015.7125014.

Bataillard P, Steffgen E, Haemmerli S, Manz A, Widmer HM (1993) An integrated silicon thermopile as biosensor for the thermal monitoring of glucose, urea and penicillin. Biosens Bioelectron 8(2):89–98. https://doi.org/10.1016/0956-5663(93)80057-V

Benam KH et al (2016) Small airway-on-a-chip enables analysis of human lung inflammation and drug responses in vitro. Nat Methods, 13(2). https://doi.org/10.1038/nmeth.3697

Bhansali S, Vasudev A (2012) Mems for biomedical applications. Elsevier

Bhavana C, Gopal J, Raghavendra P, Vanitha KM, Talasila V (2016) Techniques of measurement for Parkinson's tremor highlighting advantages of embedded IMU over EMG. In: 2016 international conference on recent trends in information technology (ICRTIT), pp 1–5. https://doi.org/10.1109/ICRTIT.2016.7569560.

Bindu DRS, Kushal, Potdar M (2014) Study of piezoelectric cantilever energy harvesters. Int J Innov Res Dev, 3(2). Accessed: Oct. 13, 2023. Available: https://www.i-scholar.in/index.php/IJIRD/article/view/144507

Bogue R (2007) MEMS sensors: past, present and future. Sens Rev 27(1):7–13. https://doi.org/10.1108/02602280710729068

Bogue R (2013) Recent developments in MEMS sensors: a review of applications, markets and technologies. Sens Rev 33(4):300–304. https://doi.org/10.1108/SR-05-2013-678

Boutry CM, Nguyen A, Lawal QO, Chortos A, Rondeau-Gagné S, Bao Z (2015) A Sensitive and Biodegradable Pressure Sensor Array for Cardiovascular Monitoring. Adv Mater 27(43):6954–6961. https://doi.org/10.1002/adma.201502535

Boutry CM et al (2018) A stretchable and biodegradable strain and pressure sensor for orthopaedic application. Nat Electron, 1(5), https://doi.org/10.1038/s41928-018-0071-7

Byrom B, McCarthy M, Schueler P, Muehlhausen W (2018) Brain Monitoring Devices in Neuroscience Clinical Research: The Potential of Remote Monitoring Using Sensors, Wearables, and Mobile Devices. Clin Pharmacol Ther 104(1):59–71. https://doi.org/10.1002/cpt.1077

Characterization and comparison of lithium niobate and quartz surface acoustic wave Devices—ProQuest. Accessed: Oct. 15, 2023. Available: https://www.proquest.com/openview/b782c2377165585c66150593533acb2e/1?pq-origsite=gscholar&cbl=18750

Cai G, Xue L, Zhang H, Lin J (2017) A review on micromixers. Micromachines, 8(9). https://doi.org/10.3390/mi8090274

Cai C, Tan J, Hua D, Qin M, Zhu N (2018) Piezoresistive temperature sensors fabricated by a surface micromachining CMOS MEMS process. Sci Rep, 8(1), https://doi.org/10.1038/s41598-018-35113-z

Cao H et al (2012) An Implantable, Batteryless, and Wireless Capsule With Integrated Impedance and pH Sensors for Gastroesophageal Reflux Monitoring. IEEE Trans Biomed Eng 59(11):3131–3139. https://doi.org/10.1109/TBME.2012.2214773

Chen A, Chatterjee S (2013) Nanomaterials based electrochemical sensors for biomedical applications. Chem Soc Rev 42(12):5425–5438. https://doi.org/10.1039/C3CS35518G

Chen C-H et al (2011) A three-dimensional flexible microprobe array for neural recording assembled through electrostatic actuation. Lab Chip 11(9):1647–1655. https://doi.org/10.1039/C0LC00718H

Chen Y, Kim Y-S, Tillman BW, Yeo W-H, Chun Y (2018) Advances in Materials for Recent Low-Profile Implantable Bioelectronics. Materials (Basel) 11(4):522. https://doi.org/10.3390/ma11040522

Chen P et al (2021) Ultrasound-driven electrical stimulation of peripheral nerves based on implantable piezoelectric thin film nanogenerators. Nano Energy 86:106123. https://doi.org/10.1016/j.nanoen.2021.106123

Chen D, Feng W, Zhang Y, Li X, Wang T (2011) A wearable wireless fall detection system with accelerators. In: 2011 IEEE international conference on robotics and biomimetics, pp 2259–2263. https://doi.org/10.1109/ROBIO.2011.6181634

Cheng R, Shao J, Gao X, Tao C, Ge J, Liu X (2016) Noninvasive assessment of early dental lesion using a Dual-Contrast photoacoustic tomography. Sci Rep. 6(1). https://doi.org/10.1038/srep21798

Chia X, Eng AYS, Ambrosi A, Tan SM, Pumera M (2015) Electrochemistry of Nanostructured Layered Transition-Metal Dichalcogenides. Chem Rev 115(21):11941–11966. https://doi.org/10.1021/acs.chemrev.5b00287

Chircov C, Grumezescu AM (2022) Microelectromechanical Systems (MEMS) for Biomedical Applications. Micromachines (Basel) 13(2):164. https://doi.org/10.3390/mi13020164

Chuang W-C, Lee H-L, Chang P-Z, Hu Y-C (2010) Review on the modeling of electrostatic MEMS. Sensors, 10(6), https://doi.org/10.3390/s100606149

Cirit M, Stokes CL (2018) Maximizing the impact of microphysiological systems with in vitro–in vivo translation. Lab Chip 18(13):1831–1837. https://doi.org/10.1039/C8LC00039E

Ciuti G, Ricotti L, Menciassi A, Dario P (2015) MEMS Sensor Technologies for Human Centred Applications in Healthcare, Physical Activities, Safety and Environmental Sensing: A Review on Research Activities in Italy. Sensors (Basel) 15(3):6441–6468. https://doi.org/10.3390/s150306441

Cobo A, Sheybani R, Meng E (2015) MEMS: Enabled Drug Delivery Systems. Adv Healthc Mater 4(7):969–982. https://doi.org/10.1002/adhm.201400772

Coffel J, Nuxoll E (2018) BioMEMS for biosensors and closed-loop drug delivery. Int J Pharm 544(2):335–349. https://doi.org/10.1016/j.ijpharm.2018.01.030

Cook D et al (2014) Lessons learned from the fate of Astra Zeneca's drug pipeline: a five-dimensional framework. Nat Rev Drug Discov, 13(6), https://doi.org/10.1038/nrd4309

Cui J et al (2022) Design and optimization of MEMS heart sound sensor based on bionic structure. Sens Actuators, A 333:113188. https://doi.org/10.1016/j.sna.2021.113188

Dagdeviren C et al (2014) Conformal piezoelectric energy harvesting and storage from motions of the heart, lung, and diaphragm. Proc Natl Acad Sci 111(5):1927–1932. https://doi.org/10.1073/pnas.1317233111

Damborský P, Švitel J, Katrlík J (2016) Optical biosensors. Essays Biochem 60(1):91–100. https://doi.org/10.1042/EBC20150010

D6T | MEMS thermal sensors | OMRON device & module Solutions—Americas. Accessed: Oct. 12, 2023. Available: https://components.omron.com/us-en/products/sensors/D6T

Digital MEMS microphones for remote monitoring system based on sound analysis—ProQuest. Accessed: Oct. 15, 2023. Available: https://www.proquest.com/openview/53b6051f95f55732f7970992ec8286c3/1?pq-origsite=gscholar&cbl=2041937

Davies HJ, Williams I, Peters NS, Mandic DP (2020) In-Ear SpO2: A tool for wearable, unobtrusive monitoring of core blood oxygen saturation. Sensors, 20(17). https://doi.org/10.3390/s20174879

De Santis M, Cacciotti I (2020) Wireless implantable and biodegradable sensors for postsurgery monitoring: current status and future perspectives. Nanotechnology 31(25):252001. https://doi.org/10.1088/1361-6528/ab7a2d

Deterre M et al (2011) Energy harvesting system for cardiac implant applications. In: 2011 Symposium on design, test, integration & packaging of MEMS/MOEMS (DTIP), pp 387–391. Accessed: Oct. 15, 2023. Available: https://ieeexplore.ieee.org/document/6108031

Di Pietrantonio F et al (2015) A surface acoustic wave bio-electronic nose for detection of volatile odorant molecules. Biosens Bioelectron 67:516–523. https://doi.org/10.1016/j.bios.2014.09.027

Duell T, Muehlbauer M, Seitzinger T, Westfall J, Rashidi R (2018) MEMS capacitive sensor for wound monitoring applications. IOP conference series: materials science and engineering, 417(1), p. 012040. https://doi.org/10.1088/1757-899X/417/1/012040.

Dugger SA, Platt A, Goldstein DB (2018) Drug development in the era of precision medicine. Nat Rev Drug Discov, 17(3). https://doi.org/10.1038/nrd.2017.226.

Dutta T, Noushin T, Tabassum S, Mishra SK (2023) Road map of semiconductor Metal-Oxide-Based sensors: a review. Sensors, 23(15), https://doi.org/10.3390/s23156849

Dziadak B, Czumak M (2022) Personal fall detector using MEMS sensors. In: 2022 23rd International conference on computational problems of electrical engineering (CPEE), pp. 1–4. https://doi.org/10.1109/CPEE56060.2022.9919414.

Encapsulated sub-millimeter piezoresistive accelerometers for biomedical applications—ProQuest. Accessed: Oct. 15, 2023. Available: https://www.proquest.com/openview/b6ae36e9f40984bb9d910962ab11d06c/1?pq-origsite=gscholar&cbl=18750&diss=y

Fatema A, Poondla S, Mishra RB, Hussain AM (2021) A Low-Cost Pressure Sensor Matrix for Activity Monitoring in Stroke Patients Using Artificial Intelligence. IEEE Sens J 21(7):9546–9552. https://doi.org/10.1109/JSEN.2021.3054406

Fernandez RE, Soma Sekhar BV, Bhattacharya E, Chadha A (2007) Studies on cantilever based triglyceride biosensor. In: 2007 International workshop on physics of semiconductor Devices, pp. 690–693. https://doi.org/10.1109/IWPSD.2007.4472613

Frommelt T, Kostur M, Wenzel-Schäfer M, Talkner P, Hänggi P, Wixforth A (2008) Microfluidic Mixing via Acoustically Driven Chaotic Advection. Phys Rev Lett 100(3):034502. https://doi.org/10.1103/PhysRevLett.100.034502

Ghanbari MM, Tsai JM, Nirmalathas A, Muller R, Gambini S (2017) An Energy-Efficient Miniaturized Intracranial Pressure Monitoring System. IEEE J Solid-State Circuits 52(3):720–734. https://doi.org/10.1109/JSSC.2016.2633510

Göreke U, Habibiabad S, Azgın K, Doğrusöz YS, Beyaz Mİ (2016) The Development and Performance Characterization of Turbine Prototypes for a MEMS Spirometer. IEEE Sens J 16(3):628–633. https://doi.org/10.1109/JSEN.2015.2488104

Gronewold TMA, Baumgartner A, Quandt E, Famulok M (2006) Discrimination of Single Mutations in Cancer-Related Gene Fragments with a Surface Acoustic Wave Sensor. Anal Chem 78(14):4865–4871. https://doi.org/10.1021/ac060296c

Guo X, Xun Q, Li Z, Du S (2019) Silicon carbide converters and MEMS devices for High-temperature power electronics: a critical review. Micromachines, 10(6). https://doi.org/10.3390/mi10060406

Hajare R, Kadam S (2021) Comparative study analysis of practical EEG sensors in medical diagnoses. Global Transitions Proceedings 2(2):467–475. https://doi.org/10.1016/j.gltp.2021.08.009

Hajare R, Reddy V, Srikanth R (2022) MEMS based sensors – A comprehensive review of commonly used fabrication techniques. Materials Today: Proceedings 49:720–730. https://doi.org/10.1016/j.matpr.2021.05.223

Han SJ, Park H, Lee JO, Choo H (2018) Effect of optical aberrations on intraocular pressure measurements using a microscale optical implant in ex vivo rabbit eyes. J Biomed Opt 23(4):1–9. https://doi.org/10.1117/1.JBO.23.4.047002

Han T, Mattinen U, Mousavi Z, Bobacka J (2021) Coulometric response of solid-contact anion-sensitive electrodes. Electrochim Acta 367:137566. https://doi.org/10.1016/j.electacta.2020.137566

Hassell BA et al (2017) Human Organ Chip Models Recapitulate Orthotopic Lung Cancer Growth, Therapeutic Responses, and Tumor Dormancy In Vitro. Cell Rep 21(2):508–516. https://doi.org/10.1016/j.celrep.2017.09.043

He C, Tan J, Jian X, Zhong G, Cheng L, Lin J (2022) A Smart Flexible Vital Signs and Sleep Monitoring Belt Based on MEMS Triaxial Accelerometer and Pressure Sensor. IEEE Internet Things J 9(15):14126–14136. https://doi.org/10.1109/JIOT.2022.3146926

He C et al (2022) A Novel Snore Detection and Suppression Method for a Flexible Patch With MEMS Microphone and Accelerometer. IEEE Internet Things J 9(24):25791–25804. https://doi.org/10.1109/JIOT.2022.3199085

He C et al (2023) A Non-Contact fall detection method for bathroom application based on MEMS infrared sensors. Micromachines, 14(1), https://doi.org/10.3390/mi14010130

Hsiao Y-P, Mukundan A, Chen W-C, Wu M-T, Hsieh S-C, Wang H-C (2022) Design of a Lab-On-Chip for cancer cell detection through impedance and photoelectrochemical response analysis. Biosensors, 12(6). https://doi.org/10.3390/bios12060405

Hu W-P, Wu Y-M, Vu C-A, Chen W-Y (2023) Ultrasensitive detection of interleukin 6 by using silicon nanowire Field-Effect transistors. Sensors, 23(2), https://doi.org/10.3390/s23020625

Huang X, Li S, Davis E, Li D, Wang Q, Lin Q (2013) A MEMS Dielectric Affinity Glucose Biosensor. J Microelectromech Syst 23(1):14–20. https://doi.org/10.1109/JMEMS.2013.2262603

Huang X, Li S, Schultz J, Wang Q, Lin Q (2009) A MEMS sensor for continuous monitoring of glucose in subcutaneous tissue. In: 2009 IEEE 22nd international conference on micro electro mechanical systems, pp 352–355. https://doi.org/10.1109/MEMSYS.2009.4805391

Huang X, Li S, Schultz J, Wang Q, Lin Q (2009) A biocompatible affinity MEMS sensor for continuous monitoring of glucose. In: 2009 4th IEEE international conference on Nano/Micro engineered and molecular systems, pp 797–802 https://doi.org/10.1109/NEMS.2009.5068698

Huang Y, Kr P. Das, Bhethanabotla VR (2021) Surface acoustic waves in biosensing applications. SensS Actuators Rep, 3, p 100041, https://doi.org/10.1016/j.snr.2021.100041

Hur Y, Han J, Seon J, Pak YE, Roh Y (2005) Development of an SH-SAW sensor for the detection of DNA hybridization. Sens Actuators, A 120(2):462–467. https://doi.org/10.1016/j.sna.2005.01.027

Hussain CM, Keçili R (2020) Chapter 8—Electrochemical techniques for environmental analysis. in: Modern environmental analysis techniques for pollutants, Hussain CM, Keçili R, Eds. Elsevier, pp. 199–222. https://doi.org/10.1016/B978-0-12-816934-6.00008-4

İlik B, Koyuncuoğlu A, Şardan-Sukas Ö, Külah H (2018) Thin film piezoelectric acoustic transducer for fully implantable cochlear implants. Sens Actuators, A 280:38–46. https://doi.org/10.1016/j.sna.2018.07.020

Jackson N, Olszewski OZ, O'Murchu C, Mathewson A (2017) Shock-induced aluminum nitride based MEMS energy harvester to power a leadless pacemaker. Sens Actuators, A 264:212–218. https://doi.org/10.1016/j.sna.2017.08.005

Jakaratanopas S, Zhuang B, Gan W, Liu P (2021) 13—Integrated microfluidic systems for genetic analysis. In: Microfluidic devices for biomedical applications (Second Edition), (James) Li X, Zhou Y, Eds. In Woodhead Publ Ser Biomater. Woodhead Publishing, pp 511–549. https://doi.org/10.1016/B978-0-12-819971-8.00004-4.

Jang K-J et al (2019) Reproducing human and cross-species drug toxicities using a Liver-Chip. Sci Transl Med, 11(517), p. eaax5516, https://doi.org/10.1126/scitranslmed.aax5516

Ji Z, Zhang M (2022) Highly sensitive and stretchable piezoelectric strain sensor enabled wearable devices for real-time monitoring of respiratory and heartbeat simultaneously. Nanotechnology and Precision Engineering (NPE) 5(1):013002. https://doi.org/10.1063/10.0009365

Jiang Y-Q, Chen J-P, Dong Y-J, Zhou F-J, Tian C-W, Chen C-Q (2022) Delivery System for Targeted Drug Therapy in Chronic Diseases. Journal of Exploratory Research in Pharmacology 7(2):112–122. https://doi.org/10.14218/JERP.2021.00052

Kaisti M et al. (2019) Clinical assessment of a non-invasive wearable MEMS pressure sensor array for monitoring of arterial pulse waveform, heart rate and detection of atrial fibrillation. npj Digit Med, 2(1), https://doi.org/10.1038/s41746-019-0117-x

Kang S-K et al (2016) Bioresorbable silicon electronic sensors for the brain. Nature, 530(7588), 7588. https://doi.org/10.1038/nature16492

Karrock T, Gerken M (2015) Pressure sensor based on flexible photonic crystal membrane. Biomed Opt Express BOE, 6(12), pp. 4901–4911, https://doi.org/10.1364/BOE.6.004901

Kattman SJ et al (2011) Stage-specific optimization of activin/nodal and BMP signaling promotes cardiac differentiation of mouse and human pluripotent stem cell lines. Cell Stem Cell 8(2):228–240. https://doi.org/10.1016/j.stem.2010.12.008

Khan HA, Nguyen TN, Shafiq G, Mirza J, Javed MA (2023) A Secure Wearable Framework for Stress Detection in Patients Affected by Communicable Diseases. IEEE Sens J 23(2):981–988. https://doi.org/10.1109/JSEN.2022.3204586

Khetani SR et al (2013) Use of micropatterned cocultures to detect compounds that cause drug-induced liver injury in humans. Toxicol Sci 132(1):107–117. https://doi.org/10.1093/toxsci/kfs326

Khetani SR, Bhatia SN (2008) Microscale culture of human liver cells for drug development. Nat Biotechnol, 26(1), https://doi.org/10.1038/nbt1361

Khoshnoud F, de Silva CW (2012) Recent advances in MEMS sensor technology-mechanical applications. IEEE Instrum Meas Mag 15(2):14–24. https://doi.org/10.1109/MIM.2012.6174574

Kim YT, Kim Y-Y, Jun C-H (1999) Needle-shaped glucose sensor with multicell electrode fabricated by surface micromachining. 3680, pp 924–930, https://doi.org/10.1117/12.341159

Kim S-G, Priya S, Kanno I (2012) Piezoelectric MEMS for energy harvesting. MRS Bull 37(11):1039–1050. https://doi.org/10.1557/mrs.2012.275

Kishore K, Kumar SS, Mukhiya R, Ali Akbar S (2021) High-resolution current mode interface for MEMS piezoresistive pressure sensor. AEU—Int J Electron Commun, 134, p 153707. https://doi.org/10.1016/j.aeue.2021.153707.

Koçak Y, Akaltun Y, Gür E (2016) Magnetron sputtered WS2; optical and structural analysis. Journal of Physics: Conference Series, 707(1), p 012028, https://doi.org/10.1088/1742-6596/707/1/012028

Koivisto T et al (2015) Automatic detection of atrial fibrillation using MEMS accelerometer. In: 2015 Computing in Cardiology Conference (CinC), pp 829–832. https://doi.org/10.1109/CIC.2015.7411039

Koo Y, Hawkins BT, Yun Y (2018) Three-dimensional (3D) tetra-culture brain on chip platform for organophosphate toxicity screening. Sci Rep, 8(1), https://doi.org/10.1038/s41598-018-20876-2

Korvink JG, Paul O, Eds (2006) MEMS: a practical guide to design, analysis, and applications. berlin, heidelberg: Springer. https://doi.org/10.1007/978-3-540-33655-6

Koyilot MC et al (2022) Breakthroughs and Applications of Organ-on-a-Chip Technology. Cells 11(11):1828. https://doi.org/10.3390/cells11111828

Kurmendra Kumar R (2019) MEMS based cantilever biosensors for cancer detection using potential bio-markers present in VOCs: a survey. Microsyst Technol, 25(9), pp 3253–3267. https://doi.org/10.1007/s00542-019-04326-1

Kwiatek MA, Pandolfino JE (2007) Prolonged reflux monitoring: Capabilities of bravo pH and impedance-pH systems. Curr GERD Rep 1(3):165–170. https://doi.org/10.1007/s12171-007-0014-0

Kyberd PJ, Poulton A (2017) Use of Accelerometers in the Control of Practical Prosthetic Arms. IEEE Trans Neural Syst Rehabil Eng 25(10):1884–1891. https://doi.org/10.1109/TNSRE.2017.2693683

Laghrouche M, Saddaoui R, Mellal I, Nachef M, Ameur S (2016) Low-cost Embedded Spirometer Based on Commercial Micro Machined Platinum Thin Film. Procedia Engineering 168:1681–1684. https://doi.org/10.1016/j.proeng.2016.11.489

Lai BFL, Huyer LD, Lu RXZ, Drecun S, Radisic M, Zhang B (2017) InVADE: Integrated vasculature for assessing dynamic events. Adv Func Mater 27(46):1703524. https://doi.org/10.1002/adfm.201703524

Lammel G (2015) The future of MEMS sensors in our connected world. In: 2015 28th IEEE International conference on micro electro mechanical systems (MEMS), pp 61–64. https://doi.org/10.1109/MEMSYS.2015.7050886

Länge K (2019) Bulk and surface acoustic wave sensor arrays for Multi-Analyte detection: a review. Sensors, 19(24). https://doi.org/10.3390/s19245382

Lange H-E, Arbeiter N, Bader R, Kluess D (2021) Performance of a Piezoelectric Energy Harvesting System for an Energy-Autonomous Instrumented Total Hip Replacement: Experimental and Numerical Evaluation. Materials (Basel) 14(18):5151. https://doi.org/10.3390/ma14185151

Lanzani G (2014) Organic electronics meets biology. Nature Mater, 13(8), https://doi.org/10.1038/nmat4021

Leary JF et al (2013) Human organ-on-a-chip BioMEMS devices for testing new diagnostic and therapeutic strategies. In: Microfluidics, BioMEMS, and Medical Microsystems XI, SPIE, pp 33–42. https://doi.org/10.1117/12.2004447.

Lee C-C, Shen I, Hume C, Cao G (2005) A Feasibility Study of PZT Thin-Film Microactuators for Hybrid Cochlear Implants. Conf Proc IEEE Eng Med Biol Soc 2005:1929–1932. https://doi.org/10.1109/IEMBS.2005.1616829

Lee HJ, Choi N, Yoon E-S, Cho I-J (2018) MEMS devices for drug delivery. Adv Drug Deliv Rev 128:132–147. https://doi.org/10.1016/j.addr.2017.11.003

LeMoyne R, Tomycz N, Mastroianni T, McCandless C, Cozza M, Peduto D (2015) Implementation of a smartphone wireless accelerometer platform for establishing deep brain stimulation treatment efficacy of essential tremor with machine learning. Annu Int Conf IEEE Eng Med Biol Soc 2015:6772–6775. https://doi.org/10.1109/EMBC.2015.7319948

Li P-Y, Givrad TK, Sheybani R, Holschneider DP, Maarek J-MI, Meng E (2010) A low power, on demand electrothermal valve for wireless drug delivery applications. Lab Chip 10(1):101–110. https://doi.org/10.1039/B910248E

Li B, Xu C, Zhang F, Zheng J, Xu C (2015) Self-polarized piezoelectric thin films: preparation, formation mechanism and application. J. Mater. Chem. C 3(34):8926–8931. https://doi.org/10.1039/C5TC01869B

Li P et al (2015) Acoustic separation of circulating tumor cells. Proc Natl Acad Sci 112(16):4970–4975. https://doi.org/10.1073/pnas.1504484112

Li S, Ma F, Bachman H, Cameron CE, Zeng X, Huang TJ (2017) Acoustofluidic bacteria separation. J Micromech Microeng 27(1):015031. https://doi.org/10.1088/1361-6439/27/1/015031

Li C, Cordovilla F, Jagdheesh R, Ocaña JL (2017) Design and optimization of a novel structural MEMS piezoresistive pressure sensor. Microsyst Technol 23(10):4531–4541. https://doi.org/10.1007/s00542-016-3187-6
Lin W-C, Hsieh C-T, Chang M-C, Liao C-H (2022) MEMS Cavity-Based pH Image Biosensor for Wound Dressings to Monitor Hard-to-Cure Wounds. IEEE Sens J 22(19):18978–18987. https://doi.org/10.1109/JSEN.2022.3197678
Liu X, Wang J-Y, Mao X-B, Ning Y, Zhang G-J (2015) Single-Shot Analytical Assay Based on Graphene-Oxide-Modified Surface Acoustic Wave Biosensor for Detection of Single-Nucleotide Polymorphisms. Anal Chem 87(18):9352–9359. https://doi.org/10.1021/acs.analchem.5b02121
Liu J, Li S, Bhethanabotla VR (2018) Integrating Metal-Enhanced Fluorescence and Surface Acoustic Waves for Sensitive and Rapid Quantification of Cancer Biomarkers from Real Matrices. ACS Sens. 3(1):222–229. https://doi.org/10.1021/acssensors.7b00876
Liu Y, Cai Y, Zhang Y, Tovstopyat A, Liu S, Sun C (2020) Materials, design, and characteristics of bulk acoustic wave eesonator: a review. Micromachines,11(7). https://doi.org/10.3390/mi11070630
Ma C, Zhao L, Zhou E-M, Xu J, Shen S, Wang J (2016) On-Chip Construction of Liver Lobule-like Microtissue and Its Application for Adverse Drug Reaction Assay. Anal Chem 88(3):1719–1727. https://doi.org/10.1021/acs.analchem.5b03869
Ma C, Peng Y, Li H, Chen W (2021) Organ-on-a-Chip: A New Paradigm for Drug Development. Trends Pharmacol Sci 42(2):119–133. https://doi.org/10.1016/j.tips.2020.11.009
Ma R-H, Wang Y-H, Lee C-Y (2011) Wireless remote weather monitoring system based on MEMS technologies. Sensors, 11(3). https://doi.org/10.3390/s110302715
Ma C et al (2020) Leukemia-on-a-chip: Dissecting the chemoresistance mechanisms in B cell acute lymphoblastic leukemia bone marrow niche, Sci Adv, 6(44), p eaba5536. https://doi.org/10.1126/sciadv.aba5536
Mallegni N et al (2022) Sensing devices for detecting and processing acoustic signals in healthcare. Biosensors,12(10). https://doi.org/10.3390/bios12100835
Maloney JM, Uhland SA, Polito BF, Sheppard NF, Pelta CM, Santini JT (2005) Electrothermally activated microchips for implantable drug delivery and biosensing. J Control Release 109(1):244–255. https://doi.org/10.1016/j.jconrel.2005.09.035
Marturano-Kruik A et al (2018) Human bone perivascular niche-on-a-chip for studying metastatic colonization. Proc Natl Acad Sci U S A 115(6):1256–1261. https://doi.org/10.1073/pnas.1714282115
Maschmeyer I et al (2015) A four-organ-chip for interconnected long-term co-culture of human intestine, liver, skin and kidney equivalents. Lab Chip 15(12):2688–2699. https://doi.org/10.1039/C5LC00392J
Mathias DN, Park J, Kim E, Joung Y-H (2016) Development of a novel noncontact ECG electrode by MEMS fabrication process. Trans Electr Electron Mater 17(3):150–154. https://doi.org/10.4313/TEEM.2016.17.3.150
Mathur A et al (2015) Human iPSC-based cardiac microphysiological system for drug screening applications. Sci Rep, 5(1). https://doi.org/10.1038/srep08883
Mazouz Z et al (2013) Discriminating DNA mismatches by electrochemical and gravimetric techniques. Biosens Bioelectron 48:293–298. https://doi.org/10.1016/j.bios.2013.04.023
Meti S, Balavald KB, Sheeparmatti BG (2016) MEMS piezoresistive pressure sensor: a survey. Accessed: Oct. 15, 2023. Available: https://www.semanticscholar.org/paper/MEMS-Piezoresistive-Pressure-Sensor%3A-A-Survey-Meti-Balavald/62e6a6031c8b61762ab490642f0d7868a2a77094
Miller PG, Shuler ML (2016) Design and demonstration of a pumpless 14 compartment microphysiological system. Biotechnol Bioeng 113(10):2213–2227. https://doi.org/10.1002/bit.25989

Mishra RB, El-Atab N, Hussain AM, Hussain MM (2021) Recent Progress on Flexible Capacitive Pressure Sensors: From Design and Materials to Applications. Advanced Materials Technologies 6(4):2001023. https://doi.org/10.1002/admt.202001023

Mishra MK, Dubey V, Mishra PM, Khan I (2019) MEMS technology: a review. J Eng Res Rep, pp 1–24, https://doi.org/10.9734/jerr/2019/v4i116891

Mitchell-Innes A, Morse R, Irving R, Begg P (2017) Implantable microphones as an alternative to external microphones for cochlear implants. Cochlear Implants Int 18(6):304–313. https://doi.org/10.1080/14670100.2017.1371974

Moradian S, Abdolvand R (2016) MEMS-based passive wireless respiration profile sensor. In 2016 IEEE Sensors, pp. 1–3 https://doi.org/10.1109/ICSENS.2016.7808422

Motamedi ME Ed (2005) MOEMS: Micro-Opto-Electro-Mechanical systems. SPIE. https://doi.org/10.1117/3.2265061

Mujahid A, Dickert FL (2017) Surface acoustic wave (SAW) for chemical sensing applications of recognition layers. Sensors, 17(12). https://doi.org/10.3390/s17122716

Nam J, Lim H, Kim D, Shin S (2011) Separation of platelets from whole blood using standing surface acoustic waves in a microchannel. Lab Chip 11(19):3361–3364. https://doi.org/10.1039/C1LC20346K

Nawroth J, Rogal J, Weiss M, Brucker SY, Loskill P (2018) Organ-on-a-Chip systems for women's health applications. Adv Healthc Mater, 7(2). https://doi.org/10.1002/adhm.201700550

Newell N, Little JP, Christou A, Adams MA, Adam CJ, Masouros SD (2017) Biomechanics of the human intervertebral disc: A review of testing techniques and results. J Mech Behav Biomed Mater 69:420–434. https://doi.org/10.1016/j.jmbbm.2017.01.037

Nguyen T-V, Ichiki M (2019) MEMS-Based sensor for simultaneous measurement of pulse wave and respiration rate. Sensors, 19(22). https://doi.org/10.3390/s19224942

Nguyen T-V, Mizuki Y, Tsukagoshi T, Takahata T, Ichiki M, Shimoyama I (2020) MEMS-Based pulse wave sensor utilizing a piezoresistive cantilever. Sensors, 20(4). https://doi.org/10.3390/s20041052

Nguyen D-HT et al. (2019) A biomimetic pancreatic cancer on-chip reveals endothelial ablation via ALK7 signaling. Sci Adv, 5(8), p eaav6789 https://doi.org/10.1126/sciadv.aav6789

Nishiyama M, Miyamoto M, Watanabe K (2010) Respiration rhythm monitoring in sleep based on weight movement using hetero-core fiber optic sensors. In: ICCAS 2010, pp 205–208. https://doi.org/10.1109/ICCAS.2010.5669885

Niu Q, Wang L, Yang H (2009) Application of a MEMS-based energy harvester for artificial heart wireless energy transmission. In: 2009 ISECS international colloquium on computing, communication, control, and management, pp 38–41. https://doi.org/10.1109/CCCM.2009.5268155

Niu W et al (2018) Summary of research status and application of MEMS accelerometers. J Comput Commun, 6(12). https://doi.org/10.4236/jcc.2018.612021

Noushin T, Tabassum S (2021) Multiplexed electrochemical sensor for Real-Time monitoring of inflammatory biomarkers. In: 2021 IEEE Sensors, pp 1–4. https://doi.org/10.1109/SENSORS47087.2021.9639859

Noushin T, Hossain NI, Tabassum S (2022) Kirigami-Patterned highly stable and strain insensitive sweat pH and temperature sensors for Long-term wearable applications. In: 2022 IEEE healthcare innovations and point of care technologies (HI-POCT), pp. 108–111. https://doi.org/10.1109/HI-POCT54491.2022.9744070

null Kshitiz et al (2012) Control of stem cell fate and function by engineering physical microenvironments. Integr Biol (Camb), 4(9), pp 1008–1018. https://doi.org/10.1039/c2ib20080e

Nunes SS et al (2013) Biowire: a platform for maturation of human pluripotent stem cell–derived cardiomyocytes. Nat Methods, 10(8). https://doi.org/10.1038/nmeth.2524

O'Hare E, Cogan JA, Dillon F, Lowery M, O'Cearbhaill ED (2022) An intraoral Non-Occlusal MEMS sensor for bruxism detection. IEEE Sens J 22(1):153–161. https://doi.org/10.1109/JSEN.2021.3128246

Oh KW, Ahn CH (2006) A review of microvalves. J Micromech Microeng 16(5):R13. https://doi.org/10.1088/0960-1317/16/5/R01

Osiadala N, Kolakowski M (2020) The use of MEMS accelerometers for remote activity and living parameters monitoring. In: 2020 Baltic URSI Symposium (URSI), pp 73–76. https://doi.org/10.23919/URSI48707.2020.9254025

Panwar LS, Kala S, Panwar V, Panwar SS, Sharma S (2017) Design of MEMS piezoelectric blood pressure sensor. In: 2017 3rd international conference on advances in computing, communication & automation (ICACCA) (Fall), pp 1–7. https://doi.org/10.1109/ICACCAF.2017.8344698

Park SW, Das PS, Chhetry A, Park JY (2017) A Flexible Capacitive Pressure Sensor for Wearable Respiration Monitoring System. IEEE Sens J 17(20):6558–6564. https://doi.org/10.1109/JSEN.2017.2749233

Pereira AN et al (2022) Flexible Sensor Suite Integrated Into Textile for Calcium Ion and Fall Detection. IEEE Sensors Letters 6(10):1–4. https://doi.org/10.1109/LSENS.2022.3210878

Pereira AN, Noushin T, Tabassum S (2021) A wearable, multiplexed sensor for Real-time and In-situ monitoring of wound biomarkers. In: 2021 IEEE Sensors, pp 1–4. https://doi.org/10.1109/SENSORS47087.2021.9639722

Podbiel D, Laermer F, Zengerle R, Hoffmann J (2020) Fusing MEMS technology with lab-on-chip: nanoliter-scale silicon microcavity arrays for digital DNA quantification and multiplex testing. Microsyst Nanoeng, 6(1). https://doi.org/10.1038/s41378-020-00187-1

Poeggel S et al (2015) Differential in vivo urodynamic measurement in a single thin catheter based on two optical fiber pressure sensors. J Biomed Opt 20(3):037005. https://doi.org/10.1117/1.JBO.20.3.037005

Potekhina A, Wang C (2019) Review of electrothermal actuators and applications. Actuators, 8(4). https://doi.org/10.3390/act8040069

Prescott JH et al (2006) Chronic, programmed polypeptide delivery from an implanted, multireservoir microchip device. Nat Biotechnol 24(4):437–438. https://doi.org/10.1038/nbt1199

Qu M et al (2021) Monitoring of physiological sounds with wearable device based on piezoelectric MEMS acoustic sensor. J Micromech Microeng 32(1):014001. https://doi.org/10.1088/1361-6439/ac371e

Reboud J et al (2012) Shaping acoustic fields as a toolset for microfluidic manipulations in diagnostic technologies. Proc Natl Acad Sci 109(38):15162–15167. https://doi.org/10.1073/pnas.1206055109

Ropmay GD, Kumar S, Rathore PK, Rangababu P, Akhtar J (2020) A MEMS based blood glucose measurement sensor using twin-cantilever structure. AIP Conf Proc 2294(1):020001. https://doi.org/10.1063/5.0031309

Rughoobur G et al (2018) Carbon nanotube isolation layer enhancing in-liquid quality-factors of thin film bulk acoustic wave resonators for gravimetric sensing. Sens Actuators, B Chem 261:398–407. https://doi.org/10.1016/j.snb.2018.01.067

Sachdeva S, Davis RW, Saha AK (2021) Microfluidic Point-of-Care testing: commercial landscape and future directions. Front Bioeng Biotechnol, 8. Accessed: Oct. 12, 2023. Available: https://www.frontiersin.org/articles/https://doi.org/10.3389/fbioe.2020.602659

Saigusa Y (2017) Chapter 5—Quartz-Based piezoelectric materials. In: Advanced piezoelectric materials (Second Edition), Uchino K, Ed. In: Woodhead Publishing in Materials. Woodhead Publishing, pp 197–233. https://doi.org/10.1016/B978-0-08-102135-4.00005-9

Saitakis M, Dellaporta A, Gizeli E (2008) Measurement of two-dimensional binding constants between cell-bound major histocompatibility complex and immobilized antibodies with an acoustic biosensor. Biophys J 95(10):4963–4971. https://doi.org/10.1529/biophysj.108.132118

Sassa F, Laghzali H, Fukuda J, Suzuki H (2010) Coulometric detection of components in liquid plugs by microfabricated flow channel and electrode structures. Anal Chem 82(20):8725–8732. https://doi.org/10.1021/ac102289a

Schmelzeisen-Redeker G, Staib A, Strasser M, Müller U, Schoemaker M (2013) Overview of a novel sensor for continuous glucose monitoring. J Diabetes Sci Technol 7(4):808–814. https://doi.org/10.1177/193229681300700402

Sharma B, Sharma A (2022) Microfluidics: Recent Advances Toward Lab-on-Chip Applications in Bioanalysis. Adv Eng Mater 24(2):2100738. https://doi.org/10.1002/adem.202100738

Sheybani R, Schober SM, Meng E (2013) 15 - Drug delivery using wireless MEMS, In Handb Mems Wirel Mob Appl, Uttamchandani D, Ed. In Woodhead Publishing Series in Electronic and Optical Materials, Woodhead Publishing, pp 489–517. https://doi.org/10.1533/9780857098610.2.489

Shi Y, Fang L, Xue Z, Qi Z (2022) Research on random drift model identification and error compensation method of MEMS sensor based on EEMD-GRNN. Sensors. 22(14), https://doi.org/10.3390/s22145225

Shlomai A et al (2014) Modeling host interactions with hepatitis B virus using primary and induced pluripotent stem cell-derived hepatocellular systems. Proc Natl Acad Sci 111(33):12193–12198. https://doi.org/10.1073/pnas.1412631111

Silemek B, Seifert F, Petzold J, Brühl R, Ittermann B, Winter L (2023) Wirelessly interfacing sensor-equipped implants and MR scanners for improved safety and imaging. Magn Reson Med 90(6):2608–2626. https://doi.org/10.1002/mrm.29818

Singhwal SS, Matekovits L, Peter I, Kanaujia BK (2022) A Study on Application of Dielectric Resonator Antenna in Implantable Medical Devices. IEEE Access 10:11846–11857. https://doi.org/10.1109/ACCESS.2022.3144664

Song P et al (2020) Recent progress of miniature MEMS pressure sensors. Micromachines, 11(1). https://doi.org/10.3390/mi11010056

St John A, Price CP (2014) Existing and emerging technologies for Point-of-Care testing. Clin Biochem Rev, 35(3), pp 155–167

Sun Y et al (2018) Wearable pulse wave monitoring system based on MEMS sensors. Micromachines (Basel) 9(2):90. https://doi.org/10.3390/mi9020090

Sun S et al (2022) MEMS Ultrasonic Fingertip Heart Rate Sensor. IEEE Sens J 22(21):20285–20292. https://doi.org/10.1109/JSEN.2022.3207355

Sun W, Chang W (2023) Bloodpressure monitoring based on flexible encapsulated sensors. Appl Sci, 13(13). https://doi.org/10.3390/app13137473

Sung JH, Shuler ML (2009) A micro cell culture analog (microCCA) with 3-D hydrogel culture of multiple cell lines to assess metabolism-dependent cytotoxicity of anti-cancer drugs. Lab Chip 9(10):1385–1394. https://doi.org/10.1039/b901377f

Sung JH, Kam C, Shuler ML (2010) A microfluidic device for a pharmacokinetic–pharmacodynamic (PK–PD) model on a chip. Lab Chip 10(4):446–455. https://doi.org/10.1039/B917763A

Tahir AM et al (2020) A Systematic Approach to the Design and Characterization of a Smart Insole for Detecting Vertical Ground Reaction Force (vGRF) in Gait Analysis. Sensors (Basel) 20(4):957. https://doi.org/10.3390/s20040957

Takahashi K, Ozawa R, Ishida M, Sawada K (2014) Estimation of diaphragm deformation induced by molecuar adsorption on mems optical interferometric biosensor. In: 2014 International conference on optical MEMS and nanophotonics, pp 5–6. https://doi.org/10.1109/OMN.2014.6924511

Taller D et al (2015) On-chip surface acoustic wave lysis and ion-exchange nanomembrane detection of exosomal RNA for pancreatic cancer study and diagnosis. Lab Chip 15(7):1656–1666. https://doi.org/10.1039/c5lc00036j

Tang N, Zhou C, Xu L, Jiang Y, Qu H, Duan X (2019) A Fully Integrated Wireless Flexible Ammonia Sensor Fabricated by Soft Nano-Lithography. ACS Sens. 4(3):726–732. https://doi.org/10.1021/acssensors.8b01690

Tetyana P, Shumbula PM, Njengele-Tetyana Z, Tetyana P, Shumbula PM, Njengele-Tetyana Z (2021) “Biosensors: Design, Development and Applications”, in *Nanopores*. IntechOpen. https://doi.org/10.5772/intechopen.97576

Tibbe MP, Leferink AM, van den Berg A, Eijkel JCT, Segerink LI (2018) Microfluidic Gel Patterning Method by Use of a Temporary Membrane for Organ-On-Chip Applications. Advanced Materials Technologies 3(3):1700200. https://doi.org/10.1002/admt.201700200

Tjulkins F et al (2016) An implantable accelerometer-Based Heart-Monitoring device with improved positional stability. J Med Devices, 10(045002). https://doi.org/10.1115/1.4034574

Toh RJ, Mayorga-Martinez CC, Han J, Sofer Z, Pumera M (2017) Group 6 Layered Transition-Metal Dichalcogenides in Lab-on-a-Chip Devices: 1T-Phase WS2 for Microfluidics Non-Enzymatic Detection of Hydrogen Peroxide. Anal Chem 89(9):4978–4985. https://doi.org/10.1021/acs.analchem.7b00302

Trani ND et al (2019) Remotely controlled nanofluidic implantable platform for tunable drug delivery. Lab Chip 19(13):2192–2204. https://doi.org/10.1039/C9LC00394K

Tseng W-K, Lin J-L, Sung W-C, Chen S-H, Lee G-B (2006) Active micro-mixers using surface acoustic waves on Y-cut 128° LiNbO3. J Micromech Microeng 16(3):539. https://doi.org/10.1088/0960-1317/16/3/009

Turutin AV et al (2023) Magnetoelectric MEMS magnetic field sensor based on a laminated heterostructure of bidomain lithium niobate and metglas. Mater, 16(2). https://doi.org/10.3390/ma16020484

Umesh S, Padma S, Srinivas T, Asokan S (2018) Fiber Bragg Grating Goniometer for Joint Angle Measurement. IEEE Sens J 18(1):216–222. https://doi.org/10.1109/JSEN.2017.2770176

Upadhyaya AM et al (2021) A comprehensive review on the optical Micro-Electromechanical sensors for the biomedical application. Front Public Health, 9. Accessed: Oct. 12, 2023. Available: https://www.frontiersin.org/articles/10.3389/fpubh.2021.759032

Unnikrishnan ASASA, Pradeep AS, Rajeev P, BSSD (2022) Design of piezoelectric heart rate monitoring sensor for wearable applications. In: 2022 6th international conference on trends in electronics and informatics (ICOEI), pp 1–6. https://doi.org/10.1109/ICOEI53556.2022.9777147.

van den Berg A, Lammerink TSJ (1998) Micro total analysis systems: microfluidic aspects, integration concept and applications. In: Microsyst Technol Chem Life Sci, Manz A, Becker H Eds. In Topics in Current Chemistry. Berlin, Heidelberg: Springer, pp 21–49. https://doi.org/10.1007/3-540-69544-3_2

Valentin NS, Hageman KN, Dai C, Della Santina CC, Fridman GY (2013) Development of a multichannel vestibular prosthesis prototype by modification of a commercially available cochlear implant. IEEE Trans Neural Syst Rehabil Eng, 21(5), pp 830–839. https://doi.org/10.1109/TNSRE.2013.2259261

Varadan VK (2000) Wireless surface acoustic wave and MEMS-based microsensors. J Acoust Soc Am, 108(5_Supplement), pp 2598–2599. https://doi.org/10.1121/1.4743672

Verhaegen K, Baert K, Simaels J, Van Driessche W (2000) A high-throughput silicon microphysiometer. Sens Actuators, A 82(1):186–190. https://doi.org/10.1016/S0924-4247(99)00365-9

Vutla SR, Regalla SP, Ramaswamy K (2019) A laminar fluid flow model for study of ventilation systems in micro-electromechanical systems (MEMS) clean room. In: Journal of physics: conference series. 1276(1), p 012057, https://doi.org/10.1088/1742-6596/1276/1/012057

Wada R, Takahashi H (2020) Time response characteristics of a highly sensitive barometric pressure change sensor based on MEMS piezoresistive cantilevers,. Jpn J Appl Phys, 59. https://doi.org/10.35848/1347-4065/ab9ba1.

Wadman M (2007) Merck settles Vioxx lawsuits for $4.85 billion. Nature 450(7168):324–326

Wang L, Sipe DM, Xu Y, Lin Q (2008) A MEMS Thermal Biosensor for Metabolic Monitoring Applications. J Microelectromech Syst 17(2):318–327. https://doi.org/10.1109/JMEMS.2008.916357

Wang C et al (2020) AuNP-Amplified Surface Acoustic Wave Sensor for the Quantification of Exosomes. ACS Sens. 5(2):362–369. https://doi.org/10.1021/acssensors.9b01869

Wang Q, Ruan T, Xu Q, Yang B, Liu J (2021) Wearable multifunctional piezoelectric MEMS device for motion monitoring, health warning, and earphone. Nano Energy 89:106324. https://doi.org/10.1016/j.nanoen.2021.106324

Wang H, Wang Y, Yan Z, Xu Z, Li J (2023) Non-contact cardiac health monitoring system based on combined ECG-SCG measurement. In: The proceedings of the 17th annual conference of China electrotechnical society, Xie K, Hu J, Yang Q, Li J, Eds. In: Lect Notes Electr Eng. Singapore: Springer Nature, pp 254–261. https://doi.org/10.1007/978-981-99-0408-2_28

Ward K, Fan ZH (2015) Mixing in microfluidic devices and enhancement methods. J Micromech Microeng 25(9):094001. https://doi.org/10.1088/0960-1317/25/9/094001

Watkins PB, Merz M, Avigan MI, Kaplowitz N, Regev A, Senior JR (2014) The clinical liver safety assessment best practices workshop: rationale, goals, accomplishments and the future. Drug Saf, 37(Suppl 1), pp S1–7, https://doi.org/10.1007/s40264-014-0181-8

Wei X, Liu J (2008) Power sources and electrical recharging strategies for implantable medical devices. Front. Energy Power Eng. China 2(1):1–13. https://doi.org/10.1007/s11708-008-0016-3

Welsch W, Klein C, von Schickfus M, Hunklinger S (1996) Development of a surface acoustic wave immunosensor. Anal Chem 68(13):2000–2004. https://doi.org/10.1021/ac960198z

Wevers NR et al (2018) A perfused human blood–brain barrier on-a-chip for high-throughput assessment of barrier function and antibody transport. Fluids and Barriers of the CNS 15(1):23. https://doi.org/10.1186/s12987-018-0108-3

Whitesides GM (2006) The origins and the future of microfluidics. Nature, 442(7101). https://doi.org/10.1038/nature05058

Wikswo JP et al (2013) Scaling and systems biology for integrating multiple organs-on-a-chip. Lab Chip 13(18):3496–3511. https://doi.org/10.1039/c3lc50243k

Wu J, Gu M (2011) Microfluidic sensing: state of the art fabrication and detection techniques. JBO 16(8):080901. https://doi.org/10.1117/1.3607430

Wu C, Du L, Wang D, Wang L, Zhao L, Wang P (2011) A novel surface acoustic wave-based biosensor for highly sensitive functional assays of olfactory receptors. Biochem Biophys Res Commun 407(1):18–22. https://doi.org/10.1016/j.bbrc.2011.02.073

Wu Q et al (2020) Organ-on-a-chip: recent breakthroughs and future prospects. Biomed Eng Online 19(1):9. https://doi.org/10.1186/s12938-020-0752-0

Xiao Y et al (2019) Ex vivo Dynamics of Human Glioblastoma Cells in a Microvasculature-on-a-Chip System Correlates with Tumor Heterogeneity and Subtypes. Adv Sci (Weinh) 6(8):1801531. https://doi.org/10.1002/advs.201801531

Xu Z et al (2016) Design and Construction of a Multi-Organ Microfluidic Chip Mimicking the in vivo Microenvironment of Lung Cancer Metastasis. ACS Appl Mater Interfaces 8(39):25840–25847. https://doi.org/10.1021/acsami.6b08746

Xu J et al (2018) Fabrication of ZnO nanorods and Chitosan@ZnO nanorods on MEMS piezoresistive self-actuating silicon microcantilever for humidity sensing. Sens Actuators, B Chem 273:276–287. https://doi.org/10.1016/j.snb.2018.06.017

Xu J, Bertke M, Wasisto HS, Peiner E (2019) Piezoresistive microcantilevers for humidity sensing. J Micromech Microeng 29(5):053003. https://doi.org/10.1088/1361-6439/ab0cf5

Xu W, Wang X, Chiu Y, Lee Y-K (2020) High Sensitivity and Wide Dynamic Range Thermoresistive Micro Calorimetric Flow Sensor With CMOS MEMS Technology. IEEE Sens J 20(8):4104–4111. https://doi.org/10.1109/JSEN.2019.2961948

Yang D, Zhu J, Zhu P (2015) SpO2 and heart rate measurement with wearable watch based on PPG. In: 2015 IET international conference on biomedical image and signal processing (ICBISP 2015), pp 1–5. https://doi.org/10.1049/cp.2015.0784

Yang X, Zhang M (2021) Review of flexible microelectromechanical system sensors and devices. Nanotechnology and Precision Engineering (NPE) 4(2):025001. https://doi.org/10.1063/10.0004301

Yarmush ML, Freedman R, Bufalo AD, Teissier S, Meunier J-R (2017) Immune system modeling devices and methods. US9535056B2, Accessed: Oct. 15, 2023. Available: https://patents.google.com/patent/US9535056B2/en

Yi N et al (2018) Fully Water-Soluble, High-Performance transient sensors on a versatile galactomannansubstrate derived from the endosperm. ACS Appl Mater Interfaces 10(43):36664–36674. https://doi.org/10.1021/acsami.8b11682

Yi Z, Zhang W, Yang B (2022) Piezoelectric approaches for wearable continuous blood pressure monitoring: a review. J Micromech Microeng 32:103003. https://doi.org/10.1088/1361-6439/ac87ba

Yoo E-H, Lee S-Y (2010) Glucose Biosensors: An Overview of Use in Clinical Practice. Sensors (Basel) 10(5):4558–4576. https://doi.org/10.3390/s100504558

Yosioka K, Kawasima Y (1955) Acoustic radiation pressure on a compressible sphere. Acta Acust Acust 5(3):167–173

You S, Zhang L, Gui J, Cui H, Guo S (2019) A Flexible piezoelectric nanogenerator based on aligned P(VDF-TrFE) nanofibers. Micromachines, 10(5). https://doi.org/10.3390/mi10050302

Yu L, Kim BJ, Meng E (2014) Chronically Implanted Pressure Sensors: Challenges and State of the Field. Sensors (Basel) 14(11):20620–20644. https://doi.org/10.3390/s141120620

Yunus S, Jonas AM, Lakard B (2013) Potentiometric biosensors. In: Encycl Biophys, Roberts GCK, Ed. Berlin, Heidelberg: Springer, pp 1941–1946. https://doi.org/10.1007/978-3-642-16712-6_714

Zerrouki C et al (2010) Biological investigation using a shear horizontal surface acoustic wave sensor: Small 'Click generated' DNA hybridization detection. Biosens Bioelectron 26(4):1759–1762. https://doi.org/10.1016/j.bios.2010.08.036

Zhang G, Liu M, Guo N, Zhang W (2016) Design of the MEMS piezoresistive electronic heart sound sensor. Sensors (Basel) 16(11):1728. https://doi.org/10.3390/s16111728

Zhang YS et al (2017) Multisensor-integrated organs-on-chips platform for automated and continual in situ monitoring of organoid behaviors. Proc Natl Acad Sci 114(12):E2293–E2302. https://doi.org/10.1073/pnas.1612906114

Zhang Y, Yang F, Sun Z, Li Y-T, Zhang G-J (2017) A surface acoustic wave biosensor synergizing DNA-mediated in situ silver nanoparticle growth for a highly specific and signal-amplified nucleic acid assay. Analyst 142(18):3468–3476. https://doi.org/10.1039/C7AN00988G

Zhang B, Korolj A, Lai BFL, Radisic M (2018) Advances in organ-on-a-chip engineering, Nat Rev Mater, 3(8). https://doi.org/10.1038/s41578-018-0034-7

Zheng YF, Gu XN, Witte F (2014) Biodegradable metals. Mater Sci Eng R Rep 77:1–34. https://doi.org/10.1016/j.mser.2014.01.001

Zheng Y et al (2019) Passive wireless measurement system based on wireless power transfer technology. Electronics, 8(9), https://doi.org/10.3390/electronics8091048

Zhou J, Khodakov DA, Ellis AV, Voelcker NH (2012) Surface modification for PDMS-based microfluidic devices. Electrophoresis 33(1):89–104. https://doi.org/10.1002/elps.201100482

Zoll RS, Schindler CB, Massey TL, Drew DS, Maharbiz MM, Pister KSJ (2019) MEMS-Actuated carbon fiber microelectrode for neural recording. IEEE Trans Nanobioscience 18(2):234–239. https://doi.org/10.1109/TNB.2019.2905505

Zschornack E et al (2013) Evaluation of the performance of a novel system for continuous glucose monitoring. J Diabetes Sci Technol 7(4):815–823. https://doi.org/10.1177/193229681300700403

Chapter 4
Analysis of Resistance and Thrombogenicity of a Composite Coating Based on Collagen and Carbon Nanotubes Using a Microfluidic Chip

K. D. Popovich, A. Yu. Gerasimenko, D. V. Telyshev, and S. V. Selishchev

Abstract This manuscript proposes a method for studying resistance to the flow and thrombogenic properties of coatings based on collagen and carbon nanotubes formed by technology spray deposition. The development of a microfluidic chip that allows tight contact between the low-volume fluid and the coating in the microchannel facilitated in vitro testing of coatings under dynamic model fluid flow conditions. The type and structure of the flow inside the microfluidic chip channel were determined. Numerical modeling made it possible to determine the field of velocities in the section at the half-height of the microchannel and the shear stress on the lower wall of the microchannel at fluid pressure of 200 kPa. The constructed microfluidic system along with the connected microfluidic units allowed evaluating the properties of the coatings by simulating the closed blood flow for several hours. The dependence of resistance on the presence of the cross-linking agent, glutaric aldehyde, in the coating composition was demonstrated. Optical profilometry revealed that the coatings have sufficient resistance to high flow shear stress and are characterized by a roughness

K. D. Popovich (✉) · A. Yu. Gerasimenko · D. V. Telyshev · S. V. Selishchev
Institute of Biomedical Systems, National Research University of Electronic Technology, Zelenograd 124498, Russia
e-mail: popovich_k@staff.sechenov.ru

A. Yu. Gerasimenko
e-mail: gerasimenko@bms.zone

D. V. Telyshev
e-mail: telyshev_d_v@staff.sechenov.ru

S. V. Selishchev
e-mail: selishchev@bms.zone

K. D. Popovich · A. Yu. Gerasimenko · D. V. Telyshev
Institute for Bionic Technologies and Engineering, I. M. Sechenov First Moscow State Medical University, Moscow 119991, Russia

V. Pachkawade and K. Guha (eds.), *Micro- and Nano-Systems in 21st-Century*, Springer Tracts in Electrical and Electronics Engineering,
https://doi.org/10.1007/978-981-96-4659-3_4

value of the same order as the titanium surface. Raman spectroscopy showed better anti-thrombogenic properties of the coatings relative to titanium.

Keywords Microfluidic chip · Composite coating · Carbon nanotubes · Collagen · Thrombogenicity · Albumin · Glutaraldehyde

Introduction

According to the Institute for Health Metrics and Evaluation in Washington, cardiovascular diseases are the leading cause of death in the world (Fig. 4.1). According to statistics from 1990 to 2019, there has been an increase in deaths related to this spectrum of diseases from 12.07 million to 18.56 million worldwide.

One of the most serious pathologies of cardiovascular diseases is heart failure. Heart failure contributes to myocardial dysfunction and further inability of the heart to provide a full blood supply to the body due to deterioration of the pumping function of the heart (Crespo-Leiro et al. 2018). Currently, the main methods of treatment of terminal heart failure are transplantation of a donor heart and implantation of a device that partially or completely replaces its function (Habal and Garan 2017). Since there is an acute shortage of donor hearts in the world, the second alternative is the most promising—the creation of ventricular support devices (VAD). One such system (VAD "Sputnik") has been developed at the National Research University "MIET" (Telyshev et al. 2019). This device has been successfully introduced into medical

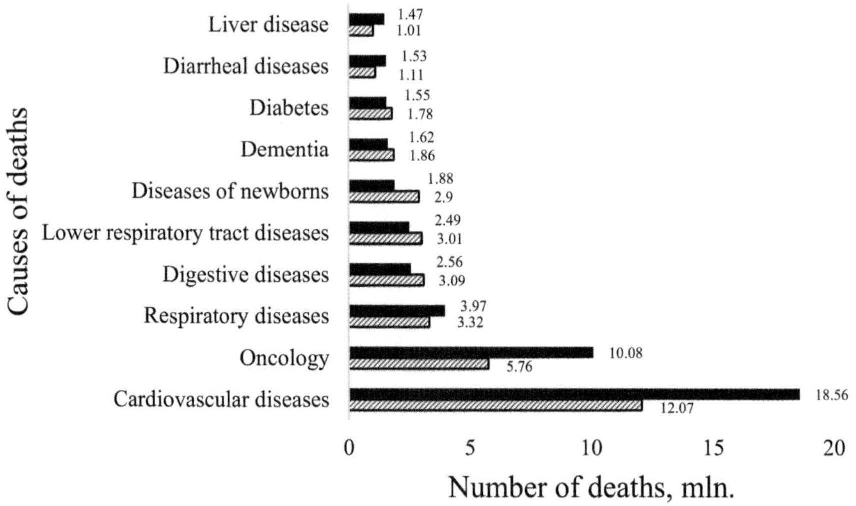

Fig. 4.1 Data on the leading causes of deaths worldwide for 2019 (black bars) and 1990 (shaded bars) from the Institute for Health Metrics and Evaluation

practice and implanted in more than 50 patients. However, despite the increased survival rate and improved quality of life of patients, in some cases, VAD implantation leads to complications in the form of thrombosis and high level of hemolysis.

Typically, the anticoagulant effect in the body is regulated by the endothelium, which is absent on artificial surfaces of medical devices. When blood flow occurs in such devices, the first thing that occurs is the adhesion of blood plasma proteins to the surface of the implanted part. This process occurs due to physical interaction forces such as Coulomb and Van der Waals forces, hydrogen bonds, and hydrophobic interactions (Hedayati et al. 2019). Protein adsorption on foreign surfaces begins with the adsorption of smaller and more abundant proteins such as human serum albumin, which is carried out by Fickian diffusion (Rabe et al. 2011). The detection of albumin is possible in the first layers of adsorption due to its lower molecular weight (66 kDa) and high concentration in blood plasma (35–53 g/l). (Kalathottukaren and Kizhakkedathu 2018; Nevídalová et al. 2020). Because of its higher affinity to the surface of foreign material, low molecular weight proteins may be partially displaced by higher molecular weight proteins, even if they are less abundant. Vroman found a sequence of changes in adsorption layers from low molecular weight proteins to high molecular weight proteins: albumin is displaced by immunoglobulins, then by fibrinogen, and finally by high molecular weight kininogen (Vroman et al. 1980). This change in the composition of the protein layer is called the Vroman effect and is confined to more hydrophilic surfaces (Vroman 2009). Platelet adhesion has been found to depend on adsorption to the surface of the above proteins (Regan-Smith et al. 2022; Liu et al. 2019; Milleret et al. 2015; Bulwan et al. 2012). In this case, receptors on the platelet surface interact with specific amino acid sequences in the adsorbed proteins. The shape of the adhered platelets changes dramatically from disc-shaped to pseudopodial. Further release of platelet contents: platelet factor 4, adenosine diphosphate, serotonin, and formation of platelet aggregates occurs. Subsequent activation of clotting factors, proteins found in platelets and plasma, is directed to the formation of thrombin. Thrombin, in turn, awakens the protein fibrin from the inactive state of the protein fibrinogen. The resulting fibrin network formed holds platelets and other blood cells together, which ensures the formation of a thrombus. The process that occurs when a foreign surface of a medical device comes into contact with blood is described (Fig. 4.2). The first step involves protein adhesion, followed by platelet activation and adhesion, and then fibrin formation. Because of these events, clots are formed on the surfaces of medical implantable devices, which adversely affect not only the performance of the device itself but also the viability of the organism in case the clot detaches from the device and enters the blood circulation of the organism.

The common anticoagulant heparin is usually administered to patients to prevent thrombosis. Its anti-clotting factor is due to the presence of a pentasaccharide sequence (bioactive center) in its molecular structure (Braune et al. 2019). This sequence has an inhibitory effect on thrombosis by neutralizing the functioning of thrombin through its binding to antithrombin. Heparin is administered to patients continuously and therapy is usually monitored by measuring the following parameters: activated clotting time, levels of anti-Factor Xa activity or activated partial

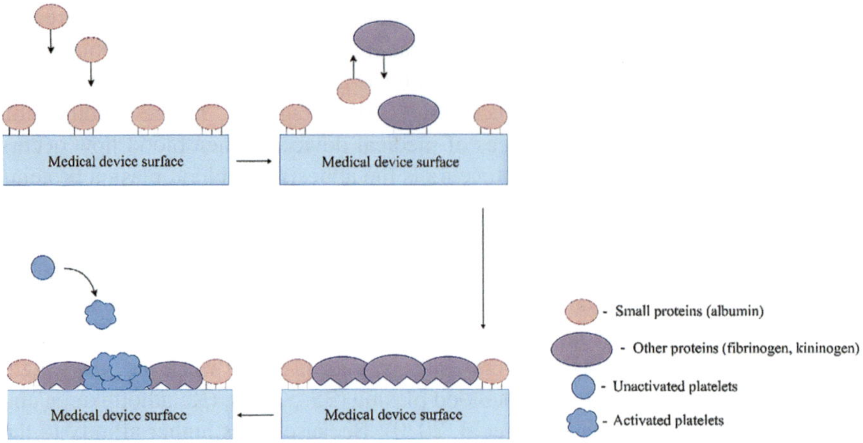

Fig. 4.2 Sequence of events leading to clot formation on blood contacting surface of medical devices

thromboplastin time (Lequier et al. 2014). However, these measurements do not always correlate correctly with the dose and effect of heparin, leading to uncertainty in monitoring the anticoagulant status of the patient (Mulder et al. 2018). Systematic heparin administration is associated with the risk of heparin-induced thrombocytopenia (HIT), which leads to blood clotting due to the production of antibodies against heparin–platelet factor 4 (TF4) complexes in the body (Cuker 2013). The mechanism of HIT involves several steps. First, heparin binds to blood platelets with the release of TF4. TF4, in turn, is able to interact with heparin, resulting in the formation of a heparin-TF4 complex, which triggers the production of antibodies in the blood. The interaction between antibodies and the heparin-TF4 complex provokes cascading reactions leading to greater platelet aggregation, causing severe thrombocytopenia and further bleeding complications. Data show that up to 8% of heparinized patients develop HIT-related antibodies and that approximately 1–5% of patients taking heparin develop HIT (Franchini 2005). Heparin use may also be ineffective if the patient has a congenital blood antithrombin deficiency (Zang et al. 2022). According to the World Health Organization, there is 1 person per 5000 people with hereditary deficiency of this protein.

Until now, the use of heparin as an anticoagulant remains a common solution to inhibit thrombosis, but the uncertainty of heparin monitoring, GIT, and congenital blood antithrombin deficiency favor the search for new anticoagulant agents.

The following solution to the problem of thrombosis when a medical device comes into contact with blood is proposed: modification of the device surface composition aimed at increasing the hemocompatible properties of the material. The fabrication of hemocompatible coatings aims at inhibiting the events that lead to the formation of blood clots. According to Ontaneda and Annich, surface modification of blood-contacting devices aimed at achieving anticoagulant effects can be divided into three main groups: endothelializing surfaces; bioactive surfaces (also called biomimetic surfaces), and biopassive surfaces (Ontaneda and Annich 2018).

The vascular endothelium (the inner surface of blood vessels) is the only surface that can truly be characterized as non-thrombotic. The non-thrombogenicity of endothelial cells is due to the presence and release of the anticoagulant heparin sulfate, thrombomodulin, and NO release. Thus, one of the engineering strategies for modifying the surface of blood pumping devices is to mimic the natural endothelium, namely surface endothelialization. In this case, the endothelial layer is formed on the surface areas of the device by seeding autologous or allogeneic cells onto the surface in vitro to achieve full haemocompatibility of the foreign material with blood (Liu et al. 2014; Figueiredo et al. 2019). In vitro seeding of the endothelial layer is a very time-consuming and expensive procedure and is challenging due to problems with cell retrieval, cell viability stability (Jaffer et al. 2015). Moreover, in case of autologous cells, there is a need for two operative procedures increasing the possibility of infection. In case allogeneic cells are used, rejection may be induced (Avci-Adali et al. 2010).

The anticoagulant effect of bioactive surfaces uses the approach of immobilizing bioactive compounds that minimize the immune response by interacting with key blood components or releasing bioactive compounds. Effective immobilization of heparin from the surface of blood contact devices has been shown to be able to inhibit thrombosis (Biran and Pond 2017). Carmeda AB has produced the first heparin-based anticoagulant biomimetic CBAS (Carmeda Bioactive Surface) coating for the surface of medical blood pumping devices. This uses covalent binding of the endpoints of heparin molecules to the surface of the device in a manner that does not interfere with the accessibility of the active center of the heparin molecules. An early study by Videm found that heparin coating has the ability to reduce complement activation by 45% (Videm et al. 1992), however, the use of heparin-coated biomaterials has been questioned as there is evidence that the clinical benefits of using heparin are not recognized (Edmunds and Colman 2006). In his study, Wöhrle compared native Jostent stents with heparin-coated stents of the same model in terms of clinical and angiographic results after 6 months. It was concluded that heparin coating does not affect the incidence of stent thrombosis or restenosis, thus confirming the ineffectiveness of heparin use (Wöhrle et al. 2001). In addition, it is still unclear whether heparin-based coatings affect HIT antibody production, as different studies have shown contradictory results (Koster et al. 2000; Pappalardo et al. 2009).

Modification strategies to make the biomaterial surface haemocompatible, i.e., resistant to protein adsorption and blood cell adhesion, are based on the modification of surface properties and aim to fabricate biopassive surfaces. In this case, polymers have shown high potential in various biotechnical applications. Collagen, a protein that forms the basis of the body's connective tissue (tendon, bone, cartilage, dermis, etc.) and provides its strength and elasticity, has high biocompatibility and low immunogenicity (Primavera et al. 2021). Collagen is actively used in cardiovascular regenerative medicine because it can be modified with drug molecules to provide gene-therapeutic effects (Zhang et al. 2018). In a research work by Tashakori-Miyanroudi, a scaffold using collagen and carbon nanofibers was fabricated for use as a cardiac patch for myocardial tissue (Tashakori-Miyanroudi et al. 2020). Medical coatings with collagen as a biocompatible polymer show high haemocompatibility properties and, in particular, anticoagulant effect and low biodegradation rate (Cherng et al. 2019).

In the case of anticoagulant coatings, rapid degradation rates result in the rapid non-functioning of the protective haemocompatible coating, so great care is taken to develop a coating with high flow resistance. It is possible to reduce the degradation rate of the material by adding a durable component to its composition. Recently, carbon nanomaterials have been used in biomedical materials to achieve such goals. One of the most common carbon nanomaterials is carbon nanotubes (CNTs). They are widely and successfully used to create biocompatible functional coatings and bulk implants with controlled surface and internal structure, as well as mechanical, electrophysical, and optical properties (Gerasimenko et al. 2021a; Demidenko et al. 2022). This is achieved due to their small size, large specific surface area, and high stability (Simon et al. 2019). Despite the wide range of potential applications in biomedical applications, CNTs have the following disadvantages: toxicity, justified by the presence of metal catalysts remaining in the synthesis process, as well as their poor solubility in water. CNTs can be functionalized with specific chemical groups to reduce toxicity and homogeneous distribution in liquid solvents and solid matrices (Kim et al. 2012). Different types of functional groups of nanotubes can be used depending on the medical application (Negri et al. 2020). Carboxyl, hydroxyl, carbonyl groups lead to water solubility and good level of biocompatibility of nanotube-based materials (Zhou et al. 2017). Also enveloping nanotubes with polymer matrix helps to reduce the thrombogenicity effect (Vakhrusheva et al. 2013). The addition of nanotubes helps to improve the haematological properties of materials for cardiovascular engineering (Gerasimenko et al. 2021b, 2022). Given these facts, carbon nanotubes can be used as fillers of biopolymer matrix from collagen polymer backbone to create haemocompatible coatings that ensure compatibility with blood (Gerasimenko et al. 2020; Savelyev et al. 2020).

Various laboratory methods are used as tools to evaluate the haemocompatibility of coatings, usually implemented under static or low shear stress conditions (Mohan et al. 2013). This approach cannot provide reliable results because the mechanical effects of blood flow significantly affect the stability of the coating and the degree of thrombosis. This problem can be investigated and solved by using a microfluidic system. Pumping fluid through a microchannel with specified dimensions at different flow rates allows the proposed coatings to be investigated in interaction with the flow. Such systems are perspective to create artificial vascular systems whose geometry is similar to natural vessels (Fenech et al. 2019; Pinto et al. 2014).

There are several known methods of microfluidic channel formation: photolithography, nanoimprint lithography, laser ablation, etc. Photolithography is a method of obtaining a certain pattern on the surface of a material due to the photosensitivity of the substance to visible or ultraviolet light. A thin photosensitive polymer film (photoresist) is applied to the substrate, after which the exposure process takes place (lighting of the photoresist through a template mask containing the desired pattern). Then, depending on the type of photoresist–positive or negative, the exposed or unexposed area of the substrate is removed, respectively. The pattern formed on the photoresist characterizes the microchannel in the microfluidic system. Structures with a minimum size of 0.5–1 μm can be obtained using this technique, but this method of MFS formation is expensive for single, non-serial use. Nanoprinting lithography (imprinting), i.e., pattern formation by mechanical deformation of the coating by a stamp includes three basic elements: a stamp formed, for example, by dry etching; a printable material in the form of a polymer layer several hundred nanometers thick; and printing equipment with temperature, pressure and parallelism control between the stamp and the substrate. The principle of such lithography is to press a stamp with a pressure of 50 to 100 bar against a thin polymer film, during which the polymer material follows the topology of the stamp surface. By means of nanoimprint lithography, it is possible to obtain microchannel smaller than 10 nm. It is worth noting that this technology works only with smooth materials because roughness promotes adhesion between the stamp and the polymer film. The disadvantage of the technology is the impossibility to change the topology of the manufactured structures, i.e., one stamp is designed for one pattern, as in the previous method of photolithography. For operative change of topology, it is necessary to produce a new stamp or mask, which is rather expensive. The third method of forming microfluidic systems is laser ablation (Fan et al. 2012; Waddell 2006). This fast and precise technology enables the removal of targeted material via laser light without direct contact. When exposed to the laser light, the substance absorbs radiation energy quickly, causing it to heat up and explosively vaporize from the targeted surface. However, uneven depth within the resonator can result in beveled sidewalls, which is a notable drawback of the technology (Masuzawa 2000). Multiple shots are necessary to reach the expected depth, and there are minor variances between shots. Achieving a small margin of error is often challenging due to the focus size, and the beam's contour is not entirely precise. Typically, the material removal rate is low, resulting in a slow process and costly equipment.

Therefore, based on the above conclusions, in the present work, we proposed a method for the formation of nanocomposite coatings based on collagen and multi-walled carbon nanotubes, providing reduced thrombogenicity, and designed a microfluidic chip to assess the impact of fluid flow on the developed coatings. Experimental results of stability and anticoagulant properties of the coatings at high shear stresses of flow were also obtained.

Materials and Methods

Materials

The coating dispersion was prepared based on the bovine collagen type II (MacMedi LLC, Russia) and carbon nanotubes (Taunit, Tambov, Russia). The nanotubes were functionalized with carboxyl groups (COOH). The average diameter of the carboxylated multiwalled carbon nanotubes (MWCNT) was 10–30 nm, length ~ 1.8–22 μm, and the specific surface area was 270 m^2/g. The degree of purity of c-MWCNT was 98%. Glutaric aldehyde (Sigma Aldrich, Missouri, MO, USA) was used to crosslink the polymer chains of the finished coatings. Phosphate-buffered solution (Sigma Aldrich, Missouri, MO, USA) and bovine serum albumin (BSA) solution (Biocloth BmbH, Munich, Germany) were used as a pumped liquid for experiments on the stability and anticoagulability of the coatings, respectively.

Composite Coating Formation

The formation of coatings consists of several stages (Fig. 4.3). At the first stage, 4 × 1.5 glass fragments were prepared as substrates. After ion cleaning, a 200 nm thick layer of titanium was formed on the glass substrates for three minutes using magnetron-sputtering technology. The sputtering power was 1000 W for 24 min. The argon pressure was 3×10^{-3} Tor In this case, the titanium layer is necessary to simulate the surface of a medical device, which is often made of titanium. For example, parts of the Sputnik-assisted circulation device that has direct contact with blood.

Next, plasma treatment of the substrates was carried out to increase the adhesion between the titanium layer and the future coating by breaking the organic bonds on the surface and functionalizing it. The substrates were plasma treated with 100 W plasma power for 10 min at 280 °C. The plasma gas included 100 cm^3 of oxygen and 100 cm^3 of argon. The pressure was 2 Tor and the reference frequency was 12.56 MHz.

The second step included the preparation of aqueous suspension (dispersion) with collagen and MWCNT. First, an aqueous suspension of collagen with a concentration of 1 wt.% in a weak solution of acetic acid (6%) was prepared. The use of an ultrasonic bath (40 W) and a magnetic stirrer for 2 and 1 h, respectively, made it possible to obtain a homogeneous solution Next, an aqueous dispersion of MWCNT with a concentration of 0.01 wt.% was prepared. An immersion ultrasonic disperser (Qsonica Sonicators, USA) was used for 40 min (20 W) to obtain a homogeneous dispersion of nanotubes. The collagen suspension was mixed with the MWCNT dispersion and exposed to magnetic stirrer and ultrasonic bath for 3 and 2 h, respectively. Then the dispersion was ready for further use.

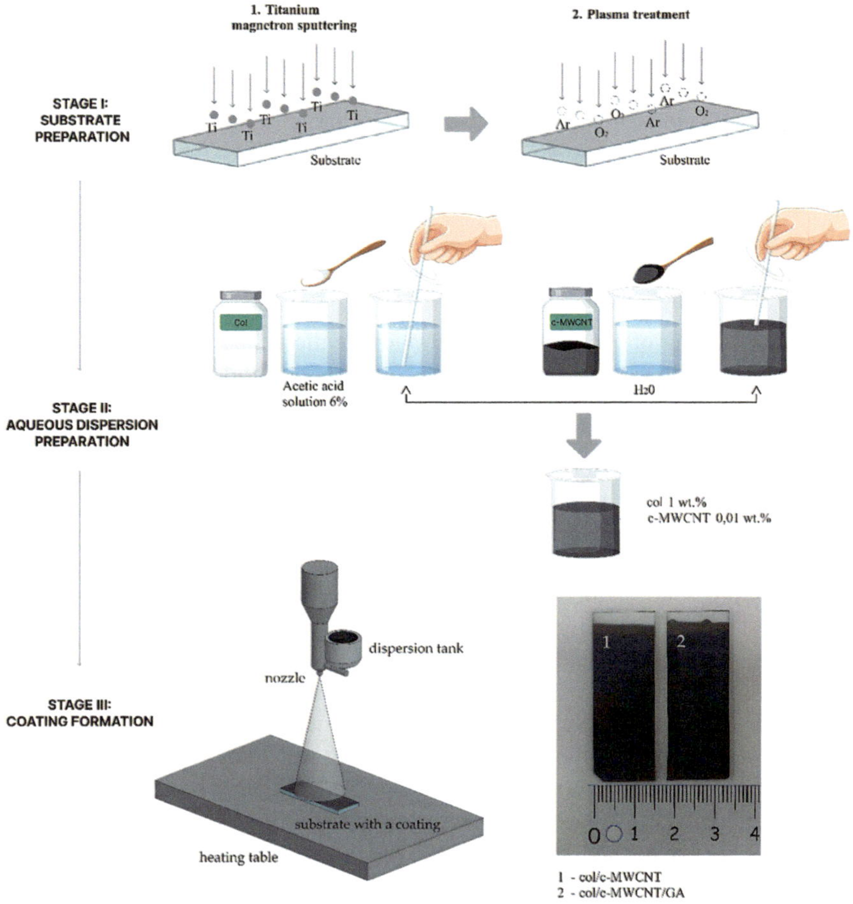

Fig. 4.3 Stages of formation of nanocomposite coatings on substrates

The coatings were formed by sputtering the prepared dispersion in the direction of the substrate followed by solvent evaporation from the sputtered layer at a constant substrate temperature of 50 °C. Repeatability of spraying was provided by the movement module, to which a nozzle with an outlet diameter of 300 nm and a dispersion reservoir with a volume of 8 ml were fixed. Automation of the composite coating spraying process with the specified characteristics of nozzle movement (time, delay, and trajectory of movement) was realized using the Arduino software platform. This system also allows programming the amount of dispersion to be sprayed, controlling the thickness of one sprayed layer. The heating table is capable of maintaining a constant substrate temperature in the range of 0–120 °C. The pressure in the compressor connected to the spraying unit is 2 kg/cm^2.

Some of the coatings were crosslinked in 50 ml glutaraldehyde (GA) solution (Sigma Aldrich, USA) for 30 min. GA is widely used as a crosslinking agent for collagen-based biomedical materials due to achieving high mechanical properties of the protein. The concentration of GA was 0.02 wt.%. After crosslinking, all samples were washed from GA in glycine solution for 5 min. Finished two types of samples are presented: collagen and MWCNT-based coating (col+MWCNT), collagen and MWCNT-based coating, crosslinked by GA (col+MWCNT+GA).

Microfluidic System Development

For in vitro testing of coatings, a special microfluidic system capable of realizing hermetic contact of the coating with a small-volume model fluid flow was developed. The assembled microfluidic chip design is shown in Fig. 4.4. This system consists of a few main elements. Element 4, made of elastic Polydimethylsiloxane (PDMS), included a formed channel for fluid transfer, which was hermetically pressed to element 5—the coated sample. The pressing was realized due to pressure from external elements 3 and 6 fixed by eight screws 2. The developed microfluidic chip was designed to be reusable, it was sufficient to rinse the channel with alcohol or water for a subsequent experiment. This system also included a recess for fixing the coating and holes for supplying and draining the liquid through the tubes 1.

The assembled microfluidic chip was 60 mm long, 45 mm wide, and 17 mm high. The weight of the system, excluding the substrate with the coating sample, was 56 g.

The channel was directly formed by casting a two-component PDMS solution into a mold made of polymer resin by 3D printing technology (Fig. 4.5). For this purpose, liquid PDMS material was mixed with crosslinking agent in a ratio of 1:10, mixed thoroughly, and poured into the prepared mold (Fig. 4.5a). The filled mold was degassed in a vacuum chamber for 30 min and cured in an oven at 70 °C for 2 h. The cured plate (Fig. 4.5b) was then left outdoors for 24 h to achieve complete stability of the material. After that, the PDMS had a stable dense and elastic state.

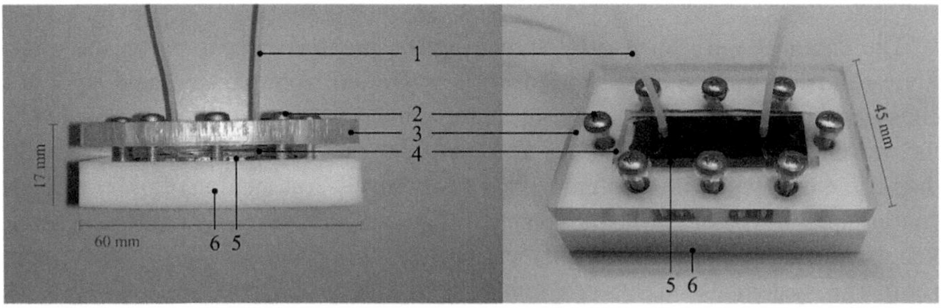

Fig. 4.4 Assembled microfluidic chip design, **a** front view and **b** side view: 1—tubes for fluid drainage and supply; 2—screws for fixing the system; 3—polymethylmethacrylate plate; 4—PDMS-based plate with channel; 5—substrate with a coating; 6—plate for inserting a coated sample

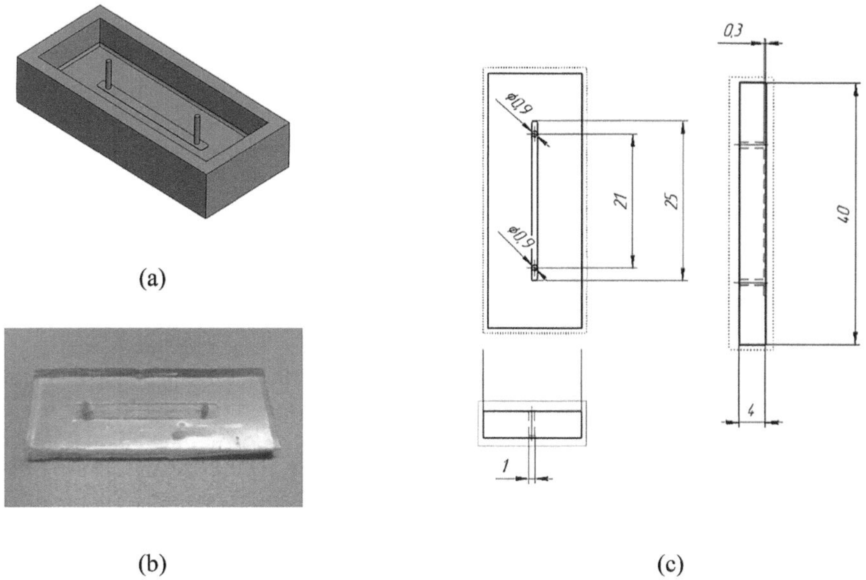

Fig. 4.5 PDMS-based plate with a formed channel: **a** mold for casting; **b** cured plate,ready for use; **c** drawing of PDMS-based plate with a channel

The geometric dimensions of the channel of the cured PDMS-based plate (Fig. 4.5c), were 25 mm in length, 1 mm in width and 300 μm in height. When pressed against the substrate, the capillary thickness was 200 μm. The immediate length of the plate was 40 mm, the width was 15 mm, and the height was 4 mm.

The developed microfluidic system had several advantages. First, the ability to provide a tight contact between the coating and the channel without leakage, achieving high shear stresses on the flow side. High adhesion of the coating and plate with the formed channel was due to the use of elastic PDMS. Secondly, the geometrical dimensions of the channel of the order of micrometer in height allowed to achieve high shear stresses of the flow, imitating the real conditions of blood-implant contact. It is known that in VAD shear stresses can reach from 0 to 150 Pa. Thirdly, the developed chip was designed for multiple use, which significantly saved costs and research time.

Numerical Simulation Methodology

To determine the type and structure of the flow inside the channel of the microfluidic chip, numerical modeling was performed. CAD-model construction was performed using the FreeCAD® package. Numerical calculation of the flow was carried out using the finite volume method implemented in OpenFOAM® package. In calculations, a hexagonal stack with the number of cells 793,952 was used. Test calculations showed that further increase in the number of cells does not lead to qualitative or quantitative change of velocity fields inside the channel and shear stresses on its walls. The calculation was performed for the maximum permissible fluid flow rate in the microfluidic system—21 ml/min.

Stand for the Study of Composite Coatings for Resistance and Hemocompatibility in the Microfluidic System

For in vitro testing, a stand was designed to determine the durability and hemocompatibility of coatings after contact with the corresponding model liquid. The stand was based on a circuit that realized the contact of coatings with model liquid under dynamic conditions (Fig. 4.6). The stand included the following units: microfluidic flow controller OB1 MK3+, liquid flow sensor MFS, developed microfluidic chip with the investigated coating, sealed flasks for input and output of model fluid and recirculation system MUX Recirculation 6/2. All of the above units were manufactured by ElveFlow. The assembled stand was fully hermetic and allowed the necessary multiple studies to test various coating formulations using a 30 ml model fluid.

Upon activation of the compressor, the generated air pressure was directed to one of the flasks, thereby driving the model liquid through the microfluidic stand. A fluid flow sensor was employed to detect the pressure within the microfluidic stand, while a flow controller facilitated the adjustment of said pressure. The microfluidic recirculation system with tubing system permitted the direction of the fluid to be

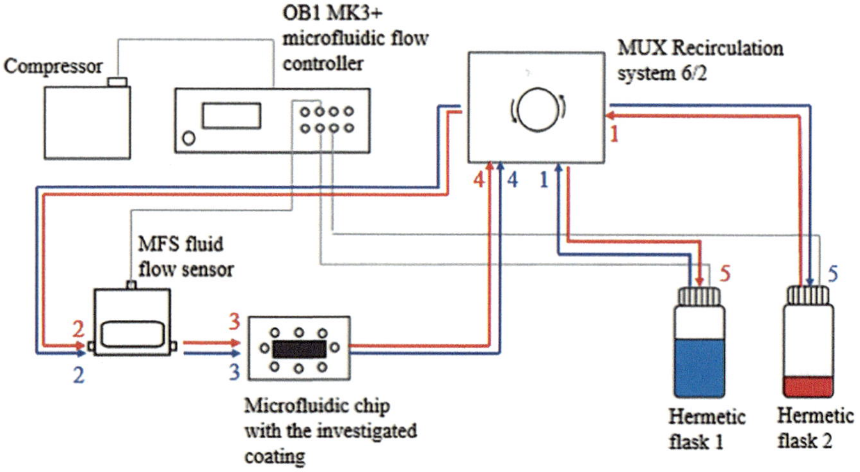

Fig. 4.6 Experimental setup of fluid pumping through the microfluidic chip, numbers indicate the sequence of flow direction

altered and the flow to be directed from one vessel to another in an alternating sequence. This technique allowed us to use a minimal volume of model fluid for long-term experimental studies without changing the direction of flow in the microfluidic chip itself, which allows us to simulate more accurately the movement of blood in vessels. In addition, this system was designed in such a way that it did not hinder the assembly and disassembly of the microfluidic chip for changing the coating under test.

Analysis of Resistance and Thrombogenicity of Composite Coatings Using a Microfluidic Stand

In vitro testing of the fabricated col+MWCNT and col+MWCNT+GA coatings was carried out in an assembled microfluidic stand. Here, the coatings were subjected to a model fluid flow of a certain shear stress and tested for stability and thrombogenicity.

In the case of coating stability tests, phosphate buffered saline (PBS) with pH equal to 7.4 was used as a model liquid. It is known that the hydrogen index can affect the degree of degradation of the coating (Cheng et al. 2012), therefore, the pH of the model fluid is equal to the pH of blood. In the microfluidic stand, the coatings were subjected to flow for 3.5 h. A fluid flow rate of 21.3 ml/min was achieved at a pressure of 2 bar (200 kPa) (Fig. 4.7).

The shear stresses (WSS) on the walls of a microchannel with a rectangular cross-section can be estimated using the following relationship (Pisapia et al. 2022):

$$\mathrm{WSS} = \frac{6\mu Q}{wh} \tag{4.1}$$

where μ—dynamic viscosity of liquid (Pa·s), Q—volumetric flow rate of liquid (m^3/s), w—channel width (m), h - channel height (m). According to relation (1), the shear

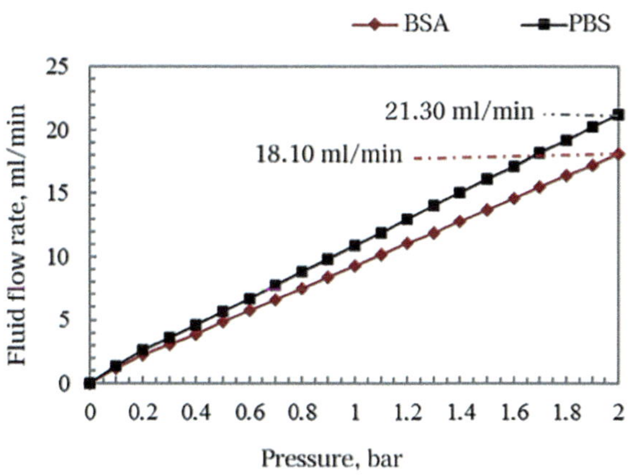

Fig. 4.7 Dependence of the flow rate value in the microfluidic system on pressure for phosphate-buffered solution (black curve) and albumin solution (red curve)

stresses on the studied coatings on the flow side were 50 Pa. As described earlier, this parameter corresponds to the range of shear stress values in VAD, which can range from 0 to 150 Pa.

Almost every half hour of pumping, the coating was removed from the chip and examined using S Neox optical profilometry for changes in the thickness of the flow contact area.

It is known that the excessive appearance of blood albumin protein molecules adhered on the surface of implantable devices is a precursor to thrombosis. Therefore, evaluation of protein adhesion to the nanocomposite coating was used to determine the degree of thrombogenicity. In this case, BSA solution was chosen as a model fluid for testing in the microfluidic stand. The solution was contacted with the coatings for 30 min at a flow rate of 18.1 ml/min.

The concentration of BSA in the solution was 4 wt.%, which corresponded to the normal concentration of albumin protein in human blood (Arne et al. 2005). After testing, the coatings were removed from the system to evaluate the adhesion of albumin protein to the coating using Raman spectroscopy. Spectra for all coatings were measured in the baseline region before and after contact with the fluid flow. Several measurements were taken for each region to obtain an average result.

Methods Results and Discussion

Numerical Simulation

The velocity field in the section at the level of half-height of the microchannel and the shear stress at the bottom wall of the microchannel were calculated (Fig. 4.8).

The shear stress was considered exactly in the region where the flow contact with the investigated coating sample took place. It was obtained that the fluid flow entering perpendicularly to the microchannel plane introduces significant fluctuations in velocity fields and shear stresses. The flow with a shear stress of 100 Pa is characterized by burgundy color. In particular, the flow pattern inside the channel can be conditionally divided into two zones—unstable flow and laminar flow. The boundary between these zones approximately coincides with the middle of the microchannel. The results showed that in order to obtain correct interaction results at given shear stresses on the flow side, the coating analysis had to be performed in the laminar flow zone to avoid all fluid fluctuations introduced by the microchannel geometry of the system.

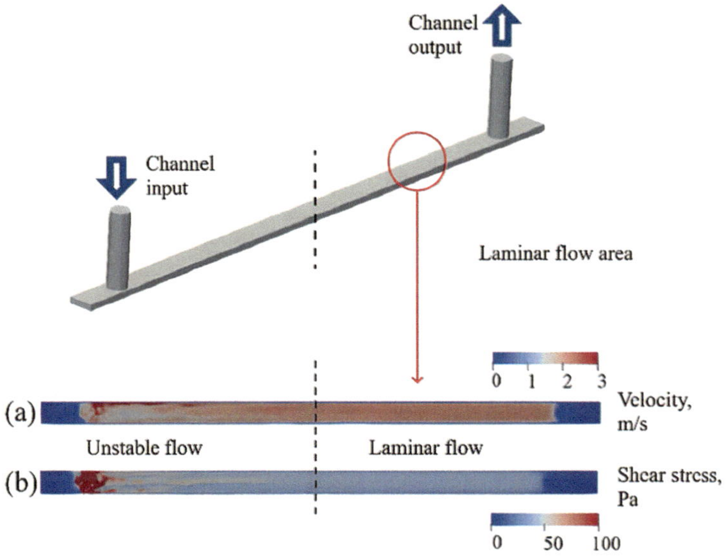

Fig. 4.8 Modeling of flow in the channel of a microfluidic system: **a** distribution of velocity field; **b** distribution of shear stresses

Optical Profilometry

Because of testing the impact of FBS flow on coatings, the stability of coatings to the flow depending on the contact time was evaluated. The study of coatings stability was carried out using a non-destructive method—optical profilometry. The measurement area included the laminar flow contact area and the area not exposed to flow contact. The surface analysis area for col+MWCNT, col+MWCNT+GA coatings was 2.77 × 3.59 mm, respectively (Fig. 4.9a, b). A total of 84 distinct fields of view (6 columns and 14 rows) were scanned over a 40-min period for each experiment cycle (Fig. 4.9c). The degree of overlap between adjacent fields of view was 10%. Prior to each individual cell scan, the autofocus function was utilized to obtain more precise data regarding the topography of the covers.

The micro-images and surface topography of the analysis area were measured during 3.5 h of flow exposure of the coatings in the microfluidic system (Figs. 4.10a–f, and 4.11a–f). The dotted line indicates the flow contact area and the blue arrows indicate the flow direction.

Optical profilometry showed that the collagen and nanotube-based coatings were uniformly sprayed on the substrate. Comparing the col +MWCNT and col+MWCNT+GA coatings, it can be concluded that there is less height difference for the coating crosslinked with GA, which indicates that this coating is more stable. The result indicates that GA stabilized the collagen-based coating.

Through optical profilometry, the roughness of the coatings was also evaluated. It is known that a high degree of roughness can contribute to coating thrombogenicity

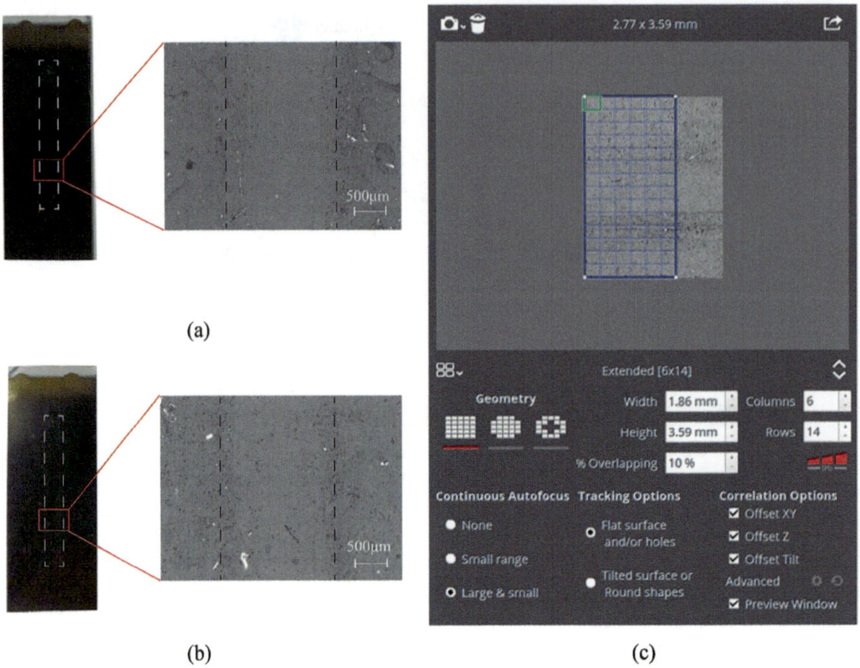

Fig. 4.9 Optical microscopy of measurement areas of 2.77 × 3.59 mm coatings after contact with PBS flow for 3.5 h: **a** coating col+MWCNT; **b** coating col+MWCNT+GA; **c** geometric parameters of coating sample scans displayed in the SensoSCAN software's advanced "Inspection Window" panel

(Hecker and Scandrett 1985). Table 4.1 demonstrates the roughness measurements of Ra coatings before in vitro testing (t = 0 h) and after contact with the flow in a microfluidic system after 3.5 h (t = 3.5 h). The surface of the VAD straightener part made of titanium was selected as a control sample.

The table shows that almost all the fabricated coatings had roughness of the order of the selected control sample (±3%). The presence of crosslinking agent as GA contributed to the reduction of roughness of all coatings within 3%.

Comparing the roughness of the fabricated coatings before and after contact with the FBS flow, the following conclusions can be drawn. The roughness decreased by 2%, 1% for col +MWCNT, col+MWCNT+GA, respectively, during 3.5 h of contact. The decrease in roughness can be explained by the degradation of coatings in the flow.

The parameter δ was introduced to quantify the stability/degree of biodegradation over time:

$$\delta = \frac{\overline{Z}_0 - \overline{Z}_k}{\overline{H}_0} 100\% \tag{4.2}$$

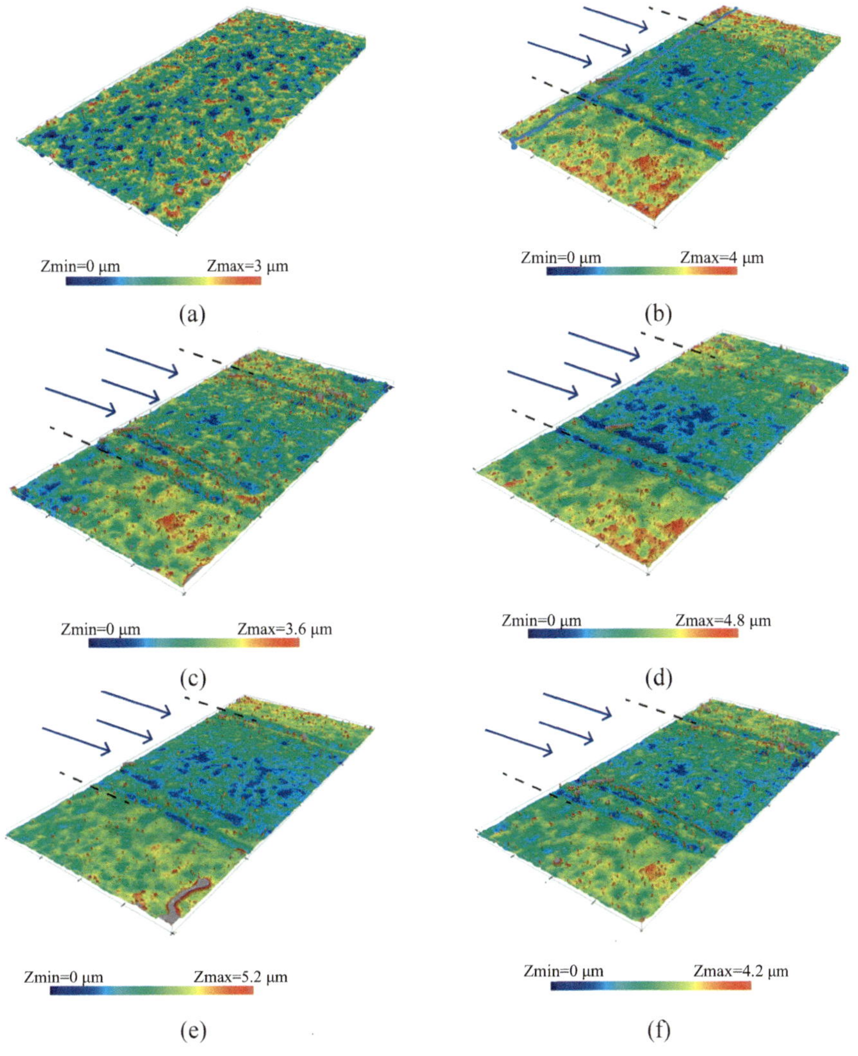

Fig. 4.10 Coating reliefs of col+MWCNT, obtained by optical profilometry, depending on the duration of contact with the FBS flow: **a** before contact; **b** 0.5 h of contact; **c** 1 h of contact; **d** 1.5 h of contact; **e** 2.5 h of contact; **f** 3.5 h of contact

where $\overline{Z}_0$—is the average profile height in the contact area not exposed to flow, $\overline{Z}_k$—is the average profile height in the contact area exposed to flow, $\overline{H}_0$—is the average thickness of the coatings, which was equal to 10 μm.

Figure 4.12 shows the results of dependence on time δ for col+MWCNT (black symbols) and col+MWCNT+GA (blue symbols) nanocomposite coatings according to formula (4.2). It was found that the col+MWCNT coating degraded by 5.5% of the total coating thickness of 10 μm after contacting with the flow for 3.5 h. The

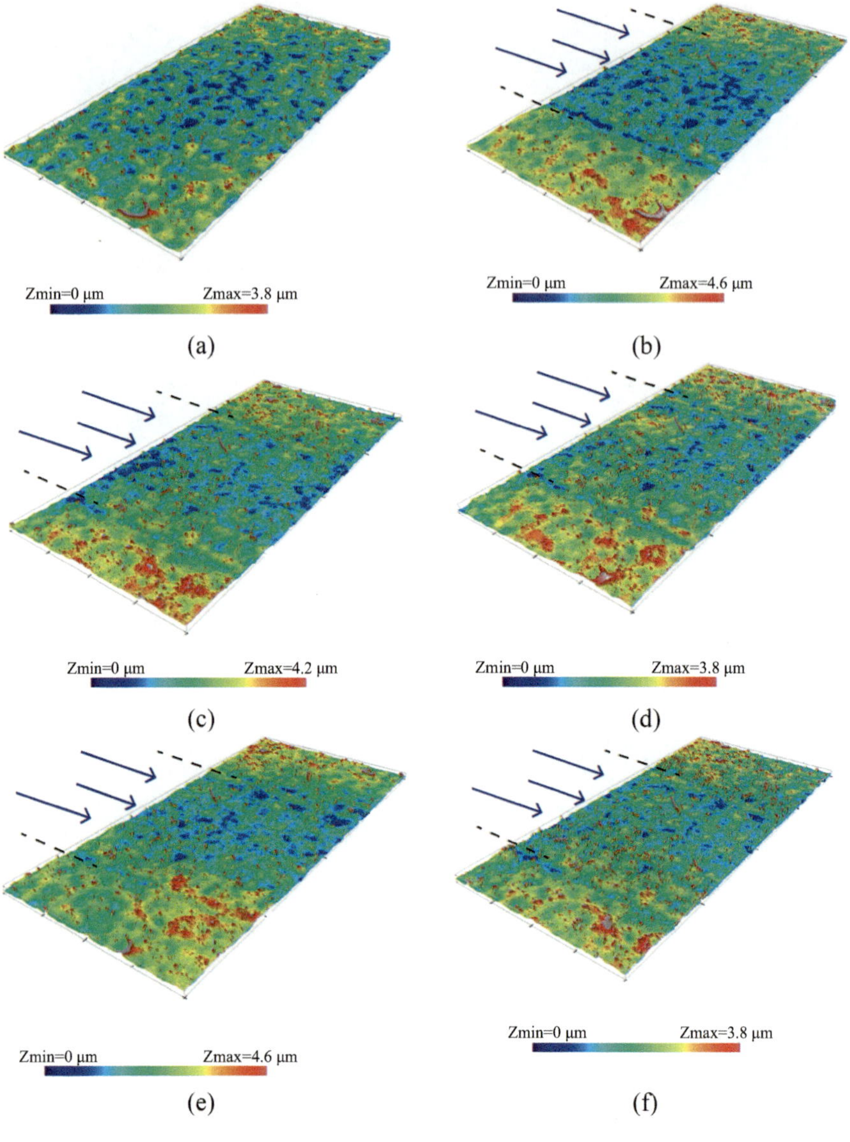

Fig. 4.11 Coating reliefs of col+MWCNT+GA, obtained by optical profilometry, depending on the duration of contact with the FBS flow: **a** before contact; **b** 0.5 h of contact; **c** 1 h of contact; **d** 1.5 h of contact; **e** 2.5 h of contact; **f** 3.5 h of contact

Table 4.1 Roughness values of fabricated coatings before and after contact with the flow of FBS solution after 3.5 h

Sample designation	t, h	Ra, μm
col+MWCNT	0.00	0.19
	3.50	0.17
col+MWCNT+GA	0.00	0.18
	3.50	0.17
Control (straightening device VAD)	0.00	0.16
	3.50	–

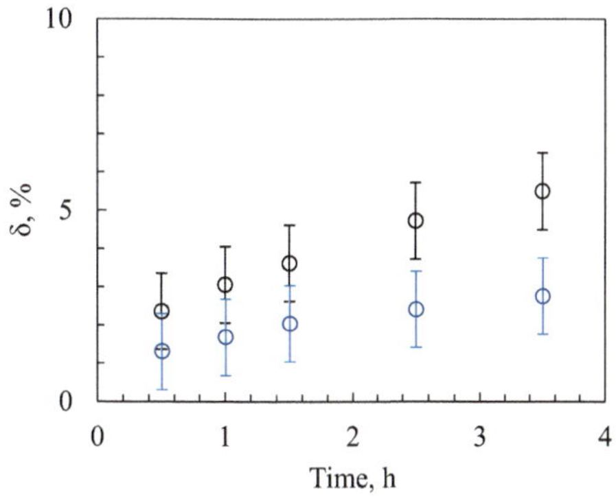

Fig. 4.12 Dependence of coating degradation thickness δ on the contact time with flow; black symbols characterize col+MWCNT coating, blue symbols—col+MWCNT+GA

coating of the same composition cross-linked with GA showed even more stable properties, degrading by 2.75% under the same conditions. Thus, cross-linking of collagen matrix by GA contributed to the increase of coating stability, practically in 2 times. Nevertheless, for both types of coatings, a rather high resistance to high shear stress flow was obtained.

Measurement of Mechanical Properties

The hardness of the coating samples were investigated by nanoindentation method using scanning nanohardness tester NanoScan-4D Compact' (NanoScan, Troitsk). The method is founded upon the principle of ascertaining a material's capacity to resist indentation, deformation, and the introduction of a harder material into it. As a harder material in the hardness tester a crystal in the indentation head (4) was used, which was pressed into the coating samples (5) to a depth of 1000 nm (Fig. 4.13a). The system incorporated a camera (1) to visualize the process of measurement, a nanohardness tester module (2) to perform parameter measurements, and a motorized table (3) to provide a measurement transition from one point on the sample surface to another.

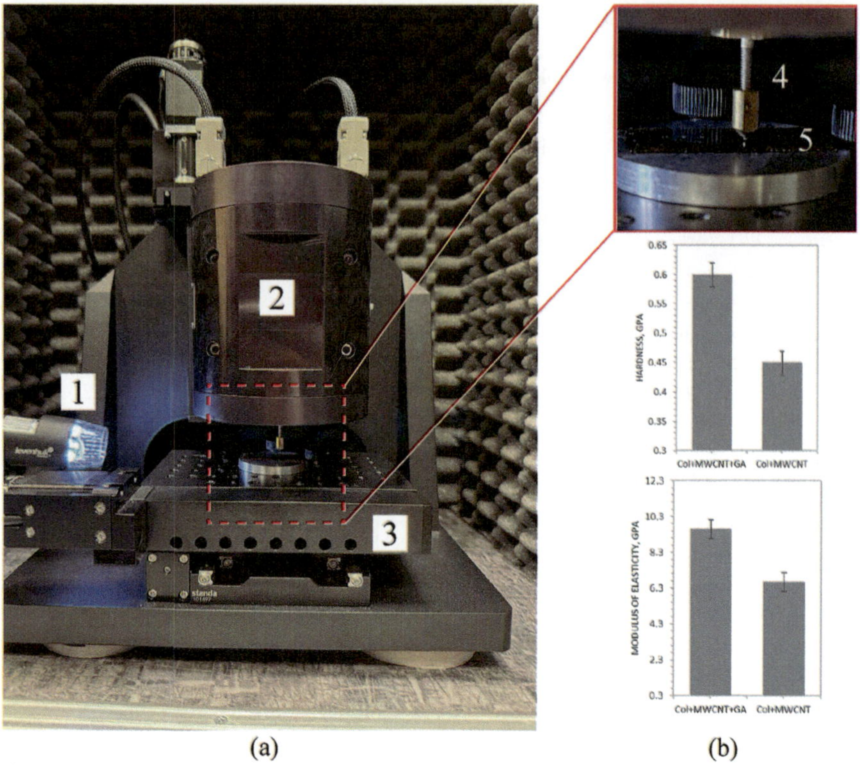

Fig. 4.13 Nanoindentation method for the study of coating samples: **a** measurement process using equipment, where 1—camera, 2—nanohardness tester module, 3—motorized table, 4—indentation head, 5—test coating sample; **b** obtained plots of hardness and modulus of elasticity of samples

For each sample, 10 cycles of measurements were performed, at different points of the surface. It was found that the hardness of col+MWCNT+GA coating is 25% higher than the hardness of col+MWCNT coating, which confirms the better mechanical stability of the sample cross-linked with GA compared to the non-cross-linked one (Fig. 4.13b). Furthermore, the modulus of elasticity of the samples was determined through the utilization of a hardness tester. The results demonstrate that coating col+MWCNT+GA exhibits greater elastic properties, with a relative increase of 30.5% in comparison to coating col +MWCNT.

Raman Spectroscopy

In the present study, the thrombogenic properties of the fabricated coatings were investigated u2sing an inVia Qontor Raman spectrometer. For each coating, 3 measurements were made in the red labeled area before and after contact with the protein flow (Fig. 4.14a). An argon laser with a wavelength of 514 nm and a power of approximately 0.125 mW was employed as the source of excitation radiation. A diffraction grating with 1800 strokes/mm provided spectral resolution of 0.5 cm^{-1}. A precision motorized stage and a built-in Olympus BX41 microscope were employed to focus the laser beam on the area under investigation.

The thrombogenic properties of the coatings were investigated by comparing the Raman spectra of the coatings in the same areas of analysis before and after contact with a flow of BSA solution for 30 min. In this case, the degree of adhesion of BSA to the nanocomposite coatings was compared to the adhesion of BSA to titanium as a control.

Spectra were recorded before (black curve) and after (red curve) contact of the BSA solution flow in the microfluidic system with the sample surface. Measuring the spectra at several different points and averaging them helped to obtain results that were more accurate. The analysis of the spectra allowed concluding the following.

First, the spectra of the titanium surface after contact with the solution of BSA compared to the original titanium surface were characterized by an increase in fluorescence in a wide range of frequencies (Fig. 4.14b). This indicated the contribution to the fluorescence response of adhered BSA molecules on the titanium surface. On average, the increase in fluorescence was 18–36%.

Secondly, the spectra of the nanocomposite coatings col+MWCNT (Fig. 4.14c) and col+MWCNT+GA (Fig. 4.14d) show characteristic modes for multi-walled carbon nanotubes (D- and G-modes), which prove the presence of carbon structures in the coatings. The D-mode with a wave number equal to 1310 cm^{-1} is due to the vibrational state of atoms perpendicular to the nanotube axis (Costa et al. 2008). The D-mode (1603 cm^{-1}) is observed in the presence of defects in the graphene/graphite structure, and its intensity is proportional to the degree of disorder in the structure of MWCNTs (Dresselhaus et al. 2005). The G-mode is characteristic of all carbon materials with sp2-hybridization. In the case of bond breakage in the graphene structure forming the tube, atoms with sp2-hybridized electrons appear outside the plane of the nanotube layers. In this case, the atoms oscillate with the frequency of the D-mode rather than the G-mode, so its intensity decreases in structures with a high level of defectivity. The intensity ratio of G- and D-mode (I_G/I_D) bands is used in analyzing the defectivity of MWCNTs (Ramos et al. 2010). The I_G/I_D is 0.53 and 0.57 for col +MWCNT and col+MWCNT+GA, respectively. Meanwhile, the I_G/I_D ratio for the initial nanotube powder is 0.54. For this case, it can be stated that neither the method of coating formation nor the interaction of nanotubes with any substance in the coating composition significantly affected the nanotube structure.

Thirdly, comparing the spectra of nanocomposite coatings col+MWCNT and col+MWCNT+GA before and after contact with BSA solution flow, the following can be observed: the total fluorescence background increased and the intensity of D- and G-mode peaks in the dependences for both coatings changed. Adhesion of BSA to the col+MWCNT coating up to 1000 cm^{-1} resulted in an increase in fluorescence up to 13%, and in the 1800–2500 cm^{-1} region up to 18%. At the same time, the intensity of D- and G-mode peaks decreased by 11% and 13%, respectively. The decrease in the intensity of D- and G-mode peaks indicates the appearance of an albumin layer

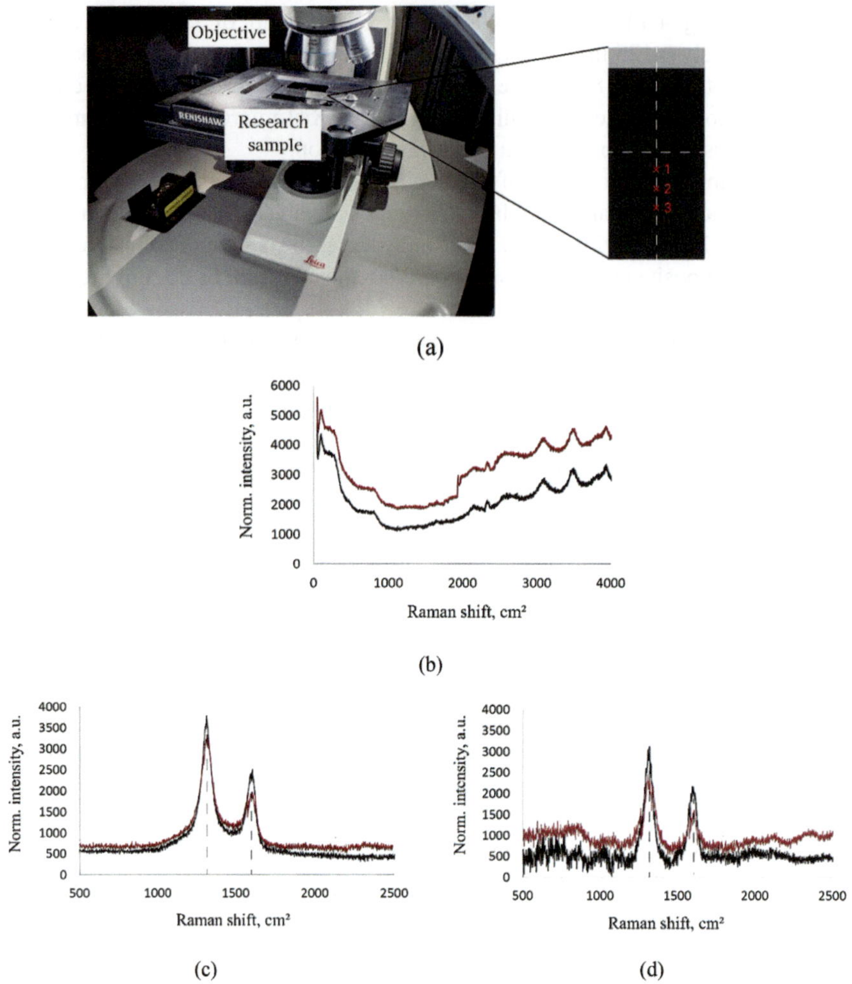

Fig. 4.14 Raman spectroscopy of obtained coating samples: **a** sample spectroscopy process; Raman spectra before (black curve) and after (red curve) contact with BSA for **b** uncoated titanium surface, **c** col+M WCNT coating and **d** col +MWCNT+GA coating

on the coating surface, which leads to attenuation of the scattered signal from the nanotubes. The amount of BSA on the col+MWCNT+GA nanocomposite coating is greater: in the region up to 1000 cm^{-1}, the increase in fluorescence reached 15%, and in the region 1800–2500 cm^{-1} up to 24%. The intensity of D- and G-mode peaks from carbon nanotubes in the coating decreased by 14% and 20%, respectively. It was obtained that GA as part of the nanocomposite coating provides slightly higher adhesion of BSA. At the same time, the highest level of adhesion was obtained for titanium surface.

From the comparison of the average fluorescence level, it can be tentatively concluded that the adhesion of BSA to col+MWCNT and col+MWCNT+GA coatings was 1.7 and 1.4 times less than that to titanium surface.

Conclusion

The contact of a medical implantable device with blood initiates a cascade of biological reactions aimed at thrombosis, which affects not only the long-term performance of the implantable device but also the life of the patient in case of thrombus detachment from the device surface. Due to limitations in systemic anticoagulation, the development of stable and antithrombogenic coatings for cardiovascular implants is an urgent task. This is especially important for circulatory assist devices, which are subjected to streamlined blood flow with high shear stress.

The composite coatings developed in this research work are a synergy of polymer (collagen) and carbon nanotubes adhered to titanium layer substrates by spraying the corresponding dispersion and subsequent solvent evaporation. Automation of the sputtering process allowed obtaining homogeneous films with a thickness of 10 μm. Some of the coatings were subjected to cross-linking of polymer chains in 0.02 wt.% GA solution to increase mechanical properties.

For hydrodynamic testing of coatings, a special microfluidic system has been developed, the function of which is to realize hermetic contact of the investigated coatings with the flow of model liquid of small volume. This system provides a plate based on elastic material polydimethylsiloxane with a formed microchannel, which under the pressure of external elements is hermetically pressed to the sample with coating. Connection of auxiliary units, such as a pressure compressor, microfluidic controller and recirculation system, allows to ensure the flow of fluid in the formed microchannel, simulating a closed blood flow with the necessary characteristics. This system is unique because, in addition to utilizing a small volume of fluid, the system is reversible, i.e., designed for multiple uses.

In the case of coating stability test, PBS was used as a model fluid. It contacted with the coatings in the system for 3.5 h at a shear stress of 50 Pa, which corresponded to the range of shear stress values in VAD. VAD can vary from 0 to 150 Pa. Numerical modeling was performed to determine the type and structure of the flow inside the channel. Numerical modeling was performed to determine the type and structure of the flow inside the channel. The velocity field in the section at the level of half-height of the microchannel and the shear stress at the bottom wall of the microchannel was calculated. It was obtained that the fluid flow entering perpendicular to the microchannel plane introduces significant fluctuations of velocity fields and shear stresses, and after half of the flow path in the channel, the flow acquires laminar character. It follows that in order to obtain correct results of interaction at given shear stresses on the flow side, the coating analysis was carried out in the laminar flow zone.

The flow resistance of coatings as a function of contact time was evaluated using optical profilometry. The measurement area included the laminar flow contact zone and the zone not exposed to flow contact. It was found that the col+MWCNT and col+MWCNT+GA coatings degraded by 5.5% and 2.75%, respectively, after contact with the flow for 3.5 h, respectively. These results indicate that cross-linking of the collagen matrix by GA contributes to an almost twofold increase in coating stability. Nevertheless, a rather high resistance to high shear stress flow was obtained for both types of coatings. In addition, the roughness of the coatings varies in the order of roughness of the titanium surface of the straightening apparatus used in the VAD.

It was found that the hardness of col +MWCNT+GA coating is 25% higher than the hardness of col +MWCNT coating, which confirms the better mechanical stability of the sample cross-linked with GA compared to the non-cross-linked one. Furthermore, the modulus of elasticity of the samples was determined through the utilization of a hardness tester. The results demonstrate that coating col + MWCNT+GA exhibits greater elastic properties, with a relative increase of 30.5% in comparison to coating col+MWCNT.

In the case of thrombogenicity tests, the coatings were contacted with a solution of bovine serum albumin protein, a precursor to thrombosis, in a microfluidic stend for 30 min. The concentration of albumin in the solution was 4 wt.%, which corresponds to the normal concentration of albumin in human blood. The attachment of albumin protein to coatings and to pure titanium was evaluated by Raman analysis before contact with the flow of albumin solution and after contact. Comparison of Raman spectra of nanocomposite coatings col+MWCNT and col+MWCNT+GA before and after contact with the flow of BSA solution allowed concluding the following. Total fluorescence background increased and the intensity of D- and G-mode peaks on the dependences for both coatings changed. From the comparison of the average fluorescence level it can be stated that the adhesion of BSA to col +MWCNT and col +MWCNT+GA coatings is 1.7 and 1.4 times less than to the titanium surface. It was obtained that GA in the composition of nanocomposite coating provides slightly higher adhesion of BSA. At the same time, the highest level of adhesion is obtained for titanium surface.

Thus, the developed collagen-based coatings can be considered good candidates for the development of blood pumping devices, having demonstrated high mechanical and antithrombogenic properties. The designed microfluidic chip is a promising tool providing full-fledged in vitro testing of coatings using small volumes of biomaterial, while simulating real conditions of blood flow in the human body.

Funding The work was supported by the Ministry of Education and Science of the Russian Federation (project FSMR-2024–0003).

Conflicts of Interest The authors have no conflicts of interest to declare that are relevant to the content of these chapters.

References

Arne HT, Sissel TE, John BE (2005) Serum albumin and blood pressure: a population-based, cross-sectional study. J Hypertens 23(4):725–730

Avci-Adali M, Ziemer G, Wendel HP (2010) Induction of EPC homing on biofunctionalized vascular grafts for rapid in vivo self-endothelialization—a review of current strategies. Biotechnol Adv 28:119–129

Biran R, Pond D (2017) Heparin coatings for improving blood compatibility of medical devices. Adv Drug Deliv Rev 112:12–23

Braune S, Latour RA, Reinthaler M, Landmesser U, Lendlein A, Jung F (2019) In vitro thrombogenicity testing of biomaterials. Adv Healthc Mater 8:1900527

Bulwan M, Wojcik K, Zapotoczny S, Nowakowska M (2012) Chitosan-based ultrathin films as antifouling, anticoagulant and antibacterial protective coatings. Biomater Sci:1963–1980

Cheng Y, Zheng YF, Zhang X, Xi TF, Wei SC (2012) Surface characteristics and corrosion behaviour of we43 magnesium alloy coated by SiC film. Appl Surf Sci 258(7):3074–3081

Cherng WJ, Pan YH, Wu TC, Chou CC, Yeh CH, Ho JJ (2019) Hemocompatibility and adhesion of heparin/dopamine and heparin/collagen self-assembly multilayers coated on a titanium substrate. Appl Surf Sci 463:732–740

Costa S, Borowiak-Palen E, Kruszyńska M, Bachmatiuk A, Kaleńczuk RJ (2008) Characterization of carbon nanotubes by Raman spectroscopy. Mater Sci Poland 26(2):433–441

Crespo-Leiro MG, Metra M, Lund LH, Milicic D, Costanzo MR, Filippatos G et al (2018) Advanced heart failure: a position statement of the heart failure association of the European society of cardiology. Eur J Heart Fail 20:1505–1535

Cuker A (2013) Clinical and laboratory diagnosis of heparin-induced thrombocytopenia: an integrated approach. Semin Thromb Hemost 40:106–114

Demidenko NA, Kuksin AV, Molodykh VV, Pyankov ES, Ichkitidze LP, Zaborovaet VA et al (2022) Flexible strain-sensitive silicone-CNT sensor for human motion detection. Bioengineering 9:36

Dresselhaus MS, Dresselhaus G, Saito R, Jorio A (2005) Raman spectroscopy of carbon nanotubes. Phys Rep 409:47–99

Edmunds LH, Colman RW (2006) Thrombin during cardiopulmonary bypass. Ann Thorac Surg 82(6):2315–2322

Fan Y, Kodzius F, Ian G (2012) Fabrication of polystyrene microfluidic devices using a pulsed co2 laser system. Foulds Microsyst Technol 18:373–379

Fenech M, Girod V, Claveria V, Meance S, Abkarian M, Charlot B (2019) Microfluidic blood vasculature replicas using backside lithography. Lab Chip 19:2096–2106

Figueiredo C, Eicke D, Yuzefovych Y, Avsar M, Hanke JS, Pflaum M et al (2019) Low immunogenic endothelial cells endothelialize the left ventricular assist device. Sci Rep 9:11318

Franchini M (2005) Heparin-induced thrombocytopenia: an update. Thromb J 3(14)

Gerasimenko AY, Zhurbina NN, Cherepanova NG, Semak AE, Zar VV, Fedorova YO et al (2020) Frame coating of single-walled carbon nanotubes in collagen on pet fibers for artificial joint ligaments. Int J Mol Sci 21:6163

Gerasimenko AY, Kurilova UE, Suetina IA, Mezentseva MV, Zubko AV, Sekacheva MI et al (2021a) Laser technology for the formation of bioelectronic nanocomposites based on single-walled carbon nanotubes and proteins with different structures, electrical conductivity and biocompatibility. Appl Sci 11:8036

Gerasimenko AY, Kurilova UE, Savelyev MS, Murashko D, Glukhova OE (2021b) Laser fabrication of composite layers from biopolymers with branched 3d networks of single-walled carbon nanotubes for cardiovascular implants. Compos Struct 260:113517

Gerasimenko AY, Kitsyuk E, Kurilova UE, Suetina IA, Russu L, Mezentseva MV et al (2022) Interfaces based on laser-structured arrays of carbon nanotubes with albumin for electrical stimulation of heart cell growth. Polymers 14:1866

Habal MV, Garan AR (2017) Long-term management of end-stage heart failure. Best Pract Res Clin Anaesthesiol 31:153–166

Hecker JF, Scandrett LA (1985) Roughness and thrombogenicity of the outer surfaces of intravascular catheters. J Biomed Mater Res 19(4):381–395

Hedayati M, Neufeld MJ, Reynoldset MM, Kipper M (2019) The quest for blood-compatible materials: recent advances and future technologies. Mater Sci Eng R Rep 138:118–152

Jaffer I, Fredenburgh J, Hirsh J, Weitz J (2015) Medical device-induced thrombosis: what causes it and how can we prevent it. J Thromb Haemost 13:72–81

Kalathottukaren M, Kizhakkedathu J (2018) Mechanisms of blood coagulation in response to biomaterials: extrinsic factors. In: Hemocompatibility of biomaterials for clinical applications. Amsterdam, pp 29–49

Kim SW, Kim T, Kim YS, Choi SH, Lim HJ, Yang SJ et al (2012) Surface modifications for the effective dispersion of carbon nanotubes in solvents and polymers. Carbon 50:3–33

Koster A, Sanger S, Hansen R, Sodian R, Mertzlufft F, Harke C et al (2000) Prevalence and persistence of heparin/platelet factor 4 antibodies in patients with heparin coated and noncoated ventricular assist devices. ASAIO J 46(319)

Lequier L, Annick G, Al-Ibrahim O, Bembea M, Brodie D (2014) The extracorporeal life support organization. ELSO

Liu T, Liu S, Zhang K, Chen J, Huang N (2014) Endothelialization of implanted cardiovascular biomaterial surfaces: the development from in vitro to in vivo. J Biomed Mater Res Part A 102:3754–3772

Liu K, Chen J, Zhang K, Wang S, Li X (2019) A Diagnostic prediction model of acute symptomatic portal vein thrombosis. Ann Vasc Surg 61:394–399

Masuzawa T (2000) State of the art of micromachining. CIRP Ann 49:473–488

Milleret M, Buzzi S, Gehrig P, Ziogas A, Grossmannc J, Schilcher K (2015) Protein adsorption steers blood contact activation on engineered cobalt chromium alloy oxide layers. Acta Biomater 24:343–351

Mohan CC, Chennazhi KP, Menon D (2013) In vitro hemocompatibility and vascular endothelial cell functionality on titania nanostructures under static and dynamic conditions for improved coronary stenting applications. Acta Biomater 9:9568–9577

Mulder MMG, Fawzy I, Lancé MD (2018) ECMO and anticoagulation: a comprehensive review. Neth J Crit Care 26:6–13

Negri V, Pacheco-Torres J, Calle D, López-Larrubia P (2020) Carbon nanotubes in biomedicine. Top Curr Chem:378

Nevídalová N, Michalcová L, Glatz Z (2020) Applicability of capillary electrophoresis-frontal analysis for displacement studies: effect of several drugs on l-tryptophan and lidocaine binding to human serum albumin. J Sep Sci 43:4225–4233

Ontaneda A, Annich GM (2018) Novel surfaces in extracorporeal membrane oxygenation circuits. Front Med 5:321

Pappalardo F, Maj G, Scandroglio A, Sampietro F, Zangrillo A, Koster A (2009) Bioline heparin-coated ECMO with bivalirudin anticoagulation in a patient with acute heparin-induced thrombocytopenia: the immune reaction appeared to continue unabated. Perfusion 24(135):7

Pinto EA, Faustino V, Rodrigues RO, Pinho D, Garcia V, Miranda J, Lima RA (2014) Rapid and low-cost nonlithographic method to fabricate biomedical microdevices for blood flow analysis. Micromachines 6:121–135

Pisapia F, Balachandran W, Rasekh M (2022) Organ-on-a-chip: design and simulation of various microfluidic channel geometries for the influence of fluid dynamic parameters. Appl Sci 12:3829

Primavera R, Razavi M, Kevadiya BD, Wang J, Vykunta A, Mascolo DD et al (2021) Enhancing islet transplantation using a biocompatible collagen-PDMS bioscaffold enriched with dexamethasone-microplates. Biofabrication 13

Rabe M, Verdes D, Seeger S (2011) Understanding protein adsorption phenomena at solid surfaces. Adv Colloid Interface Sci 162:87–106

Ramos SC, Vasconcelos G, Antunes EF, Lobo AO, Trava-Airoldi VJ (2010) Wettability control on vertically-aligned multi-walled carbon nanotube surfaces with oxygen pulsed dc plasma and co2 laser treatments. Diam Relat Mater 19:752–775

Regan-Smith S, Fritzen R, Hierons SJ, Ajjan AA, Blindauer CA, Stewart AJ (2022) Strategies for therapeutic amelioration of aberrant plasma ZN^{2+} handling in thrombotic disease: targeting fatty acid/serum albumin-mediated effects. Int J Mol Sci 23:10302

Savelyev MS, Gerasimenko AY, Vasilevsky PN, Fedorova YO, Groth T, Ten GN et al (2020) Spectral analysis combined with nonlinear optical measurement of laser printed biopolymer composites comprising chitosan/swcnt. Anal Biochem 598:113710

Simon J, Flahaut E, Golzio M (2019) Overview of carbon nanotubes for biomedical applications. J Mater 12:624

Tashakori-Miyanroudi M, Rakhshan K, Ramez M, Asgarian S, Janzadeh A, Azizi Y et al (2020) Conductive carbon nanofibers incorporated into collagen bio-scaffold assists myocardial injury repair. Int J Biol Macromol 163:1136–1146

Telyshev D, Denisov M, Markov A, Fresiello L, Verbelen T, Selishchev S (2019) Energetics of blood flow in Fontan circulation under VAD support. Artif Organs 44:50–57

Vakhrusheva TV, Gusev AA, Gusev SA, Vlasova II (2013) Albumin reduces thrombogenic potential of single-walled carbon nanotubes. Toxicol Lett 221:137–145

Videm V, Svennevig JL, Fosse E, Semb G, Osterud A, Mollnes TE (1992) Reduced complement activation with heparin-coated oxygenator and tubings in coronary bypass operations. J Thorac Cardiovasc Surg 103:806–813

Vroman L (2009) When blood is touched. Materials 2:1547–1557

Vroman L, Adams A, Fischer G, Munoz P (1980) Interaction of high molecular weight kininogen, factor XII, and fibrinogen in plasma at interfaces. Blood 55:156–159

Waddell EA (2006) Laser ablation as a fabrication technique for microfluidic devices. Methods Mol Biol 321

Wöhrle J, Al-Khayer E, Grötzinger U, Schindler C, Kochs M, Hombach V et al (2001) Comparison of the heparin coated vs the uncoated jostent—no influence on restenosis or clinical outcome. Eur Heart 22(19):1808

Zang L, Zhu H, Wang K, Liu Y, Yu F, Zhao W (2022) Not just anticoagulation—new and old applications of heparin. Molecules 27:6968

Zhang D, Wu X, Chen J, Lin K (2018) The development of collagen based composite scaffolds for bone regeneration. Bioact Mater 3:129–138

Zhou L, Forman H, Yi G, Lunec G (2017) Multi-walled carbon nanotubes: a cytotoxicity study in relation to functionalization, dose and dispersion. Toxicol In Vitro 42:292–298

Part III
Energy Harvesting

Chapter 5
Advancements in MEMS Vibration Energy Harvesters

Seyedfakhreddin Nabavi

Abstract This chapter provides a comprehensive overview about design, optimization, fabrication, and application of MEMS vibration energy harvesters. The content is structured to initially introduce various transduction methods applicable at the MEMS/NEMS scale for converting vibration into electricity. Thereafter, microfabrication methods for fabrication of each standalone transduction method are discussed. The inherent shortcomings of each group of MEMS harvesters such as narrow bandwidth or low power efficiency are explained in detail and up to date employed proposed solutions to overcome these obstacles are highlighted. In addition, the chapter delves into the crucial aspect of MEMS harvesters, power management systems, where various circuitry architectures designed to rectify, regulate, and store the harvested energy are studied. To provide practical relevance of the MEMS harvesters, their application in diverse fields such as Internet of Things (IoT), and wearable devices is explained.

Keywords Micro-electromechanical systems (MEMS) · Transduction · Piezoelectric · Electrostatic · Electromagnetic · Power management systems · Wearable · IoT

Introduction

The functionality of consumer electronics is dependent on the power supply, i.e., without a power supply, the intended electronic devices will not function, and if there are fluctuations in the power supply, the performance of the electronic consumers will be affected (Nabavi and Bhadra 2021a). A reliable power supply is therefore necessary for the secure operation of electronic circuits and systems. To give electronic

S. Nabavi (✉)
Nditive Technologies Inc, St. John's, NL, Canada
e-mail: snabavi@mun.ca

École de Technologie Supérieure (ÉTS), Montréal, QC, Canada

V. Pachkawade and K. Guha (eds.), *Micro- and Nano-Systems in 21st-Century*, Springer Tracts in Electrical and Electronics Engineering,
https://doi.org/10.1007/978-981-96-4659-3_5

devices the feature of portability (Nabavi and Bhadra 2021b), the use of batteries has been recognized as a vibrant solution. As the lifespan of the battery is limited, it needs to be replaced or recharged at regular intervals. If the battery has to be used in an inaccessible or hard-to-reach location, such as an implanted body, frequent battery replacement is not an ideal practice. Therefore, harvesting electrical energy from the environment is a favorable approach. Figure 5.1 illustrates the energy sources available in the environment, namely light, heat, radio frequency, and vibration. Indeed, each energy source has its own merits depending on the situation. However, it should be emphasized that vibrations can originate from various sources, including human activity. And thanks to MEMS (microelectromechanical system) technology, miniaturized vibrometry harvesters can be offered with an adequate level of efficiency (Nabavi et al. 1052).

Since MEMS technology invention in the early 1950s, it has made significant progress, fueled by the demand for smaller devices with better performance and lower cost. MEMS technology has proven its effectiveness in various applications such as resonators (Kolahdouz-Moghaddam et al. 2023), switches (Sharma et al. 2023), optical aligners (Nabavi et al. 2022a), micropositioners (Nabavi et al. 2021),

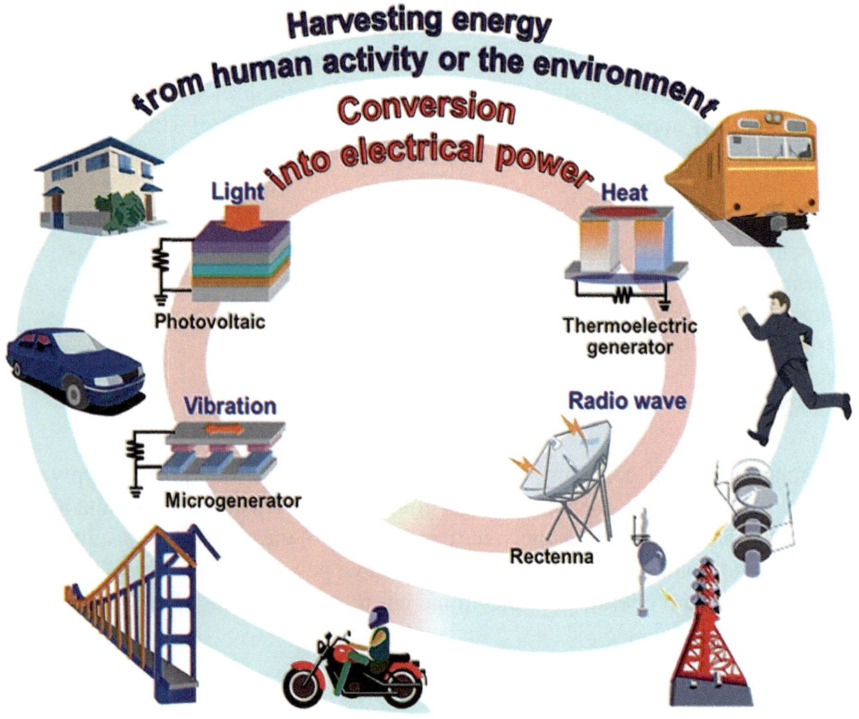

Fig. 5.1 Overview of energy resources and harvesting techniques, adopted from (Fujitsu Global 2023)

and beyond, highlighting its vast advantages. MEMS technology is an approach that can provide a deformable mechanical structure under external excitation such as vibration, while the overall physical aspects of the system do not exceed a few hundred micrometers. Improving the efficiency of vibration-based MEMS energy harvesters mainly consists of using efficient materials or optimizing the mechanical configuration of the harvesters (Nabavi and Zhang 2018a).

This chapter first explains the transduction approaches used at the MEMS scale to convert vibration into electricity. It then discusses previous efforts to optimize the mechanical structures to overcome the inherent shortcomings of MEMS-based vibration energy harvesters. It also provides a brief explanation about a necessary component of any type of MEMS vibration energy harvesting device, the power management circuitry. In addition, the applicability of MEMS vibration energy harvesters in real-world settings is outlined.

Conversion of Vibration into Electricity

At the MEMS scale, typically vibration can be converted to electricity through three main conversion approaches, which are known as electrostatic, electromagnetic, and piezoelectric-based vibration-to-electricity conversion. In this subsection, working principle of each conversion technique is described in detail.

Electrostatic

Electrostatic vibration energy harvesters operate on the basis of the variable capacitor. Thus, the mechanical structure of the electrostatic harvester is devised to form a variable capacitor whose capacitance can be changed in response to an external excitation. In this context, the harvester comprises a series of fixed and movable mechanical structures in the form of fingers (comb-fingers) (Nabavi et al. 2022b). In Fig. 5.2a, the general configuration of an electrostatic-based vibration energy harvester is illustrated. According to this illustration, it can be seen that the movable part of the harvester undergoes motion, causing variation in capacitance between each fixed and movable fingers. Figure 5.2b represents the implantation of the electrostatic MEMS vibration harvester in order to be able to produce power. As shown, such an energy harvester requires a bias voltage (relatively high amplitude) to create electrical charge on the surface of electrodes for polarization.

The energy, E, scavenged by an electrostatic energy harvester according to capacitance variation can be expressed as follows (Warshel and Aqvist 1991):

$$E = \frac{1}{2} V_{bias}^2 \Delta C, \tag{1}$$

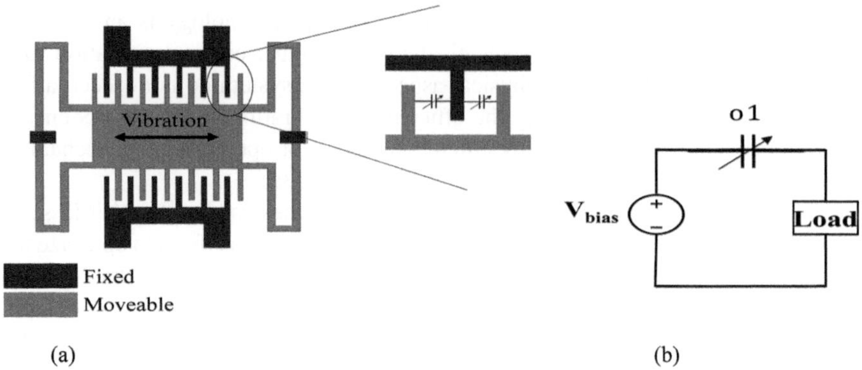

Fig. 5.2 **a** General configuration of electrostatic based vibration energy harvester, and **b** its implementation

where V_{bias} is the bias voltage and ΔC denotes the changes in the capacitance. It can be inferred that the harvested energy in such an energy conversion technique is highly dependent on the bias voltage. However, in practical energy harvesting systems, where an external supply voltage is unavailable, the applicability of this energy conversion technique for the creation of MEMS vibration energy harvesters is relatively low.

To eliminate the need for an external supply voltage in the electrostatic-based energy harvesters, the use of electrets has been proposed (Tao et al. 2015). Electrets are dielectric materials that can store electrical charges in their structure, similar to permanent magnets in magnetostatics, for an extended period of time. Figure 5.3 represents a schematic of the electret-based electrostatic energy harvesters and its equivalent circuit (Nabavi and Zhang 2016a). In this figure, Q_e is the electret charge, Q is the total charge between movable electrode and fixed electrode, R is the load resistance, $C(t)$ is the capacitance of the electrostatic energy harvester as a function time, and U_e is the surface voltage on electret layer. The relative movement between the fixed and moving electrodes leads to a change in capacitance, which ultimately generates an electrical load, $i = dQ/dt$.

Electromagnetic

Electromagnetic vibration energy harvesters work on the basis of Faraday's law of induction, in which a conductor (coil) cuts through the magnetic flux, hence an electric current is induced in the coil. Similar to electrostatic harvesters, a fixed (stator) and a moving (rotor) part are required for such vibration to electricity energy conversion method (Tan et al. 2016). Figure 5.4 shows a general configuration of the electromagnetic energy harvester (Costanzo et al. 2022). In this figure, a permanent magnet is suspended by means of a spring, and a coil larger than the magnet is

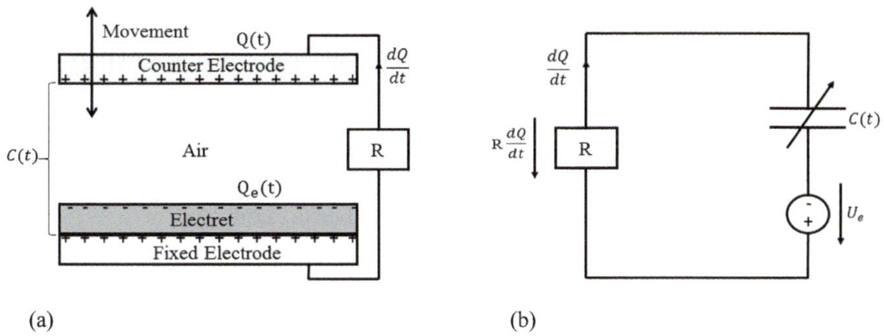

Fig. 5.3 **a** General configuration of electret-based electrostatic energy harvester, and **b** its equivalent circuit (Nabavi and Zhang 2016a)

mounted under the magnet. The vibration causes the magnet to move forwards and backwards toward the coil, which induces an electric current/voltage in the coil. It is important to note that the positioning of the coil and magnet can be determined as either fixed or moving components, depending on the manufacturing capacity.

The amount of induced voltage in the electromagnetic coil can be computed by (Norhisam et al. 2008).

$$U = -N\frac{d\phi}{dt}, \tag{2}$$

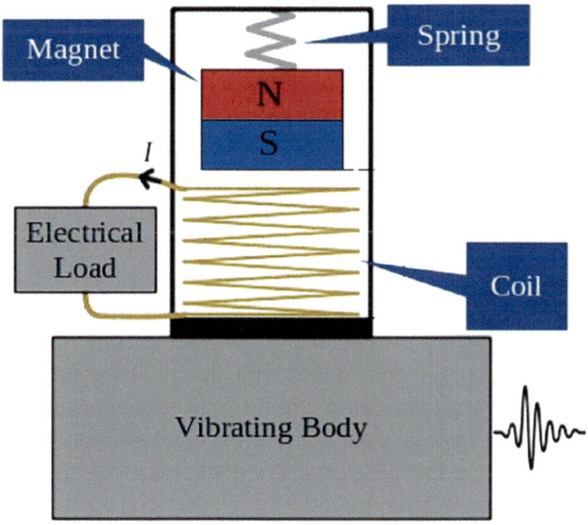

Fig. 5.4 General configuration of the electromagnetic energy harvester in which the magnet is considered to be movable and the coil to be fixed (Costanzo et al. 2022)

where U is the generated voltage, N is the number of turns, and ϕ is the total flux. The flux density is highly dependent on the magnetic field, which is given by:

$$\phi = BA\sin(\theta), \tag{3}$$

where B is the magnetic field, A denotes the perpendicular area to the magnetic field, and θ is the angle at which the magnetic field reaches the coil.

Piezoelectric

Piezoelectric materials generate electrical signals when they are subjected to a mechanical load. The amount of energy generated in piezoelectric materials is directly proportional to the mechanical force applied. A wide range of materials exhibit the piezoelectric property such as zinc oxide (ZnO) (Xu and Li 2021), aluminum nitride (AlN) (Nabavi and Zhang 2016b), Aluminum Scandium Nitride (AlScN) (Wang et al. 2020), and $Pb(Zr_xTi_1 - xPb(Zr_xTi_{1-x})O_3$ (PZT) (Lei et al. 2011). The energy conversion efficiency of any piezoelectric material depends on its electromechanical coupling factor, K, which is defined as follows:

$$K^2 = \frac{Electrical\ energy\ converted}{Mechanichal\ energy\ applied} \tag{4}$$

With piezoelectric materials, energy conversion can occur in two different modes, namely d_{31} and d_{33} (Kim et al. 2012a). Figure 5.5 illustrates the applied stress, the generated electric field, and the electrode configuration for the operating modes d_{31} and d_{33}. In mode d_{31}, the applied stress and the generated electric field are perpendicular to each other, necessitating electrode placement at the top and bottom of the piezoelectric material. In contrast, in operating mode d_{33}, the applied stress and the generated electric field are in the same direction, which requires an interdigital electrode to be used on one side of the piezoelectric layer.

A vibration energy harvester based on the piezoelectric material can simply be constructed as a cantilever beam. In this configuration, the piezoelectric material can be deposited on one side (unimorph) or on both sides (bimorph) of the cantilever (Kim et al. 2012b). When the cantilever oscillates, the piezoelectric material is mechanically stressed, resulting in the generation of an electrical voltage. Figure 5.6 shows a typical structure for the piezoelectric vibration energy harvester. Regardless of the operating mode, the energy harvester consists of three different layers: the base, the piezoelectric material, and the electrodes. All materials are perfectly stacked on top of each other and anchored from one side.

The open circuit voltage generated by the piezoelectric materiel can be estimated as (Jeon et al. 2005):

$$V_{oc} = \sigma g_{3i} D \tag{5}$$

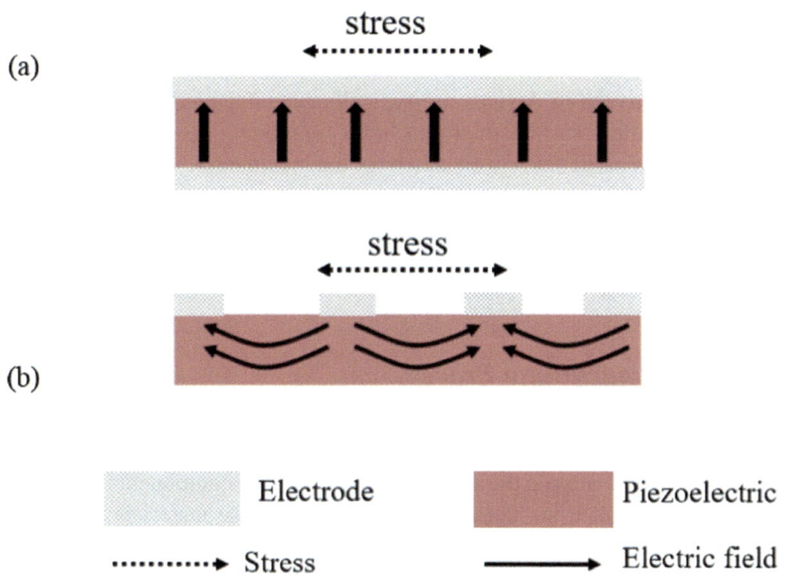

Fig. 5.5 Illustration of piezoelectric operating modes, **a** mode d_{31} and **b** mode d_{33}. Arrows inside the piezoelectric material indicate the generated electric fields

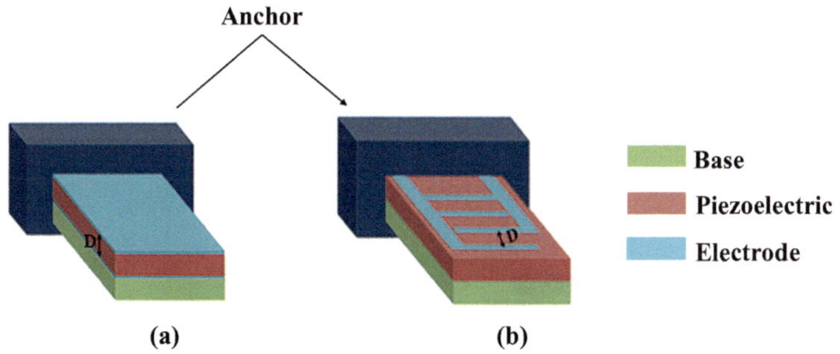

Fig. 5.6 General configurations of the piezoelectric based vibration energy harvester, **a** d_{31} mode and **b** d_{33} mode

where σ is the applied mechanical stress, g_{3i} is the piezoelectric coefficient while the index i determines the operating mode, either 1 or 3. D is the distance between electrode, which is the space between the top and bottom electrodes in mode $d_{31,}$ and spacing between the interdigitated electrodes in mode d_{33}, as shown in Fig. 5.6. It can be inferred that the amplitude of the harvested voltage by the piezoelectric-based vibration energy harvesters is influenced by the magnitude of the applied mechanical stress, the piezoelectric coefficient, and the electrode configuration.

Fabrication

MEMS energy harvesting devices are typically manufactured using two primary fabrication techniques: surface micromachining and bulk micromachining. Surface micromachining is particularly well suited for the manufacture of capacitive sensors and actuators, such as electrostatically operated energy harvesting devices. Bulk micromachining, on the other hand, is used for the manufacture of electrostatic, electromagnetic, and piezoelectric energy harvesting devices.

Surface Micromachining Process

In surface micromachining process, movable structures are created by depositing and structuring different layers on the substrate. These moving components of the devices are made from so-called structural (device) layers. These materials are deposited on sacrificial layers that have been previously deposited or grown to support the structural layers. The sacrificial layers are then removed as part of the manufacturing process, a step known as the release process. Figure 5.7 shows a simplified standard process for surface micromachining process (Nabavi et al. 2021). As shown in this figure, nitride and oxide layers serve as insulator and sacrificial layers, respectively, while two polysilicon layers, namely *Ply 0* and *Ply 1*, are used to fabricate both the fixed and moving parts of the MEMS device.

Figure 5.8 shows a schematic of the energy harvester based on the electrostatic energy conversion, which is manufactured using a surface micromachining process (Zhu et al. 2015). In this energy harvester, four serpentine springs support the moving plate. These springs are connected to the silicon substrate solely at anchor points,

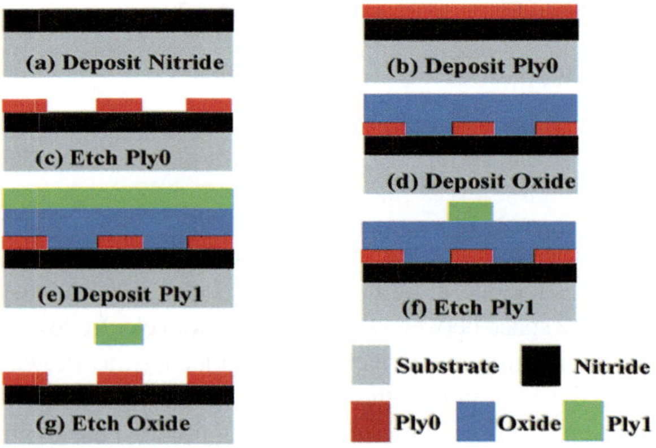

Fig. 5.7 Flow diagram of a standard surface micromachining surface for the manufacturing of an electrostatic based energy harvester, reprinted with permission from Nabavi et al. (2021)

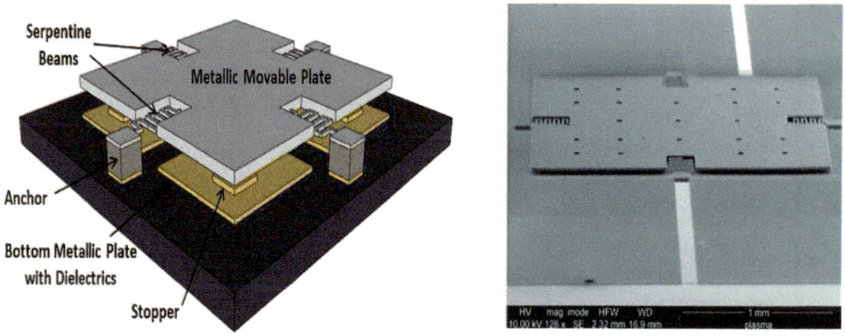

(a) (b)

Fig. 5.8 **a** Schematic of the electrostatic vibration energy harvester designed based on the surface micromachining process, and **b** micrograph of the fabricated prototype, reprinted with permission from Zhu et al. (2015)

which are made of electroplated nickel. This creates an air gap between the moving metal plate and the metal on the surface.

Bulk Micromachining Process

In contrast to surface micromachining, bulk micromachining is about producing MEMS structures inside the substrate and incorporating a thicker structural layer. This approach is based on the deep etching of materials such as silicon or glass. Bulk micromachining fabrication processes, which involve a combination of anisotropic and isotropic etching, have found wide application in the fabrication of various mechanical structures.

Figure 5.9 shows a visual representation of a standard micromachining process. This process uses silicon wafers (SOI wafers) (Nabavi and Zhang 2018b). It involves several sequential steps, including deposition of materials (e.g., piezoelectric and conductive layers), subsequent patterning (lithography), and removal of unwanted areas, including full etching of the substrate using the Deep Reactive Ion Etch (DRIE) technique.

Figure 5.10 illustrates piezoelectric energy harvesters manufactured using the standard bulk micromachining process (Nabavi and Zhang 2019a). Complete etching of the substrate leads to the formation of a cantilever without a proof mass, while partial etching of the substrate introduces a proof mass into the structure of the harvester.

An example of an electromagnetic-based energy harvester fabricated by bulk micromachining process is shown in Fig. 5.11 (Wang et al. 2009). In this energy harvester, a planar coil is formed on the silicon substrate. On a separate wafer, a serpentine-shaped planar spring made of copper with an opening in the center for

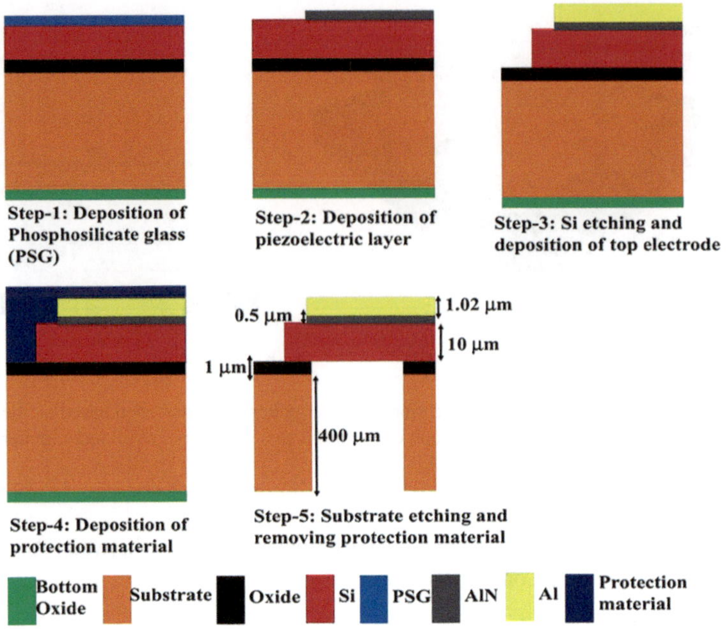

Fig. 5.9 Flow diagram of a standard bulk micromachining for the manufacturing of a piezoelectric energy harvester with a proof mass at the tip of the cantilever, reprinted with permission from Nabavi and Zhang (2018b)

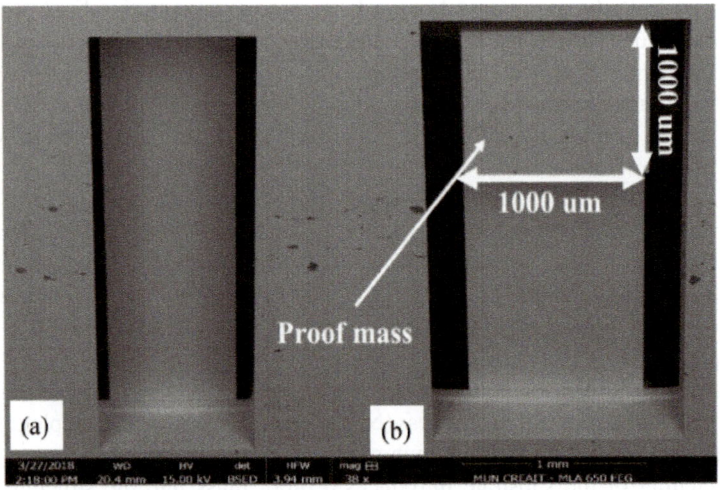

Fig. 5.10 Micrographs of the piezoelectric-based energy harvesters: **a** without a proof mass and **b** with a proof mass, both fabricated using the standard bulk micromachining process, reprinted with permission from Nabavi and Zhang (2019a)

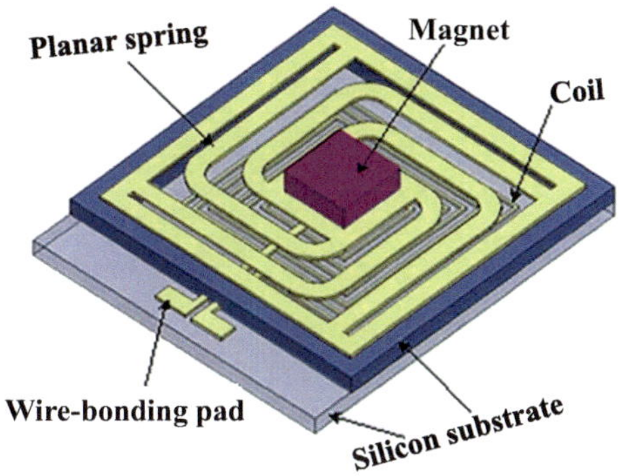

Fig. 5.11 Schematic of the electromagnetic-based energy harvester, showcasing the assembly of the coil and magnet, reprinted with permission from Wang et al. (2009)

positioning the permanent magnet is manufactured. The two wafers, i.e., the coil and the movable magnet, are then attached together using the micro-assembly. It is obvious that the production of electromagnetic energy harvesters requires more steps compared to their counterparts, the electrostatic and piezoelectric energy harvesters.

Design of MEMS Energy Harvesters

MEMS harvesters are mechanical devices that perform best at their fundamental resonant frequency. Since the majority of vibration sources in the environment operate at low frequencies, it is critical to design MEMS harvesters with resonant frequencies that match the vibration frequencies. The resonant frequency of a vibrating mechanical structure is determined by its physical dimensions. This means that the fundamental resonant frequency is usually high for devices with a size of a few hundred micrometers (e.g., MEMS devices).

An effective approach to lowering the resonant frequency, such as in piezoelectric energy harvesters with a cantilever design, is to incorporate a proof mass. This increases the overall mass of the device, which leads to a reduction in its oscillation frequency. Figure 5.12 shows an electrostatic energy harvester in which the use of a relatively heavy microball in the structure of the harvester results in an operating resonant frequency of less than 20 Hz.

The frequency and magnitude of the vibration sources in the environment are not constant, i.e., they can vary over time. On the other hand, the MEMS vibration devices have a relatively narrow frequency bandwidth, i.e., in the order of a few Hz. To increase the effectiveness of the MEMS harvesters, it is therefore necessary to increase their operating bandwidth (Nabavi and Zhang 2018).

An effective approach to extending the operating frequency range of MEMS energy harvesters is to introduce non-linearity into the system. This is achieved by integrating a pair of permanent magnets into the structure of the harvester (Harne and Wang 2013). For example, in a cantilever-based piezoelectric harvester, one magnet is positioned at the tip of the cantilever, while another magnet with the same polarity

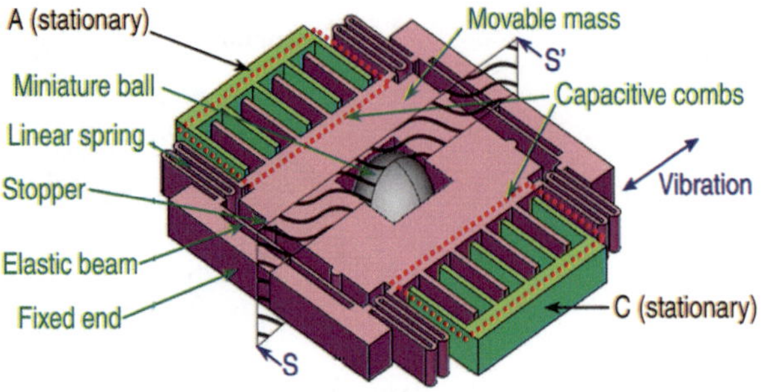

Fig. 5.12 Schematic representation of the electrostatic energy harvester, where a moving mass in the center of the energy harvester serves to reduce the operating resonant frequency (Lu et al. 2018)

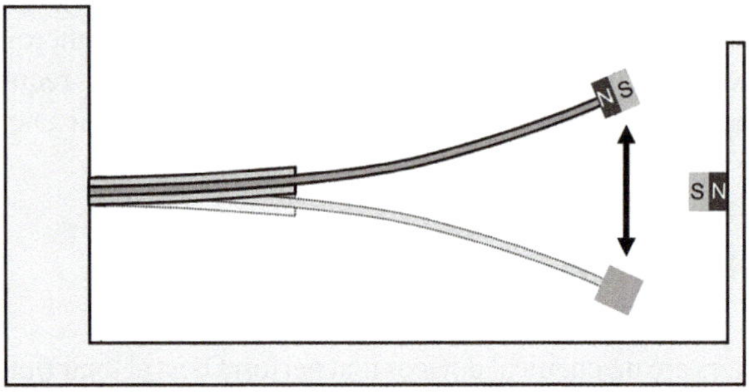

Fig. 5.13 Schematic representation of a bistable piezoelectric energy harvester

is located near the cantilever tip in the rest position, as shown in Fig. 5.13. This magnet configuration generates a repulsive force that makes the rest position of the cantilever unstable.

Figure 5.14 shows an example of the piezoelectric vibration energy harvester with a two-stage vibratory structure coupled with permanent magnets (Tang and Li 2014). The first stage captures low-frequency ambient vibrations. When the magnetic repulsive force pushes the second stage away, its spring deforms. If the restoring force exceeds the magnetic force, the second stage is released and vibrates freely at its own resonant frequency. In this way, the harvester can convert low-frequency ambient vibrations into high-frequency vibrations and thus achieve an upward frequency conversion.

It is obvious that extending the operating frequency bandwidth by using magnetic force has a significant impact on the dimensions of the harvester. In other words, the size of such an energy harvester cannot be reduced due to the need for micro-assembly and careful placement of the magnet pairs. In contrast, the use of a multimode geometry, which can be achieved by incorporating multi degrees of freedom (DOF), can extend the operating frequency of MEMS-based energy harvesters. In such a geometry, the resonant modes are close to each other. To achieve this, the harvester usually

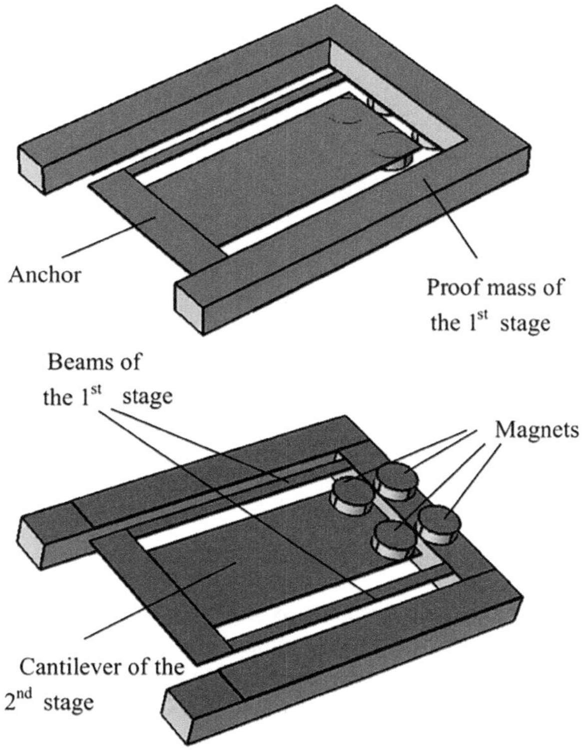

Fig. 5.14 Structure of the 2-stage piezoelectric vibration energy harvester with frequency up-conversion capability, reprinted with permission from Tang and Li (2014)

consists of several sections that can oscillate freely at different frequencies, so that multi DOF are realized. Figure 5.15, for example, shows a piezoelectric multimode energy harvester (Toyabur et al. 2017). As can be seen, the harvester consists of 4 free-standing cantilevers with proof masses attached together so that multiple resonant mode shapes can be observed at frequencies close to each other.

One technique for extending the operating frequency of MEMS vibration harvesters is to use a nonlinear multimode structure, which means that the harvester structure has multi DOF in addition to nonlinearity behavior. Such an energy harvester requires a fairly complex structure. Therefore, the use of design automation techniques, such as genetic algorithm (GA) and artificial intelligence, can greatly facilitate the design of these particular energy harvesters (Nabavi and Zhang 2017).

Figure 5.16 shows a multi-mode nonlinear piezoelectric MEMS vibration harvester that has been optimized by an automated design and optimization methodology based on the GA (Nabavi and Zhang 2019b). This energy harvester features a distinctive geometry anchored on two opposite sides and can be considered as a double clamped beam. As shown in Fig. 5.16, this energy harvester consists of three

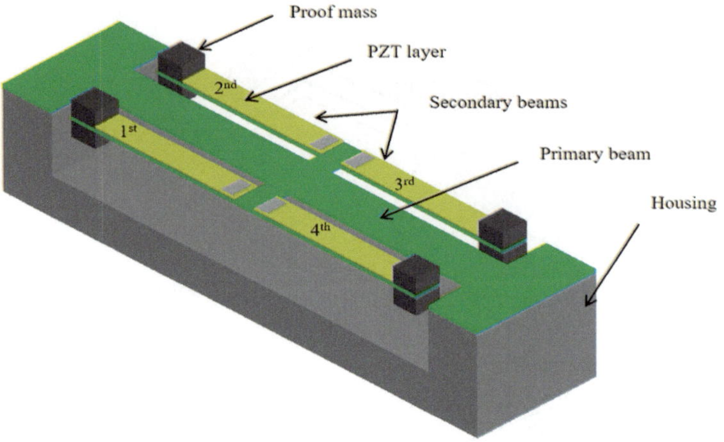

Fig. 5.15 Schematic of the multimodal piezoelectric energy harvester, where four cantilevers are connected to construct a multi degrees of freedom structure, reprinted with permission from Toyabur et al. (2017)

proof masses, including a central main mass and two lateral masses at different locations, allowing a uniform mass distribution throughout the MEMS harvester. These three proof masses are suspended from the serpentine beams, which have piezoelectric material deposited on their surface. Therefore, the GA, known as the evolutionary optimization method, is used to optimize the size of the proof masses as well as the length of the beams to minimize the frequency gap between all three modes.

Figure 5.17 shows the measurement of the frequency spectrum for the nonlinear multimode energy harvester. It can be seen that the first three modes of the MEMS harvester are close to each other, whereas the second mode, which occurs at around 261.8 Hz, exhibits nonlinear behavior due to its sharp jump.

The phenomenon of nonlinearity is influenced by the excitation amplitude. In other words, the frequency at which the jump phenomenon occurs is shifted according to the excitation amplitude (vibration). Figure 5.18 shows the measured velocity of the center of the nonlinear multimode energy collector shown in Fig. 5.16 in the time and frequency domain. It can be seen that the jump phenomenon shifted by 14 Hz when the acceleration was increased from 0.2 m/s^2 to 1 m/s^2.

Power Management System

It is essential to point out that the voltage generated by MEMS harvesters is inherently an AC signal, with an amplitude that can be as modest as a few hundred millivolts, due to the compact size of MEMS scavenging devices. Consequently, power management systems capable of converting the harvested AC signal to DC and delivering a constant voltage to the load are considered essential component of MEMS energy harvesters.

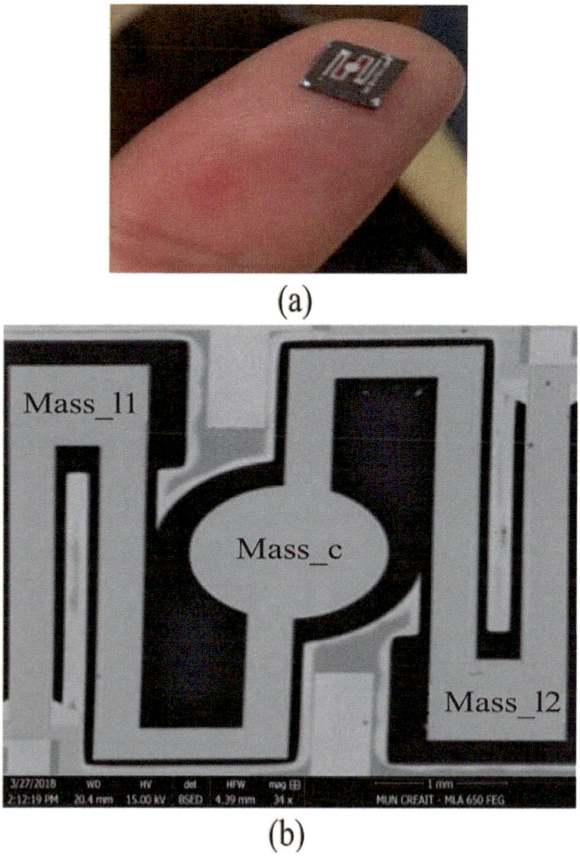

Fig. 5.16 Images of **a** the finger-top and **b** the top view of a nonlinear piezoelectric MEMS multi-mode vibration harvester designed based on the GA, reprinted with permission from Nabavi and Zhang (2019b)

A typical power management system consists of a rectifier that converts the AC signal to DC and a regulator or DC-DC convertor that ensures delivering a steady voltage to the load (Aktakka and Najafi 2014).

A well-known approach for converting an AC signal to DC is using a full-wave rectifier. In such a circuitry two pairs silicon diodes are implemented. The forward voltage drop of silicon diodes is about 0.7 V, which means during each phase almost the amplitude of the DC signal dropped by 1.4 V, although the voltage drop can be reduced by using Schottky diodes, however, such an AC to DC conversion circuitry is efficient for MEMS-based energy harvesters.

An effective method of implementing a full-wave rectifier, even in a CMOS technology, is to use MOSFET transistors instead of diodes. The circuit diagram of the full-wave rectifier based on MOSFETs is shown in Fig. 5.19 (Nabavi and Bhadra 2023). In this circuit, transistors *M2* and *M3* are activated during the positive phase

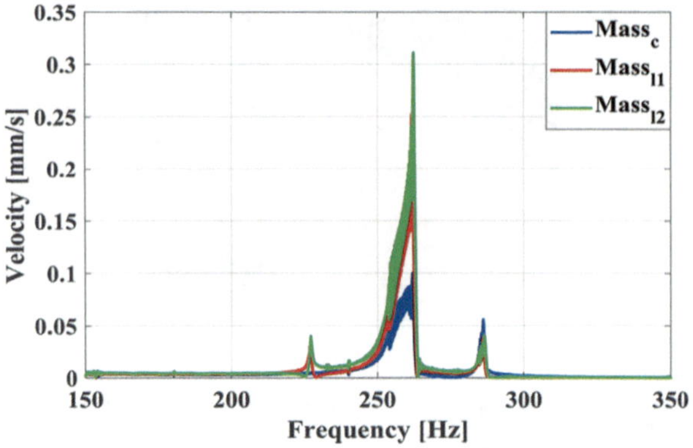

Fig. 5.17 Frequency spectrum of the piezoelectric non-liner multimode energy harvester, where the first three modes are close to each other and the second mode has a non-linear behavior, reprinted with permission from Nabavi and Zhang (2019b)

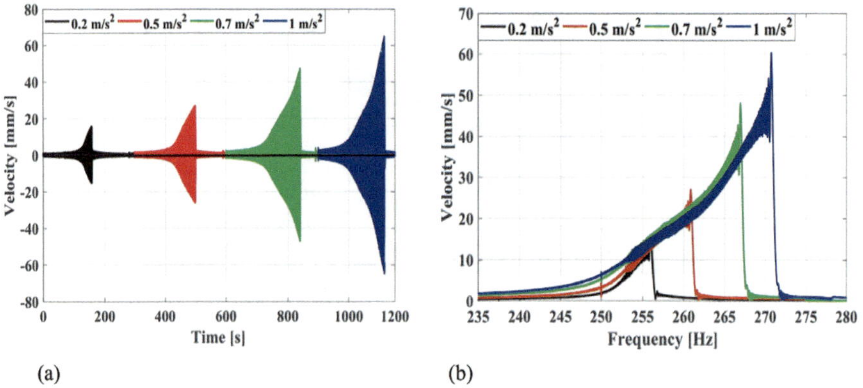

Fig. 5.18 Measured velocity of the nonlinear multimode energy harvester when the energy harvester was excited by forward frequency sweeping in its nonlinear (second) mode, with data shown **a** in the time domain and **b** in the frequency domain, reprinted with permission from Nabavi and Zhang (2019b)

of the incoming AC signal, while *M1* and *M4* are deactivated. This activation is controlled by the respective negative and positive voltage levels at their gates. Consequently, *Vn* is connected to ground, creating a current path through *M3* that delivers *Vp* to C_{start}, as shown in red in Fig. 5.19. Conversely, during the negative phase of the AC input signal, shown in blue in Fig. 5.19, M1 and M4 are activated and *M2* and *M3* are deactivated. This results in the reverse polarity of *Vn* being applied to C_{start} via *M4*.

Another interface circuit known for its proven effectiveness in piezoelectric energy harvesters is the Synchronized Switch Harvesting on Inductor (SSHI) (Lallart and

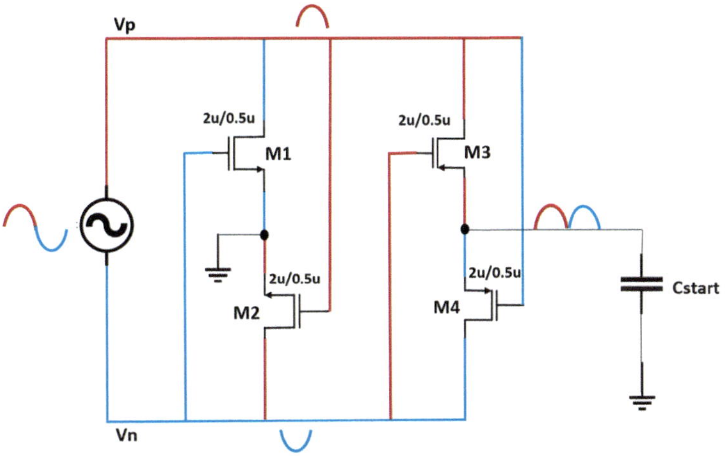

Fig. 5.19 Schematic of a full-wave rectifier utilizing MOSFET transistors (Nabavi and Bhadra 2023)

Lombardi 2020). This AC to DC converter significantly increases the energy extraction efficiency from MEMS harvesters. The circuit includes an inductor and a switch. The switch remains open except during peak deflection when it is activated for a short period, creating an oscillating electrical network (RLC) involving the inductor and the internal piezoelectric capacitance. After half an electrical period, the switch is deactivated, resulting in a quasi-instantaneous reversal of the harvested voltage. Figure 5.20a represents the schematic of SSHI circuit and Fig. 5.20b demonstrates a prototype of piezoelectric-based energy harvester integrated with CMOS power management system (Du et al. 2019).

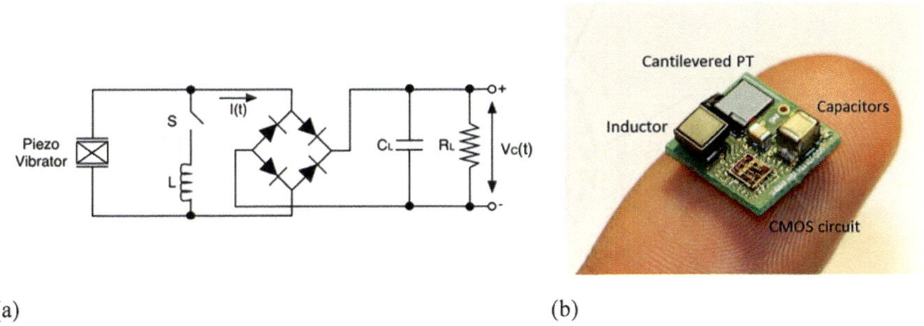

Fig. 5.20 **a** Schematic of the SSHI interface circuit and **b** fabricated piezoelectric-based MEMS harvester integrated with the SSHI, reprinted with permission from Du et al. (2019)

Applications of MEMS Energy Harvesters

Internet of Things (IoT) can manage a large number of smart wireless devices and form a network infrastructure connected to the Internet. With the development of IoT, wireless sensor nodes (WSN) have been used in various fields, such as environmental and conditional monitoring. WSNs consist of a collaborative network of embedded sensors called sensor nodes, which have sensing, computation, and communication functions. The main purpose of WSNs is to capture and collect data for specific analysis (Tang et al. 2018). A major challenge in maintaining the effectiveness of sensor nodes lies in the power supply. Usually, sensor nodes are powered by batteries, but the large number of installed nodes and their occasional placement in inaccessible locations make frequent battery replacement impractical (Sun et al. 2018).

A practical approach to overcoming the limited lifetime of wireless sensor nodes is the use of MEMS energy harvesters. These devices are a viable solution to extend the lifetime of wireless sensor nodes as they are compact and can be easily integrated into sensor systems. The integration of MEMS energy harvesters into wireless sensor nodes can advance the approach to deploying self-powered systems and overcome the challenges associated with the widespread use of sensor networks in diverse and often challenging environments.

The power consumption of portable devices is generally low. Therefore, the required energy can be provided by a MEMS energy harvesting system. MEMS energy harvester can be designed to harvest energy from human movement. Thus far, deploying MEMS energy harvesters has shown great potential for the development of wearable devices without batteries. Figure 5.21 represents a harvester utilizing a rotational proof mass and magnetically plucked piezoelectric beams, which specifically cater to harvesting energy efficiently from human motion (Xue et al. 2018).

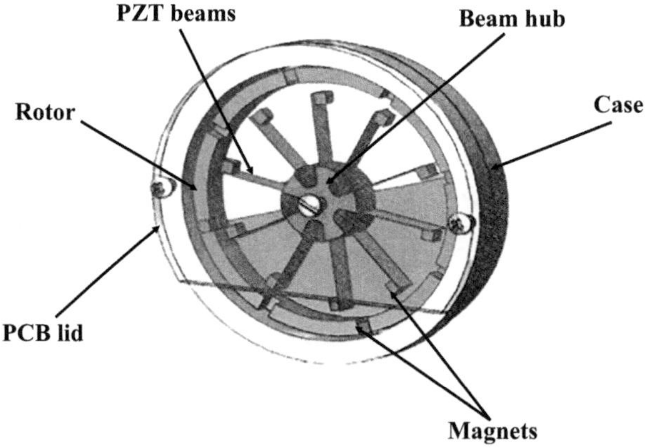

Fig. 5.21 Schematic of the rotational piezoelectric based energy harvester that can be worn on the human wrist. The energy harvester converts low and irregular frequencies of human motion into electricity, adopted from Xue et al. (2018)

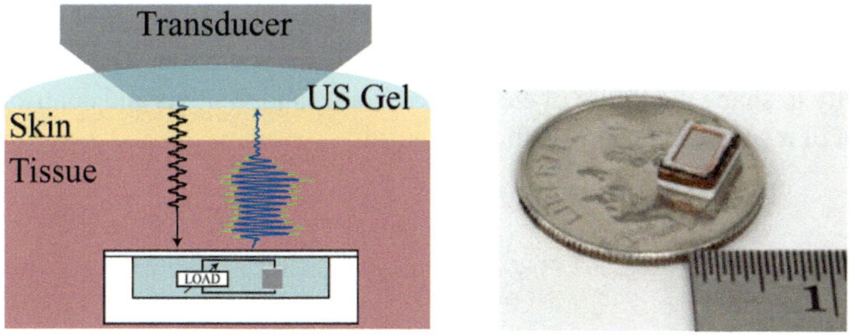

(a) (b)

Fig. 5.22 **a** Overview of an ultrasound-coupled IMD, in which the transmitted ultrasonic wave generates harvestable vibrations, and **b** prototype of an implantable piezoelectric energy harvester for energy harvesting from ultrasonic waves, adapted from Shen and Maharbiz (2019)

Implantable biomedical devices (IMDs) represent another category of low-power electronic devices that can be directly powered by MEMS energy harvesters. These devices are surgically implanted in the body, primarily serving functions related to sensing and actuation. Given the anticipated extended duration of IMDs within the body (in vivo), the utilization of MEMS energy harvesters becomes an attractive option. Figure 5.22a provides an overview of ultrasound-coupled IMDs, where ultrasonic waves induce vibrations in the IMD that are integrated with a vibration harvester and ultimately lead to the charging of the IMD (Shen and Maharbiz 2019). An example of a vibration energy harvester with the capability of harnessing power from ultrasonic waves is shown in Fig. 5.22b.

Conclusion and Future Perspective

This chapter explains the design, optimization, fabrication, and applications of MEMS vibration energy harvesters. It discusses the different conversion methods used to convert vibrations into electricity, including electrostatic, electromagnetic, and piezoelectric techniques. The manufacturing processes, whether surface micromachining or bulk micromachining, are outlined. The importance of design elements such as resonant frequency matching, nonlinearity, and multimode geometries to improve the efficiency of MEMS energy harvesters has proven to be an effective way to increase the versatility of MEMS vibration harvesters.

A critical focus is given to power management systems, indispensable for converting, regulating, and storing harvested energy. Notably, these circuitries can be seamlessly implemented using CMOS technology without compromising the compact size of MEMS harvesters. The chapter highlights the significant roles of MEMS energy harvesters in diverse applications such as wearable devices and IoT.

As technology advances, the potential of MEMS energy harvesters is expected to grow, opening up new possibilities for autonomous operation. These devices have the ability to shape the future of electronic circuits and systems and contribute to the creation of a smart and energy-efficient world.

References

Aktakka EE, Najafi K (2014) A micro inertial energy harvesting platform with self-supplied power management circuit for autonomous wireless sensor nodes. IEEE J Solid-State Circuits 49:2017–2029

Costanzo L, Lo Schiavo A, Sarracino A, Vitelli M (2022) Stochastic thermodynamics of an electromagnetic energy harvester. Entropy 24:1222

Du S, Jia Y, Zhao C, Amaratunga GAJ, Seshia AA (2019) A Nail-Size piezoelectric energy harvesting system integrating a MEMS transducer and a CMOS SSHI circuit. IEEE Sens J 20:277–285

Fujitsu Global Available online: https://www.fujitsu.com/global/ (accessed on 11 November 2023)

Harne RL, Wang KW (2013) A review of the recent research on vibration energy harvesting via bistable systems. Smart Mater Struct 22:23001

Jeon YB, Sood R, Jeong J-H, Kim S-G (2005) MEMS power generator with transverse mode Thin Film PZT. Sensors Actuators A Phys 122:16–22

Kim S-B, Park H, Kim S-H, Wikle HC, Park J-H, Kim D-J (2012a) Comparison of MEMS PZT cantilevers based on d31 and d33 modes for vibration energy harvesting. J Microelectromechanical Syst 22:26–33

Kim S-G, Priya S, Kanno I (2012b) Piezoelectric MEMS for energy harvesting. MRS Bull 37:1039–1050

Kolahdouz-Moghaddam A-R, Nabavi S, Nabki F (2023) Lamé resonator integrated with chevron-shaped thermal actuators to improve motional resistance and temperature stability. J Microelectromechanical Syst

Lallart M, Lombardi G (2020) Synchronized switch harvesting on electromagnetic system: a nonlinear technique for hybrid energy harvesting based on active inductance. Energy Convers Manag 203:112135

Lei A, Xu R, Thyssen A, Stoot AC, Christiansen TL, Hansen K, Lou-Moeller R, Thomsen EV, Birkelund K (2011) MEMS-Based thick film PZT vibrational energy harvester. In proceedings of the 2011 IEEE 24th international conference on micro electro mechanical systems, pp 125–128

Lu Y, Marty F, Galayko D, Laheurte J-M, Basset P (2018) A power supply module for autonomous portable electronics: Ultralow-Frequency MEMS electrostatic kinetic energy harvester with a comb structure reducing air damping. Microsystems & Nanoeng. 4, 28

Nabavi S, Bhadra S (2021a) Smart mandibular advancement device for intraoral monitoring of cardiorespiratory parameters and sleeping postures. IEEE Trans Biomed Circuits Syst 15:248–258

Nabavi S. Bhadra S (2021b) Design and development of a wristband for continuous vital signs monitoring of COVID-19 patients. In proceedings of the 2021 43rd annual international conference of the IEEE engineering in medicine & biology society (EMBC), pp 6845–6850

Nabavi S, Bhadra S (2023) A 65-Nm CMOS Self-Supplied power management system for near-field wirelessly powered biomedical devices. Electron 12:4622

Nabavi S, Zhang L (2016a) Portable wind energy harvesters for low-power applications: a survey. Sensors 16:1101

Nabavi S, Zhang L (2016b) MEMS piezoelectric energy harvester design and optimization based on genetic algorithm. In proceedings of the 2016 IEEE international ultrasonics symposium (IUS); pp 1–4

Nabavi S, Zhang L (2017) Design and optimization of piezoelectric mems vibration energy harvesters based on genetic algorithm. IEEE Sens J 17:7372–7382
Nabavi S, Zhang L (2018a) Frequency tuning and efficiency improvement of piezoelectric mems vibration energy harvesters. J Microelectromechanical Syst 28:77–87
Nabavi S, Zhang L (2018b) Curve-Shaped anchor for durability and efficiency improvement of piezoelectric MEMS energy harvesters. In proceedings of the 2018 IEEE SENSORS, pp 1–4
Nabavi S, Zhang L (2018c) Design and optimization of a low-resonant-frequency piezoelectric MEMS energy harvester based on artificial intelligence. In proceedings of the proceedings; 2, p 930
Nabavi S, Zhang L (2019a) T-Shaped piezoelectric structure for High-Performance MEMS vibration energy harvesting. J Microelectromechanical Syst 28:1100–1112
Nabavi S, Zhang L (2019b) Nonlinear Multi-Mode wideband piezoelectric mems vibration energy harvester. IEEE Sens J 19:4837–4848
Nabavi S, Aljaroudi A, Zhang L (2018) T-Shaped piezoelectric vibratory MEMS harvester with integration of highly efficient power management system. In proceedings of the journal of physics: conference series, 1052, p 12102
Nabavi S, Ménard M, Nabki F (2021) Surface micromachined Out-of-Plane electrostatic MEMS actuator integrated with displacement sensor. In proceedings of the 2021 IEEE sensors; pp 1–4
Nabavi S, Ménard M, Nabki F (2022) Vertical electrostatic MEMS aligner with integrated silicon nitride optical waveguides. In proceedings of the 2022 IEEE sensors, pp 1–4
Nabavi S, Ménard M, Nabki F (2022) A SOI Out-of-Plane electrostatic MEMS actuator based on in-plane motion. J Microelectromechanical Syst
Norhisam M, Norrimah A, Wagiran R, Sidek RM, Mariun N, Wakiwaka H (2008) Consideration of theoretical equation for output voltage of linear displacement sensor using meander coil and pattern guide. Sensors Actuators A Phys 147:470–473
Sharma S, Nabavi S, Rabih AAS, Ménard M, Nabki F (2023) Hybrid MEMS actuator with 3 degrees-of-freedom for efficient planar optical switching. J Microelectromechanical Syst
Shen K, Maharbiz MM (2019) Design of ceramic packages for ultrasonically coupled implantable medical devices. IEEE Trans Biomed Eng
Sun H, Yin M, Wei W, Li J, Wang H, Jin X (2018) MEMS based energy harvesting for the internet of things: a survey. Microsyst Technol 24:2853–2869
Tan Y, Dong Y, Wang X (2016) Review of MEMS electromagnetic vibration energy harvester. J Microelectromechanical Syst 26:1–16
Tang Q, Li X (2014) Two-Stage wideband energy harvester driven by multimode coupled vibration. IEEE/ASME Trans Mechatronics 20:115–121
Tang X, Wang X, Cattley R, Gu F, Ball AD (2018) Energy harvesting technologies for achieving Self-Powered wireless sensor networks in machine condition monitoring: a review. Sensors 18:4113
Tao K, Lye SW, Miao J, Tang L, Hu X (2015) Out-of-Plane Electret-Based MEMS energy harvester with the combined nonlinear effect from electrostatic force and a mechanical elastic stopper. J Micromechanics Microengineering 25:104014
Toyabur RM, Salauddin M, Park JY (2017) Design and experiment of piezoelectric multimodal energy harvester for low frequency vibration. Ceram Int 43:S675–S681
Wang P, Tanaka K, Sugiyama S, Dai X, Zhao X, Liu J (2009) A micro electromagnetic low level vibration energy harvester based on MEMS technology. Microsyst Technol 15:941–951
Wang J, Park M, Mertin S, Pensala T, Ayazi F, Ansari A (2020) A film bulk acoustic resonator based on ferroelectric aluminum scandium nitride films. J Microelectromechanical Syst 29:741–747
Warshel A, Aqvist J (1991) Electrostatic energy and macromolecular function. Annu Rev Biophys Biophys Chem 20:267–298
Xu Z, Li Z (2021) Design and fabrication of ZnO-Based SAW sensor using low power homo-buffer layer for enhanced humidity sensing. IEEE Sens J 21:7428–7433
Xue T, Yeo HG, Trolier-McKinstry S, Roundy S (2018) Wearable inertial energy harvester with sputtered bimorph lead zirconate titanate (PZT) Thin-Film Beams. Smart Mater Struct 27:85026

Zhu JX, Lin J, Yuksek NS, Almasri M, Feng ZC (2015) Dynamic phenomena and analysis of MEMS capacitive power harvester subjected to low-frequency excitations. Nonlinear Dyn 79:673–688

Part IV
CMOS MEMS

Chapter 6
CMOS Amalgamated MEMS/NEMS: Pioneering Technology for the Modern Semiconductor Cosmos and Future World

Soumya Sen, Abhishek Chauhan, Ashish Raman, and Mamta Khosla

Abstract This chapter explores the integration of MEMS/NEMS with CMOS technology, highlighting its potential to revolutionize various fields by merging micro/nano-electro-mechanical systems with CMOS circuits on a single chip. This integration allows for simultaneous use of mechanical, electrical, and nanoscale components, enabling the development of sophisticated systems. MEMS/NEMS with CMOS enhances sensitivity, precision, miniaturization, and power efficiency while improving signal processing. The fabrication process involves CMOS circuitry creation, followed by the addition of MEMS/NEMS structures, leading to compact and high-performance systems. Applications span biosensors, RF MEMS for wireless communication, nano-optics, and nanomechanical resonators for high-frequency oscillators. Additionally, CMOS-based MEMS/NEMS systems have gained prominence in sensor applications such as gas sensors, biosensors, and pressure sensors. While CMOS technology remains the most commercialized for its scalability and performance, new technologies like CNTFET, Nanowire FET, Nanosheet FET, and TFET are being explored to address the limitations of MOS technology at smaller scales. MEMS sensors stand out for their lightweight, portability, low cost, low power consumption, and high sensitivity, making them vital for modern systems and advanced analysis. This integration is set to play a crucial role in the development of beyond-Moore devices and future technological innovations.

S. Sen (✉) · A. Chauhan · A. Raman · M. Khosla
Department of ECE, Dr. B.R. Ambedkar, National Institute of Technology, Jalandhar, India
e-mail: sensoumya8730@gmail.com

A. Chauhan
e-mail: abhishekc.ec.21@nitj.ac.in

A. Raman
e-mail: ramana@nitj.ac.in

M. Khosla
e-mail: khoslam@nitj.ac.in

V. Pachkawade and K. Guha (eds.), *Micro- and Nano-Systems in 21st-Century*, Springer Tracts in Electrical and Electronics Engineering,
https://doi.org/10.1007/978-981-96-4659-3_6

Keywords MEMS/NEMS · Moore's law · CMOS · Biosensor · Nanowire

Introduction

Micro-Electro-Mechanical Systems (MEMS) and Nano-Electro-Mechanical Systems (NEMS) integrated with Complementary Metal–Oxide–Semiconductor (CMOS) technology represent a transformative and pivotal combination that is poised to shape the landscape of the future world across various industries. This convergence of disciplines is driving innovation and enabling the creation of incredibly compact, efficient, and powerful devices that have the potential to revolutionize diverse sectors, ranging from healthcare and communications to consumer electronics and beyond.

At the heart of this integration is the interlink between traditional semiconductor technology (CMOS) and the world of micro and nano-scale mechanical systems. CMOS technology forms the basis of modern digital electronics, facilitating the creation of transistors and integrated circuits that power our computing devices. MEMS and NEMS, on the other hand, introduce the dimension of motion and mechanical manipulation at incredibly small scales Fig. 6.1. depicts the integrated CMOS-MEMS hierarchical fusion. These systems can comprise sensors, actuators, switches, resonators, and more, all on the scale of micrometers or even nanometers. The synergy between MEMS/NEMS and CMOS offers several key advantages:

Miniaturization: By combining MEMS/NEMS with CMOS, it becomes possible to create highly miniaturized devices that can perform complex functions. This is crucial for applications where size and portability are essential, such as wearable devices, implantable medical devices, and sensor networks.

Integration: CMOS technology allows for the seamless integration of MEMS/NEMS components with electronic circuitry on a single chip. This integration streamlines manufacturing processes and enhances overall device performance, while reducing the need for external connections.

Efficiency: CMOS provides low-power electronics, and when coupled with MEMS/NEMS devices, this combination leads to energy-efficient systems. Such efficiency is vital for applications like Internet of Things (IoT) devices that need to operate on limited power sources for extended periods.

Sensing and Actuation: MEMS/NEMS devices excel at sensing various physical parameters like pressure, acceleration, temperature, and more. They can also actuate mechanical motion in response to electrical signals. When integrated with CMOS, these capabilities open up avenues for advanced sensors and actuators that drive innovations in robotics, automotive safety, environmental monitoring, and medical diagnostics.

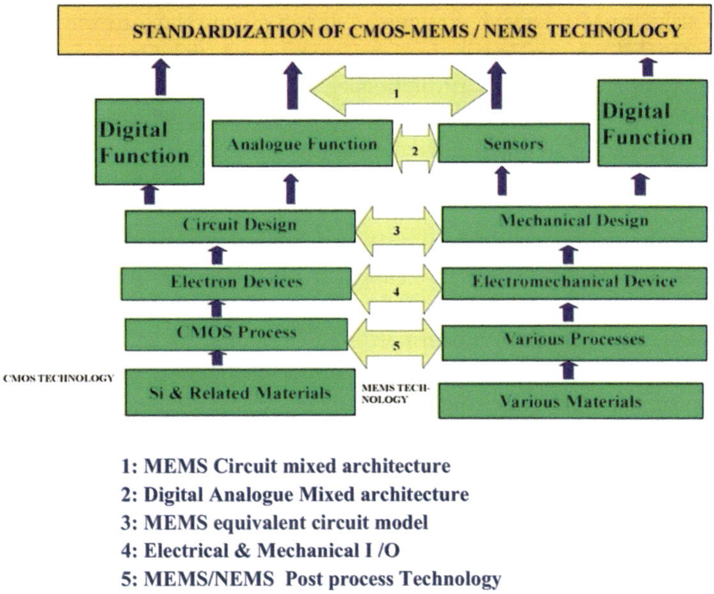

Fig. 6.1 Integration and Fabrication steps for CMOS-MEMS technology

Communication: The amalgam of MEMS/NEMS with CMOS can enable the development of novel communication devices. For instance, micro-mirrors can be used to modulate light for optical communication, leading to faster and more efficient data transmission.

Biomedical Applications: This technology fusion is transforming healthcare through the creation of lab-on-a-chip devices for rapid diagnostics, implantable devices for precise drug delivery, and wearable health monitors that can continuously gather vital signs.

Consumer Electronics: The integration of MEMS/NEMS with CMOS has contributed to the development of high-performance, compact microphones, accelerometers, gyroscopes, and more, enhancing the capabilities of smartphones, tablets, and other portable gadgets.

Automotive Industry: This combination is pivotal in the advancement of automotive safety systems, such as airbag deployment sensors, tire pressure monitoring systems, and advanced driver-assistance systems (ADAS).

As we move toward an era of pervasive connectivity, automation, and data-driven decision-making, the synergistic relationship between MEMS/NEMS and CMOS stands as a key enabler of this future world. Its influence extends across industries, driving advancements that promise to make our lives safer, more efficient, and more connected than ever before.

Micro-Electro-Mechanical Systems (MEMS) and Nano-Electro-Mechanical Systems (NEMS) integrated with Complementary Metal–Oxide–Semiconductor (CMOS) technology herald a revolutionary convergence that stands as a cornerstone in shaping the imminent future across a multitude of industries. This fusion of precision engineering at micro and nano scales with the efficiency and functionality of CMOS ushers in an era of unprecedented possibilities, propelling innovation to new horizons.

MEMS, miniature mechanical systems imbued with electronic capabilities, have already demonstrated their prowess in applications ranging from accelerometers and gyroscopes in smartphones to pressure sensors in medical devices. Their synergy with CMOS, the bedrock of modern electronic circuits, engenders a symphony of miniaturization, energy efficiency, and advanced functionality. By allowing MEMS devices to be intricately fabricated alongside electronic circuits on a single silicon wafer, the integration becomes seamless, reducing both size and power consumption while augmenting overall system performance.

NEMS, an evolution beyond MEMS, delves into the realm of nanoscale dimensions, where individual atoms and molecules orchestrate astonishing feats. Collaborating NEMS with CMOS fabricates a paradigm where infinitesimal mechanical structures interact harmoniously with nanoscale electronics. This fusion not only accentuates the sensitivity and precision of sensors but also begets transistors and logic elements operating at previously unfathomable scales. Consequently, this transformative marriage of NEMS and CMOS opens vistas for ultra-high-density data storage, ultra-sensitive biosensors, and even computation at the quantum scale.

Industries spanning from healthcare to telecommunications to automotive and beyond stand to reap the rewards of this synergistic convergence. In healthcare, the integration of MEMS and CMOS gives rise to an array of point-of-care diagnostic tools, revolutionizing patient monitoring and disease detection. In telecommunications, the amalgamation of NEMS and CMOS promises communication devices that exploit the full potential of the electromagnetic spectrum, leading to enhanced connectivity and data transfer rates.

Transportation, another beneficiary, witnesses the emergence of self-monitoring vehicles with MEMS-based sensors and CMOS-based control systems, driving us closer to the realization of autonomous driving. Moreover, the aerospace sector sees a transformation with miniaturized sensors that withstand extreme conditions, safeguarding aircraft and satellites alike. In the energy sector, MEMS-NEMS-CMOS integration nurtures efficient energy harvesting, optimizing power consumption and fostering sustainable practices.

However, this convergence is not without its challenges, as issues of power management, heat dissipation, and fabrication precision necessitate careful consideration (Failed 2019; Verd and Segura 2019; Nasrollahpour et al. 2020; Frey et al. 2009; Baltes et al. 2002; Qu 2016).

In summation, the intertwining of MEMS and NEMS with CMOS technology marks a pivotal juncture in the tapestry of technological advancement. It propels us toward a future defined by compactness, efficiency, and ingenuity across myriad sectors, heralding a new era where the inconceivable becomes reality and reshapes the very fabric of our existence (Table 6.1).

By integrating nanomaterials with MEMS, NEMS, and CMOS technology, a new class of devices has emerged with unprecedented capabilities and performance.

The integration of these technologies enables the development of sensors, actuators, and transducers that are not only incredibly small but also highly sensitive and responsive. For example, nanomaterial-enhanced MEMS/NEMS devices can detect minuscule changes in temperature, pressure, chemical composition, or even biological entities, making them invaluable in applications like environmental monitoring, medical diagnostics, and industrial process control.

One area where this integration has shown promise is in the field of biomedical devices. Nanomaterial-enhanced MEMS/NEMS devices integrated with CMOS technology can be used to create highly sensitive biosensors capable of detecting biomolecules at ultralow concentrations. These devices have the potential to revolutionize disease detection and monitoring by providing rapid, accurate, and cost-effective diagnostic tools.

Furthermore, the integration of nanomaterials with MEMS/NEMS and CMOS technology has opened up new avenues for energy harvesting and storage. Nanomaterials can enhance the efficiency of energy conversion processes, leading to the development of miniature power sources that can be integrated into wearable devices, wireless sensors, and even implantable medical devices.

Table 6.1 Recent Advances and Significance for different application area in MEMS/NEMS and CMOS hybrid technology

Serial number	Application area	Significance	Recent advances
1	Internet of Things (IoT)	Enabling smaller, more efficient sensors and actuators for ubiquitous connectivity and smart systems	Development of ultra-low-power MEMS sensors with on-chip CMOS readout circuits
2	Biomedical devices	Enhancing healthcare through miniaturized, implantable devices and lab-on-chip systems	Integration of MEMS-based sensors for monitoring vital signs. NEMS-based drug delivery systems
3	Consumer electronics	Enabling miniaturized, high-performance components for portable devices	MEMS-based microphones, accelerometers, and gyroscopes for smartphones
6	Energy harvesting	Converting ambient energy into usable electrical power	Integration of MEMS/ NEMS-based devices for vibration and thermal energy harvesting
7	Environmental monitoring	Providing real-time data on pollution, weather, and natural disasters	MEMS-based gas and chemical sensors for environmental sensing networks

In the realm of communication technology, the integration of nanomaterials with MEMS/NEMS and CMOS has led to the development of advanced optical and photonic devices. These devices can manipulate light at the nanoscale, enabling ultra-fast data transmission, high-resolution imaging, and optical computing.

In conclusion, the integration of Micro-Electro-Mechanical Systems (MEMS) and Nano-Electro-Mechanical Systems (NEMS) with Complementary Metal–Oxide–Semiconductor (CMOS) technology has provided a platform for the seamless integration of nanomaterials into a wide range of applications. This convergence of technologies has led to the creation of highly sensitive sensors, efficient energy devices, advanced communication tools, and innovative biomedical instruments that have the potential to reshape industries and improve the quality of life for individuals around the world.

CMOS-MEMS/NEMS Hybrid Systems

CMOS MEMS/NEMS hybrid systems epitomize a remarkable and visionary fusion, seamlessly amalgamating two frontiers of modern technology: the highly sophisticated Complementary Metal–Oxide–Semiconductor (CMOS) and the intricately intricate world of Micro/Nanoelectromechanical Systems (MEMS/NEMS). This union engenders a novel paradigm that unites the precision of electronic circuitry with the tangible and dynamic capabilities of micro and nanoscale mechanical structures. In essence, these systems embody a symphony of innovation, where the delicate precision of CMOS technology harmonizes with the exquisite artistry of MEMS/NEMS fabrication. CMOS, celebrated for its ubiquity in integrated circuits and microprocessors, empowers these hybrid systems with the computational prowess and electronic finesse that the modern world relies upon. Conversely, MEMS/NEMS introduces a realm of infinitesimal mechanical wonders, where structures as minuscule as a fraction of a human hair come to life with movement, resonance, and mechanical metamorphosis.

The convergence of CMOS and MEMS/NEMS is akin to orchestrating a grand collaboration between maestros from two disparate domains. This partnership unlocks possibilities that surpass the sum of its parts: it allows for the coexistence of intricate electronic pathways and delicate mechanical components on a single chip, akin to harmonizing an ensemble of diverse instruments in a symphony of technological marvels. His symbiosis gives birth to a host of applications, where sensors can detect the faintest nuances of the environment, actuation can bring about precise mechanical motion under digital command, and communication devices can transcend limitations in processing speed and signal quality. These applications span industries and domains, ranging from medical diagnostics that delve into the minutiae of the human body, to the intricacies of industrial automation, and the captivating cosmos of advanced communication networks. The cress section schematic is shown in Fig. 6.2.

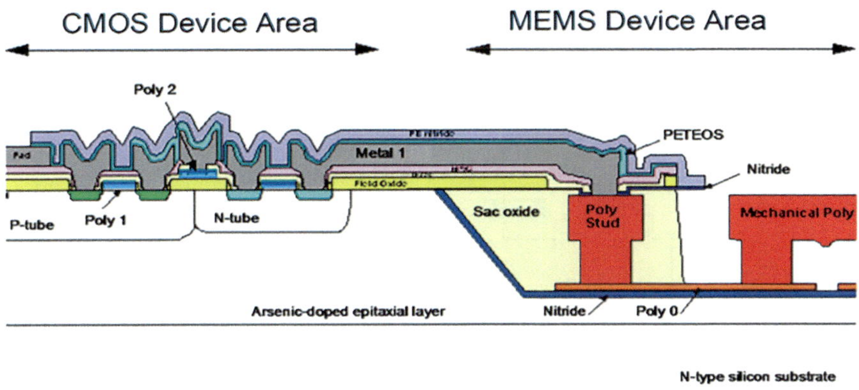

Fig. 6.2 Cross-section schematic of pre CMOS MEMS (Frey et al. 2009)

Furthermore, these hybrid systems address fundamental challenges in the technological landscape. By marrying the energy efficiency of CMOS with the potential for motion and adaptation inherent in MEMS/NEMS, a new era of power-conscious yet highly responsive devices emerges. This, coupled with the inherent miniaturization that the hybridization facilitates, redefines the boundaries of what can be achieved within the confines of a diminutive chip.

One of the key advantages of CMOS MEMS/NEMS hybrids is the ability to have electronic and mechanical components coexisting on a single chip. This physical proximity facilitates seamless communication between electronic and mechanical subsystems, enabling real-time interaction and feedback loops. For instance, a microsensor built on the same chip as CMOS circuitry can detect a physical parameter, such as temperature or pressure, and instantly feed that information into the electronic circuitry for further processing.

This integration opens the door to a plethora of applications. In the realm of sensors, CMOS MEMS/NEMS hybrids can create highly sensitive, miniature sensors capable of detecting even subtle changes in the environment. These sensors find utility in areas like environmental monitoring, medical diagnostics, and industrial automation.

Actuators, which convert electronic signals into mechanical motion, also benefit from this hybrid approach. These integrated devices can exhibit precise control over movement, enabling applications in robotics, microfluidics, and optical systems.

Communication devices, too, experience a transformation. The integration of MEMS/NEMS elements with CMOS technology can lead to more efficient signal processing, faster data transmission, and improved signal quality. This is particularly valuable for high-frequency communication systems, such as wireless communication and radar applications.

Moreover, these hybrid systems address challenges like power consumption and miniaturization. By integrating MEMS/NEMS components with CMOS, power-efficient designs become achievable. Figure 6.3 depicts the SRAM circuit with the

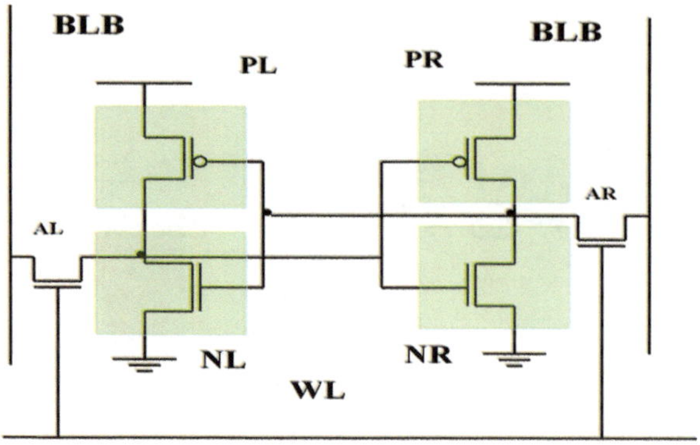

Fig. 6.3 Integrated CMOS MEMS/NEMS SRAM circuit

hybrid technology of MEMS/NEMS with CMOS, possessing the benefits of one chip for enhanced memory functionalities. MEMS/NEMS elements could interact with memory cells in novel ways, yielding unique architectures, better sensing, or energy-efficient designs. The compact nature of these systems lends itself to miniaturized devices that can be deployed in constrained spaces. In conclusion, CMOS MEMS/NEMS hybrid systems embody a captivating convergence of electronics and mechanics. This synergy not only unlocks novel functionalities but also paves the way for innovative solutions across various domains, from healthcare to telecommunications, revolutionizing the way we perceive and interact with technology.

CMOS MEMS/NEMS hybrid systems stand as a testament to an extraordinary and forward-looking synthesis, effortlessly harmonizing two frontiers of contemporary technology: the exquisitely refined Complementary Metal–Oxide–Semiconductor (CMOS) realm and the mesmerizingly intricate domain of Micro/Nanoelectromechanical Systems (MEMS/NEMS). This fusion not only underscores the marriage of precision electronics and mechanical prowess but also encapsulates the embodiment of innovation itself, transcending the boundaries of what was once thought achievable.

The intricate dance between CMOS and MEMS/NEMS engenders a synergy that transforms possibilities into realities. CMOS, a stalwart in the realm of integrated circuits, harnesses semiconductor properties to fashion intricate electronic pathways that enable computation, signal processing, and communication. Conversely, MEMS/NEMS traverse the delicate micro and nano domains, crafting mechanical structures that respond to forces, vibrations, and even the faintest touch. When these two dimensions interweave, they spark a symphony that orchestrates capabilities far beyond the sum of their parts.

CMOS, with its precision manufacturing and electronic elegance, becomes the canvas upon which MEMS/NEMS paint their art. These mechanical masterpieces, carved with the precision of nanometer scales, impart unprecedented functionalities to the electronic landscape. Sensors and actuators, once tethered to standalone devices, now coexist seamlessly on a single chip, birthing a generation of integrated systems that redefine compactness and efficiency.

The innovation unfurls further as CMOS MEMS/NEMS hybrid systems venture into the uncharted territories of quantum effects and emergent phenomena. Quantum dots integrated into CMOS circuitry blur the lines between classical and quantum computing, promising exponential leaps in processing power. Simultaneously, the harmonious dance of MEMS/NEMS with quantum bits (qubits) orchestrates a choreography where quantum mechanics waltz with mechanical vibrations, opening doors to quantum sensing and information processing paradigms that challenge convention.

Industries across the spectrum stand to benefit profoundly from this convergence. Consumer electronics embrace more responsive touchscreens, immersive virtual reality experiences, and augmented reality applications that meld seamlessly with the physical world. Biomedical engineering embraces personalized diagnostics, ushering in an era where wearable health-monitoring devices interweave with our daily lives, seamlessly transmitting vital data to medical professionals. The automotive sector embraces enhanced safety through intelligent collision detection and predictive maintenance systems, all nested within the compact space of a CMOS MEMS/NEMS hybrid chip.

Nevertheless, this union of CMOS and MEMS/NEMS does not come without its share of challenges. Managing power distribution, mitigating thermal effects, and navigating the intricacies of nanoscale fabrication demand an orchestra of expertise across disciplines. Yet, it is precisely this multidisciplinary symphony that fuels the momentum of innovation, inviting collaborations that redefine the boundaries of human ingenuity.

The convergence of CMOS and MEMS/NEMS into hybrid systems is a grand opus of technology, composing a narrative of integration, elegance, and limitless potential. It signifies not only the culmination of scientific progress but also the inception of a new chapter where intricate mechanics meld seamlessly with elegant electronics, redefining the landscape of human interaction, computation, and exploration.

At the heart of the technological revolution, CMOS MEMS/NEMS hybrid systems emerge as a magnificent embodiment of visionary convergence, seamlessly uniting two distinct frontiers of modern innovation: the meticulously refined domain of Complementary Metal–Oxide–Semiconductor (CMOS) technology and the captivatingly intricate realm of Micro/Nanoelectromechanical Systems (MEMS/NEMS). This harmonious fusion doesn't merely represent a mere junction of technologies; it encapsulates the very essence of progress, transcending conventional boundaries and igniting a beacon of possibilities that illuminate the path toward an awe-inspiring future.

CMOS, the cornerstone of digital electronics, wields silicon's transformative capabilities to orchestrate the intricate symphony of transistors and logic gates. It enables the art of computation, data processing, and communication to flourish, birthing the digital age that shapes our world. Meanwhile, MEMS/NEMS ventures into the realm of the infinitesimally small, crafting mechanical systems with the precision of a watchmaker at the nanoscale. These structures resonate with the energy of the cosmos and respond to the gentlest touch, introducing a realm of sensors, actuators, and mechanical marvels that elevate technology to an art form.

In this union, CMOS MEMS/NEMS hybrid systems illuminate a pathway toward innovation that resonates with both finesse and intricacy. CMOS acts as the stage where MEMS/NEMS unveil their brilliance, providing a canvas on which mechanical artistry finds its electronic counterpart. Miniaturized sensors and micro-actuators seamlessly coexist with intricate electronic circuits, leading to devices that effortlessly straddle multiple dimensions of functionality and utility.

This harmonious marriage extends beyond the visible spectrum, delving into the ethereal domain of quantum phenomena. Quantum dots integrated within CMOS components blur the lines between classical and quantum computing, promising computational capabilities that defy conventional limits. Simultaneously, the delicate dance of MEMS/NEMS with quantum states unlocks new dimensions in sensing, enabling the detection of minute changes in physical properties that were once beyond the realm of detection.

Industries, from healthcare to communication, from aerospace to energy, are poised to reap transformative benefits from this synergy. Wearable health monitors gain newfound accuracy, as MEMS/NEMS sensors integrated with CMOS technology provide real-time data streams for early diagnosis and precision treatment. Telecommunication devices become more versatile, combining MEMS/NEMS-based filters with CMOS amplifiers to enhance signal processing and network efficiency. Moreover, space exploration benefits from miniaturized, resilient MEMS/NEMS components that withstand the rigors of outer space, paired with CMOS control systems that navigate and communicate with unparalleled accuracy.

In the midst of these soaring opportunities, the amalgamation of CMOS and MEMS/NEMS isn't without its challenges. Power management, thermal dissipation, and nanoscale fabrication intricacies demand careful orchestration. However, these obstacles are the crucibles in which innovation thrives, fostering collaborative efforts and interdisciplinary solutions that redefine technological boundaries.

In summation, the nexus of CMOS and MEMS/NEMS within hybrid systems is an orchestration of the future, a tapestry woven from threads of electronic sophistication and mechanical ingenuity. It symbolizes not just a convergence of technologies, but a celebration of human creativity, curiosity, and resilience. As CMOS MEMS/NEMS hybrid systems continue to shape the landscape of possibility, they beckon us to journey deeper into the uncharted territories of the technological cosmos, inspiring us to create, explore, and imagine beyond the limits of today (Verd and Segura 2019; Hagleitner and Kirstein 2005; Riverola et al. 2016; Jo and Choi 2018; Riverola Borreguero 2018; Islam 2020; Lin and Dai 2018; Li et al. 2020; Session 2005).

More Than Moore Era

The "More than Moore" era represents a paradigm shift in semiconductor technology beyond the traditional scaling of complementary metal–oxide–semiconductor

(CMOS) devices. In this era, the focus shifts from merely reducing transistor size to exploring new ways to integrate various functionalities, materials, and technologies on a single chip. The integration of CMOS technology with MEMS (Micro-Electro-Mechanical Systems) and NEMS (Nano-Electro-Mechanical Systems) in hybrid systems embodies the essence of the More than Moore approach. Let's delve into how the More than Moore era is shaping CMOS-MEMS/NEMS hybrid systems:

Diversification of Integration

In the More than Moore era, the goal is to integrate diverse technologies and functionalities beyond just digital logic. The integration of CMOS with MEMS/NEMS components introduces mechanical and sensing capabilities, enabling chips to perform tasks beyond traditional computation. This diversification opens up new opportunities for applications ranging from healthcare to the Internet of Things (IoT).

Enhanced Functionality

CMOS-MEMS/NEMS hybrid systems offer enhanced functionality by combining electronic, mechanical, and sensing components. For example, a single chip can integrate sensors, actuators, and control electronics, enabling real-time feedback and dynamic control in applications like robotics and wearable devices.

Sensing and Actuation Synergy

MEMS/NEMS components offer the ability to sense and actuate the physical world, while CMOS provides the processing power to interpret and respond to these inputs. This synergy leads to real-time closed-loop control systems that are capable of making rapid decisions based on sensor data.

Miniaturization and Integration

The More than Moore era encourages the miniaturization and integration of multiple functionalities on a single chip. CMOS-MEMS/NEMS hybrid systems achieve this by combining electronic, mechanical, and sensing components on the same chip, resulting in smaller, more compact devices with improved performance.

Energy Efficiency and Autonomous Systems

The integration of MEMS/NEMS components allows for energy-efficient operations. For instance, sensors can be optimized for low-power consumption, and energy harvesting from the environment can power both CMOS and MEMS/NEMS components. This is critical for the development of autonomous systems, wearables, and IoT devices.

Novel Materials and Fabrication Techniques

The More than Moore era encourages the exploration of novel materials and fabrication techniques to realize advanced devices. In CMOS-MEMS/NEMS hybrid systems, this could involve integrating nanomaterials, such as carbon nanotubes or graphene, as sensing elements or for mechanical reinforcement.

Advanced Sensing and Control Applications

CMOS-MEMS/NEMS hybrid systems enable advanced sensing and control applications. For instance, in healthcare, these systems can provide real-time monitoring of physiological parameters and deliver targeted therapies. In industrial settings, they can enable precise control of manufacturing processes and automation.

Multidisciplinary Collaboration

The success of CMOS-MEMS/NEMS hybrid systems in the More than Moore era relies on collaboration between experts from various fields, including electronics, mechanics, material science, and software development. Integrating such diverse expertise is essential for addressing the challenges associated with these complex systems.

The More than Moore era represents a shift from solely focusing on transistor scaling to exploring new avenues for integration and functionality. CMOS-MEMS/NEMS hybrid systems exemplify this shift by combining electronic and mechanical components to create multifunctional devices with enhanced capabilities. This approach is driving innovation in various industries and paving the way for new applications that were previously unattainable with traditional CMOS technology alone.

It signifies a pivotal juncture in the evolution of Complementary Metal–Oxide–Semiconductor (CMOS) technology, where the trajectory of innovation shifts from simply scaling down transistor dimensions, as per Moore's Law, to encompass a broader range of advancements that extend beyond conventional miniaturization. In this era, the focus pivots toward enhancing chip functionalities, performance, and capabilities by integrating diverse components and technologies alongside traditional CMOS processes.

At its core, the "More than Moore" philosophy acknowledges the physical limitations and challenges associated with shrinking transistors to nanoscale dimensions. It recognizes that while transistor scaling remains crucial, there are inherent constraints imposed by quantum effects, power dissipation, and other physical phenomena. As a result, the emphasis expands to encompass a multitude of strategies that harness new materials, architectures, and technologies to enhance chip performance and capabilities in diverse ways. One of the key pillars of the "More than Moore" era is the integration of heterogeneous components and systems on a single chip. This involves combining CMOS technology with a diverse array of functionalities, such as sensors, actuators, radio frequency (RF) devices, and photonics, to create hybrid systems that offer enhanced features and novel capabilities. This integration allows for synergistic interactions between disparate technologies, enabling advancements in areas like healthcare, automotive, telecommunications, and beyond. The integration of MEMS (Microelectromechanical Systems) and NEMS (Nanoelectromechanical Systems) components with CMOS technology exemplifies this paradigm. By embedding mechanical structures within the same chip as electronic circuitry, these hybrid systems can enable a new class of sensors, actuators, and devices that harness both electronic and mechanical functionalities. This synergy leads to improved performance, reduced power consumption, and innovative applications such as microsensors, microfluidics, and wearable devices. The pursuit of higher levels of integration also brings forth the concept of System-on-Chip (SoC) and System-in-Package (SiP) solutions, where multiple functionalities are combined into compact arrangements.

This leads to improved system performance, reduced power consumption, and optimized chip area utilization, fostering the development of highly efficient and compact electronic systems. In conclusion, the "More than Moore" era embodies a holistic shift in CMOS technology, acknowledging that the trajectory of progress extends beyond traditional transistor scaling. By integrating diverse technologies, materials, and components alongside CMOS processes, this era opens doors to unprecedented advancements in a myriad of applications. From enabling new generations of sensors and actuators to facilitating innovative packaging solutions and specialized devices, the "More than Moore" era presents a rich landscape of possibilities that redefine how we harness the power of CMOS technology for the challenges and opportunities of the modern world.

The emergence of the "More than Moore" era heralds a remarkable transformation akin to the mythical rebirth of the phoenix, breathing new life into the trajectory of CMOS technology. Figure 6.4 shows the variation of the beyond Moore era with technology nodes. In this epochal shift, the conventional boundaries of innovation are shattered, propelling CMOS technology into a dimension where creativity flourishes beyond the confines of silicon's atomic structure. This metaphorical phoenix rises from the ashes of Moore's Law, transcending the well-trodden path of linear scaling and ascending to a realm where imagination takes flight, and possibilities are as boundless as the sky.

With renewed vitality, the CMOS technology of the "More than Moore" era undergoes a renaissance. The constraints of silicon's atomic lattice, once seen as insurmountable, are now transcended through ingenious feats of engineering and materials science. Exotic materials, such as gallium nitride and silicon carbide, infuse new life into transistors, amplifying their efficiency and performance. These materials serve

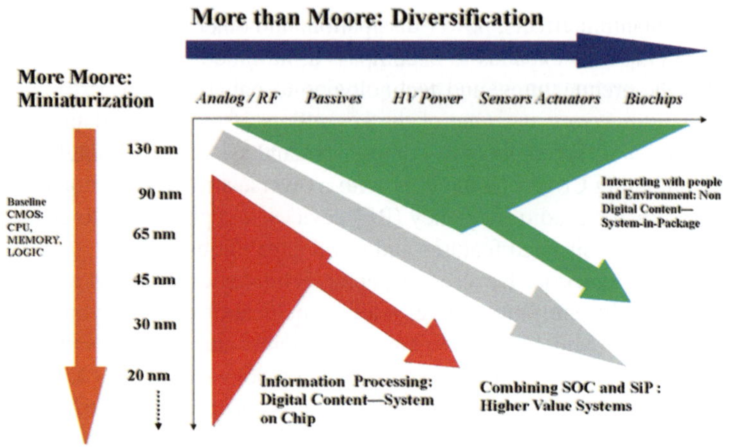

Fig. 6.4 Beyond Moore's law on the basis of technology nodes

as the phoenix's feathers, enabling devices to soar to new heights previously deemed unattainable.

This transcends the atomic structure of silicon by embracing heterogeneity and collaboration. Interdisciplinary partnerships between electronics, photonics, mechanics, and more converge, akin to the diverse elements that compose the phoenix's majestic form. The amalgamation of disparate technologies empowers CMOS devices with multifunctional capabilities, enabling them to simultaneously compute, sense, and communicate, much like the phoenix embodies myriad qualities in its graceful flight.

The innovations of the era stand as rare gems of human ingenuity. Quantum effects, which were once considered esoteric phenomena, now entwined with CMOS circuits, bestowing them with the ethereal qualities of quantum computation and sensing. This integration defies the conventional boundaries of silicon's behavior, reminiscent of the phoenix's ability to transcend the natural cycle of life and death.

The "More than Moore" symbolizes the expansion of CMOS technology into new dimensions. Advanced packaging techniques create three-dimensional landscapes, stacking chips upon chips like the layers of the phoenix's plumage, defying the flat confines of traditional scaling. This vertical integration not only maximizes space but also rejuvenates the potential for innovation.

As the CMOS technology soars, it illuminates the horizon with a spectrum of possibilities that extend beyond the limits of imagination. Devices tailored for specific industries emerge like the phoenix's unique plumage, each feather a distinct solution (Roy et al. 2013; Fang 2013; Maleville 2015; Wise 2010; Simon 2021; Menon et al. 2023).

In sum, the "More than Moore" era orchestrates a transformation, rekindling the fire of innovation within CMOS technology. CMOS technology ascend to unprecedented heights, embracing diversity, interdisciplinary, and quantum enchantment.

Just as the phoenix's flight ignites wonder and admiration, so too does the evolution of CMOS technology inspire awe, propelling us toward a future where creativity knows no bounds and innovation reigns supreme.

CMOS-Compatible Fabrication Processes

A CMOS-compatible fabrication process is a sophisticated and meticulously designed manufacturing approach that adheres to the principles and standards of Complementary Metal–Oxide–Semiconductor (CMOS) technology. This process is specifically engineered to create integrated circuits and devices that seamlessly integrate with existing CMOS infrastructures, ensuring compatibility, reliability, and optimal performance. At its core, a CMOS-compatible fabrication process aligns with the foundational elements of CMOS technology, which is the bedrock of modern digital electronics. It involves intricate steps that encompass both the creation of active electronic components, such as transistors and diodes, and the passive components like interconnects, capacitors, and resistors that form the intricate web of a functional integrated circuit. Starting with a substrate, often made of silicon, the CMOS-compatible process involves the formation of intricate structures using a combination of photolithography, chemical vapor deposition, etching, and doping techniques. Each layer of the integrated circuit is meticulously patterned and crafted, with each step conforming to the stringent requirements of CMOS standards.

The term "compatible" is critical here. The term "compatible" assumes paramount significance within the context of the integration between Complementary Metal–Oxide–Semiconductor (CMOS) technology and Micro/Nanoelectromechanical Systems (MEMS/NEMS). This compatibility serves as the cornerstone, ensuring a harmonious convergence of two distinct realms, while safeguarding the integrity and performance of both CMOS and MEMS/NEMS components.

A "CMOS-compatible" process acts as a vigilant custodian, meticulously orchestrating the steps of integration to prevent any potential disruption to the underlying CMOS structures. The intricacies of material deposition, etching, and patterning are choreographed with meticulous precision to mitigate any adverse effects that could compromise the functionality of CMOS elements. This compatibility extends beyond mere coexistence; it signifies a seamless collaboration where the sum of the parts is greater than its individual components.

Central to this compatibility is the vigilant effort to eliminate cross-contamination, a threat that could arise due to the introduction of MEMS/NEMS materials or processing steps into the CMOS fabrication environment. Specialized protocols and equipment are employed to ensure that particles, contaminants, or unwanted materials do not infiltrate the CMOS ecosystem, preserving its purity and performance.

Furthermore, the compatibility encompasses thermal considerations. The delicately balanced thermal profiles that govern CMOS fabrication must not be disrupted by the introduction of MEMS/NEMS processes. Temperature-sensitive CMOS components require a harmonious environment, shielding them from potential thermal damage that could arise from unfamiliar fabrication steps. This is akin to orchestrating a symphony where different instruments must harmonize without overpowering one another.

The meticulous design of a CMOS-compatible process takes into account potential interferences that could arise from the integration itself. These interferences might manifest as unintended electrical interactions, mechanical disturbances, or even changes in material properties that result from the convergence of the two domains. Rigorous testing, simulation, and optimization become imperative to ensure that the end product maintains the performance and reliability standards expected of both CMOS and MEMS/NEMS technologies.

In essence, the notion of compatibility serves as the guiding principle, preserving the integrity of CMOS components while forging a partnership with MEMS/NEMS structures. This compatibility ensures that the integration is not a compromise, but an enhancement, where the innovative capabilities of both realms are harnessed without sacrificing performance or reliability. As the dance of technology advances, "CMOS-compatible" becomes the assurance of harmonious coexistence, ushering in a new era where the lines between electronics and mechanics blur, and the synergy between the two elevates the realm of what is achievable. Figure 6.5 depicts the CMOS-compatible fabrication process. A CMOS-compatible process ensures that the steps involved, including material deposition and patterning, do not adversely affect the underlying CMOS structures. The process is designed to avoid cross-contamination, thermal damage, or other interference that could compromise the functionality of the CMOS components. Moreover, a CMOS-compatible process extends its reach beyond just electronic components. It accommodates the integration of other functionalities, such as Microelectromechanical systems (MEMS) and Nanoelectromechanical systems (NEMS), as well as sensors, actuators, and communication devices. This integration is achieved through careful consideration of the materials used, the thermal and mechanical stresses they might induce, and the potential for interference with the CMOS circuits. In addition, it takes into account factors such as power consumption, heat dissipation, and device performance to ensure that the resulting devices are not only functional but also optimized for their intended applications. The process often involves stringent quality control and testing methodologies to verify the reliability and consistency of the fabricated devices. In the grand tapestry of modern electronics, a CMOS-compatible fabrication process serves as the canvas upon which intricate and diverse functionalities are woven together. It empowers the creation of integrated circuits that blend the power of CMOS electronics with additional features, fostering a landscape of innovation that addresses a wide array of applications, from communication to healthcare, from computing to automation (Ray et al. 2008; Gong and Hogan 2014; Izhaky et al. 2006; Ginet et al. 2011; Schropfer et al. 2004).

This process operates as a meticulous custodian, assuming the role of a skilled conductor who orchestrates a complex symphony of integration without allowing a single note of discord to disrupt the harmonious composition of the underlying CMOS structures. This orchestration is not merely a mechanical exercise; it's a

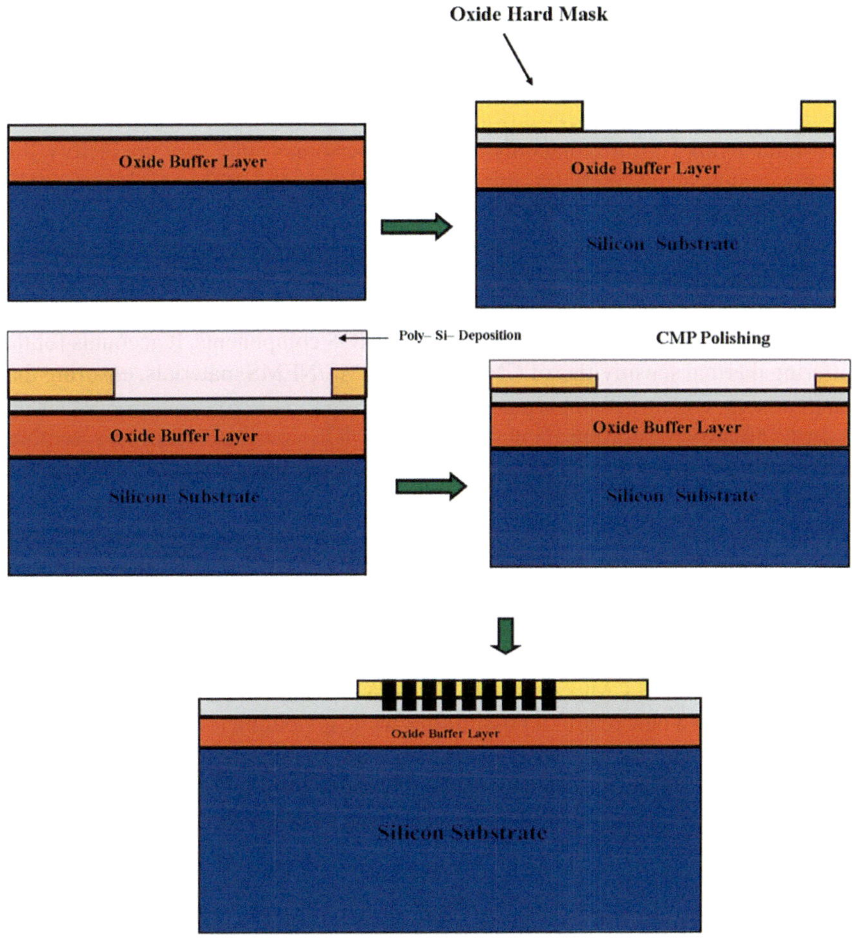

Fig. 6.5 CMOS compatible fabrication process flow

carefully choreographed dance that ensures the seamless integration of disparate elements while safeguarding the delicate balance of CMOS technology.

Imagine this process as a virtuoso conductor who leads an orchestra of diverse instruments, each with its unique timbre and character. Just as a conductor's baton guides musicians, the CMOS-compatible process directs the integration steps, guiding the intricate dance of materials and processes. Every deposition, etching, and patterning maneuver is executed with surgical precision, akin to an ensemble's synchronization in a grand symphony.

Like an astute custodian of a historical masterpiece, the CMOS-compatible process anticipates potential disruptions before they can manifest. It employs a meticulous screening process, examining each fabrication step under a microscope of scrutiny. It assesses the impact of introducing new materials or processes into the hallowed CMOS environment, ensuring that no element could potentially undermine the structural or functional integrity of the CMOS components.

In this role, the CMOS-compatible process stands as a gatekeeper against cross-contamination, ensuring that no unwanted elements infiltrate the realm of CMOS. Just as a vigilant guard ensures that only authorized personnel enter a secure facility, this process guards against particles, impurities, or alien materials that could potentially compromise the sanctity of CMOS structures.

Furthermore, the custodial nature of a "CMOS-compatible" process extends to the realm of thermal management. Just as a conservator monitors the environment of a delicate artifact, the process scrutinizes thermal conditions to prevent any disruptions that could jeopardize the nuanced balance of CMOS components. It accounts for the differing thermal sensitivities of CMOS and MEMS/NEMS materials, ensuring that temperatures remain within permissible bounds.

In the grand orchestration of integration, the CMOS-compatible process resembles a seasoned conductor who doesn't just guide but also anticipates. It navigates the terrain of potential interferences, adjusting the tempo, and refining the nuances to ensure that the symphony plays out flawlessly. Rigorous testing, simulation, and validation become the tools of this custodial craft, validating that the integration adheres to the stringent standards of both CMOS and MEMS/NEMS technologies.

Eventually, a "CMOS-compatible" process is a testament to meticulousness, an exemplar of precision, and a guarantor of harmony. Its vigilant custodianship ensures that the integration of CMOS with MEMS/NEMS doesn't merely coexist but flourishes, unlocking a new realm of possibilities while upholding the pristine essence of CMOS technology. Just as a conductor's art elevates a musical composition, so does the CMOS-compatible process elevate the technological symphony, orchestrating a performance where innovation and integration are perfectly in tune.

Real-Time Sensing and Control

Real-time sensing and control in CMOS-MEMS/NEMS (Micro/Nanoelectromechanical Systems) hybrid systems represent a dynamic synergy that capitalizes on the convergence of electronics and mechanical elements. This integration enables the instantaneous detection, analysis, and response to various physical phenomena, delivering a new dimension of functionality and efficiency in a multitude of applications. At the heart of this concept lies the capability to sense changes in the environment or parameters in real time. Sensors integrated within the CMOS-MEMS/NEMS hybrid system can detect factors such as temperature, pressure, acceleration, humidity, and more. The captured data is rapidly converted into electronic signals by the CMOS circuitry, enabling instantaneous analysis and interpretation. The interaction doesn't stop at mere sensing; the integration of MEMS/NEMS components provides the means to actuate mechanical responses based on the sensed data. For instance, if a CMOS-MEMS/NEMS hybrid system detects an increase in temperature, it can trigger a mechanical actuator to adjust a cooling mechanism or change the state of a microvalve for precise temperature control. This closed-loop system, where sensing and actuation occur seamlessly, lends itself to applications like smart thermostats, environmental monitoring, and even medical devices. The real-time aspect of CMOS-MEMS/NEMS systems is pivotal in time-sensitive applications. Consider a scenario where an autonomous vehicle equipped with these systems detects an obstacle in its path. The real-time sensing and processing enable instant decision-making, triggering appropriate mechanical responses to avoid collisions. In communication systems, real-time control mechanisms optimize signal processing, leading to enhanced data throughput and reliability.

The coexistence of CMOS and MEMS/NEMS elements also enables advanced control strategies. Algorithms embedded in the CMOS circuitry can analyze the sensed data and make decisions based on predefined conditions. These decisions then guide the mechanical actuators to execute precise movements or adjustments. This intricate dance between electronics and mechanics can be seen in applications like image stabilization in cameras, where CMOS-MEMS/NEMS systems work together to counteract unwanted vibrations and movements. Furthermore, real-time sensing and control enhance the efficiency and accuracy of medical devices. For example, in insulin pumps for diabetes management, these hybrid systems can sense blood glucose levels in real time and automatically adjust the insulin delivery rate, ensuring optimal blood sugar control. In sum, real-time sensing and control in CMOS-MEMS/NEMS hybrid systems amplify the capabilities of both electronics and mechanics, resulting in a dynamic interplay that responds instantaneously to changes in the environment or desired parameters. This fusion extends across a vast spectrum of applications, revolutionizing industries and enhancing the way we interact with technology in the modern world Figure 6.6.

The dynamic synergy achieved through real-time sensing and control in CMOS-MEMS/NEMS (Micro/Nanoelectromechanical Systems) hybrid systems unveils a realm of technological marvel that harnesses the convergence of electronics and mechanical intricacies. This symbiotic relationship gives rise to a harmonious interplay, where the finesse of microelectronics and the tangible precision of mechanical structures blend seamlessly, opening doors to a universe of unprecedented possibilities and applications.

At the core of this synergy lies the capacity for real-time sensing. CMOS technology, with its precision fabrication and signal processing prowess, acts as the neural network of the system, enabling the rapid acquisition, interpretation, and transmission of data. MEMS/NEMS components, on the other hand, serve as the system's sensory organs, translating physical phenomena into measurable signals. Be it the gentlest vibration, a minute pressure change, or a subtle temperature shift, these mechanical structures capture the essence of the physical world and channel it into a language comprehensible to the digital domain.

Micro-Electro mechanical system is a technique that is used to create micro scale integrated devices like a chip, in which we can combine both mechanical & Electrical Components. The existence of features smaller than 100 nm that are not produced using conventional machining but rather employing different methods known as

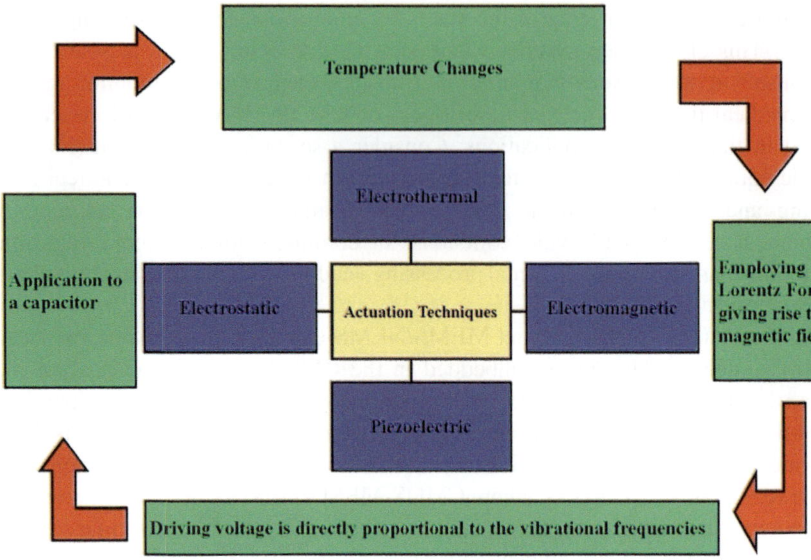

Fig. 6.6 Real time sensing scenario for CMOS-MEMS hybrid devices

micro-fabrication technology appears to be a common trait among these devices. This straight forward definition would also apply to microelectronics, but MEMS have a property that electronic circuits do not. MEMS feature holes, cavities, channels, cantilevers, membranes, etc., and in some ways resemble 'mechanical' parts, whereas electrical circuits are by nature solid and compact structures.

This amalgamation extends beyond mere sensing, delving into the realm of real-time control. As the CMOS circuits assimilate the sensory input, they orchestrate instantaneous responses, orchestrating a dance of precise actions guided by the mechanical realm's cues. This control mechanism drives actuators, tuning parameters, and manipulating physical processes with a finesse that was once inconceivable. The result is a symphony of feedback loops, where the digital and mechanical components harmonize to maintain stability, optimize efficiency, and even correct deviations in real time.

The implications of this synergy span across industries with profound implications. In the realm of healthcare, CMOS-MEMS/NEMS systems yield exquisitely accurate diagnostics and personalized treatment options. Microsensors delicately measure biological indicators, while CMOS circuits analyze the data and transmit real-time information to medical professionals. Moreover, this integration is instrumental in advanced prosthetics, where seamless interactions with the human body are enabled, restoring mobility and sensory perception.

In the automotive sector, CMOS-MEMS/NEMS hybrid systems drive the evolution of autonomous vehicles. Sensors monitor road conditions, traffic patterns, and vehicle parameters, while CMOS technology processes the data, enabling the vehicle to respond to complex scenarios in real time. This synthesis of sensing and control underpins not only safety mechanisms but also the dream of fully self-driving cars.

Energy optimization is another beneficiary, as CMOS-MEMS/NEMS systems enable real-time adjustments in power generation and consumption. Smart grids, empowered by this synergy, efficiently balance demand and supply, reacting promptly to changes in usage patterns or renewable energy availability.

As this synergy evolves, it gives birth to an ecosystem where the once disparate domains of electronics and mechanics merge into a seamless continuum. Challenges such as power management, integration complexities, and ensuring robustness against external factors become the focal points of innovation. Collaborations across disciplines flourish, nurturing a landscape where engineers, physicists, biologists, and more unite to create systems that embrace the intricacies of both the digital and physical worlds.

In essence, the real-time sensing and control realized within CMOS-MEMS/NEMS hybrid systems encapsulate the pinnacle of harmonious collaboration between electronics and mechanics. This union redefines the boundaries of human achievement, fostering a future where technologies anticipate, adapt, and respond in ways that mirror the elegance of the natural world.

The harmonious coexistence of Complementary Metal–Oxide–Semiconductor (CMOS) and Micro/Nanoelectromechanical Systems (MEMS/NEMS) elements within integrated systems doesn't just culminate in seamless integration; it extends its influence into the realm of advanced control strategies, imbuing devices with a level of intelligence and autonomy that fundamentally transforms their capabilities. This convergence signifies the dawn of a new era where algorithms embedded in the CMOS circuitry acquire, analyze, and act upon the sensed data in a proactive and informed manner.

At the heart of this transformation lies the inherent computational power of CMOS technology. This powerful computing platform integrates digital intelligence with the precision of MEMS/NEMS sensors and actuators. Algorithms within the CMOS circuitry serve as the neural pathways, processing the continuous stream of data garnered by MEMS/NEMS elements with unparalleled speed and accuracy. By dissecting patterns, identifying trends, and discerning anomalies, these algorithms make real-time decisions that optimize device performance or trigger responsive actions.

This ability for data analysis is not limited to standalone devices but also extends to networked systems. In the context of the Internet of Things (IoT), CMOS-MEMS/NEMS hybrid systems become nodes in an interconnected web, sharing information and insights across devices. Algorithms, once relegated to data centers, now inhabit the very heart of these devices, enabling distributed decision-making that enhances the efficiency, reliability, and overall functionality of the entire network.

The marriage of algorithms with MEMS/NEMS-CMOS systems finds applications across industries. In healthcare, these systems facilitate early disease detection and treatment optimization. For instance, MEMS/NEMS sensors could monitor vital signs, while CMOS-based algorithms continuously analyze the data to identify deviations from normal health parameters. This timely analysis could trigger alerts to medical professionals, leading to swift interventions and improved patient outcomes.

The automotive sector leverages this synergy for safety and efficiency gains. MEMS/NEMS sensors capture real-time information about road conditions, weather, and vehicle performance, while CMOS-embedded algorithms decipher this data to inform adaptive cruise control, collision avoidance, and even predictive maintenance systems.

Furthermore, energy management benefits from this integration. Smart grids, powered by CMOS-MEMS/NEMS devices, can respond dynamically to fluctuations in energy demand and supply.

The integration of CMOS-MEMS/NEMS devices presents a transformative opportunity that extends its positive influence to the realm of energy management, particularly within the dynamic context of smart grids. As the world transitions toward more sustainable and efficient energy consumption, this convergence emerges as a pivotal enabler, propelling the evolution of energy distribution and consumption into a new era of responsiveness and optimization.

Smart grids, often envisioned as the cornerstone of the modern energy landscape, emerge as the ideal canvas for the integration of CMOS-MEMS/NEMS devices. These grids represent a departure from the traditional, unidirectional energy flow, instead fostering a bidirectional exchange of information and power between producers and consumers. The infusion of CMOS-MEMS/NEMS technologies into this framework injects a newfound intelligence and adaptability that revolutionizes energy management.

Considering these smart grids as a living organism, with CMOS-MEMS/NEMS devices functioning as sensory receptors and decision-making centers. These devices serve as eyes and ears, capturing real-time data on energy demand, supply, and grid conditions. Whether it's a surge in consumption during peak hours, fluctuations due to renewable energy sources, or unexpected technical issues, these devices are poised to sense and quantify these changes with remarkable precision.

The integration goes beyond mere sensing; it empowers these devices to act as intelligent decision-makers. The CMOS circuitry embedded within the system processes the sensory data with astounding speed, leveraging advanced algorithms to analyze patterns and trends. This analysis fuels informed decisions, optimizing energy distribution, predicting demand spikes, and adapting the grid to varying conditions. It's akin to a neural network, where data flows, computations are performed, and decisions are made in real time to ensure the grid operates efficiently and reliably.

This level of responsiveness manifests as a two-fold benefit. First, the ability to respond dynamically to energy fluctuations ensures efficient utilization of resources. Surplus energy generated by solar panels, wind turbines, or other renewable sources can be swiftly directed to areas experiencing high demand. Conversely, during low-demand periods, energy can be stored or redirected to ensure minimal wastage. This optimization not only reduces costs but also minimizes the environmental footprint associated with energy production.

Secondly, the integration contributes to grid stability. In cases of unexpected spikes in demand or unforeseen disruptions, CMOS-MEMS/NEMS devices can trigger immediate responses. Faults can be isolated, and power rerouted to minimize the impact of outages. The grid becomes self-healing and adaptive, ensuring an uninterrupted energy supply to consumers.

However, achieving this integration is not without its complexities. Coordinating the communication between a multitude of CMOS-MEMS/NEMS devices, ensuring their reliability and accuracy, and managing potential data privacy and security concerns necessitate robust design and implementation strategies. Algorithms analyze consumption patterns, anticipate peaks, and optimize energy distribution, reducing waste and promoting sustainable practices.

The integration of algorithms within CMOS-MEMS/NEMS systems isn't without its complexities. Ensuring low-power consumption and efficient memory utilization, while also accounting for real-time processing demands, requires innovative design strategies and optimization techniques. Moreover, data privacy and security must be paramount in these systems, warranting robust encryption and protection mechanisms to prevent unauthorized access to sensitive information.

In essence, the coexistence of CMOS and MEMS/NEMS elements, empowered by embedded algorithms, transforms devices into intelligent entities capable of data-driven decision-making. This synergy births a new breed of devices that anticipate, adapt, and respond to the world around them with a level of sophistication that ushers in a future where technology becomes not just a tool, but a partner in human endeavors.

The integration of CMOS-MEMS/NEMS devices within smart grids introduces a paradigm shift in energy management. These devices become the neural centers of an intelligent grid, sensing, analyzing, and adapting to dynamic energy demand and supply. This symbiosis transforms grids from static infrastructures to living, dynamic networks that optimize efficiency, promote sustainability, and ensure reliability. As we navigate the path toward a greener and more energy-efficient future, the integration of CMOS-MEMS/NEMS devices into smart grids emerges as a beacon of innovation and progress.

Real-time sensing and control in CMOS-MEMS/NEMS hybrid systems, specifically incorporating nanomaterials like nanotubes and tunnel FETs (Field-Effect Transistors), offer a transformative approach to engineering systems with unprecedented capabilities. This integration leverages the strengths of both micro/nanoelectromechanical systems (MEMS/NEMS) and complementary metal–oxide–semiconductor (CMOS) technology to create highly integrated, multifunctional systems with enhanced sensitivity, responsiveness, and efficiency (Ho and Tai 1996; Glueck et al. 2013; Chung et al. 2004; Židek et al. 2012).

Nanomaterials in CMOS-MEMS/NEMS Hybrid Systems

Nanomaterials, such as carbon nanotubes (CNTs), graphene, and other nanoscale structures, exhibit unique mechanical, electrical, and chemical properties. Integrating these materials into CMOS-MEMS/NEMS hybrid systems presents several advantages:

Enhanced Sensing Capabilities: Nanomaterials' large surface area and high aspect ratios make them ideal candidates for sensing applications. They can be functionalized to selectively interact with target molecules, enabling highly sensitive and specific detection in real time.

Mechanical Strength: Nanomaterials like CNTs possess exceptional mechanical strength and flexibility. Incorporating them into MEMS/NEMS devices enhances structural integrity and durability, crucial for robust real-time applications.

Electrical Properties: Nanomaterials can serve as conductive paths, allowing for efficient electrical signal transmission between components. They can also be integrated into electrodes for improved sensing and actuation.

Energy Efficiency: Nanomaterials can contribute to energy-efficient operation due to their unique electrical characteristics and the potential for low-power signal processing.

Tunnel FETs in CMOS-MEMS/NEMS Hybrid Systems

Tunnel FETs are an emerging transistor technology that exploits quantum tunneling for charge carrier transport. Integrating tunnel FETs into CMOS-MEMS/NEMS hybrid systems offers several advantages:

Ultra-Low Power Consumption: Tunnel FETs exhibit subthreshold slopes that are steeper than conventional MOSFETs, leading to lower power consumption during switching. This is especially advantageous for battery-operated devices.

Fast Switching Speeds: The tunneling mechanism enables rapid switching, facilitating real-time response in control applications.

Enhanced Energy Efficiency: The combination of tunnel FETs with nanomaterials and MEMS/NEMS devices can lead to highly energy-efficient systems suitable for continuous monitoring and control.

Reduced Heat Dissipation: The lower power dissipation of tunnel FETs results in reduced heat generation, which can contribute to the reliability and longevity of the integrated system.

Applications of CMOS-MEMS/NEMS Hybrid Systems with Nanomaterials and Tunnel FETs:

The integration of nanomaterials and tunnel FETs into CMOS-MEMS/NEMS hybrid systems has the potential to revolutionize various domains:

Biomedical Devices: Real-time health monitoring devices with ultra-sensitive biosensors for early disease detection and personalized medicine.

Environmental Sensing: Highly responsive sensors for real-time monitoring of pollutants, gases, and contaminants in the environment.

Internet of Things (IoT): Battery-efficient, real-time sensing nodes for IoT applications in smart cities, agriculture, and industrial automation.

Autonomous Systems: Enhanced control systems for autonomous vehicles, drones, and robotics with faster decision-making based on real-time sensor data.

Energy Harvesting: Integration of energy harvesters with real-time monitoring and control circuits, enabling self-powered systems.

Quantum Technologies: Integration of nanomaterial-based quantum sensors and tunnel FETs for advanced quantum computing and sensing applications.

It's important to acknowledge that while the potential benefits are significant, there are also challenges related to nanomaterial integration, fabrication techniques, scalability, and reliability that need to be addressed (Table 6.2).

Table 6.2 Various aspects of MEMS/NEMS and CMOS hybrid technology

Aspect	Description	Recent advances
Integration methods	Techniques for combining MEMS/NEMS devices with CMOS electronics	Monolithic Integration, Wafer-Level Bonding, Flip-Chip Bonding,
Device types	MEMS/NEMS devices that benefit from CMOS integration	Accelerometers, Gyroscopes, RF MEMS Switches, Microphones, Resonators, Biosensors, Energy Harvesters
Monolithic integration	Fabricating MEMS/NEMS and CMOS components on the same silicon substrate	MEMS capacitive accelerometers with integrated CMOS readout circuits, Piezoelectric energy harvesters with CMOS power management
Through-Silicon vias	Vertical connections through silicon wafers for 3D integration of MEMS/NEMS and CMOS	Stacked MEMS capacitive accelerometers with CMOS circuits, 3D integrated inertial sensors
materials	Substrate and materials used for MEMS/NEMS and CMOS integration	Silicon, Silicon Carbide (SiC), Silicon-on-Insulator (SOI), Polymers
Applications	Areas where integrated MEMS/NEMS-CMOS systems are making an impact	5GCommunication (RF MEMS), Wearables (Inertial Sensors), IoT (Gas Sensors),

Challenges in CMOS-MEMS/NEMS Hybrid Systems

Integrating CMOS (Complementary Metal–Oxide–Semiconductor) technology with MEMS (Micro-Electro-Mechanical Systems) and NEMS (Nano-Electro-Mechanical Systems) in hybrid systems offers remarkable potential, but it also poses several challenges that need to be addressed for successful implementation. These challenges stem from the inherent differences between electronic and mechanical components, as well as the complexities of combining these technologies. Here, we'll delve into some of the key challenges in CMOS-MEMS/NEMS hybrid systems:

Process Compatibility and Integration

Materials Compatibility: MEMS/NEMS structures often require different materials than traditional CMOS electronics, leading to compatibility issues during fabrication. Finding materials that can withstand CMOS processing conditions without affecting performance is crucial.

Process Compatibility: The fabrication processes for CMOS, MEMS, and NEMS may have conflicting requirements, such as temperature and chemical compatibility. Developing processes that can accommodate the diverse needs of both electronics and mechanical structures can be challenging.

Fabrication Complexity

Multi-Technology Integration: Integrating MEMS/NEMS with CMOS often involves combining multiple fabrication technologies on the same chip. This complexity increases the chances of defects and yield loss during manufacturing.

Alignment and Registration: Precise alignment of MEMS/NEMS structures with CMOS components is critical. Achieving accurate alignment across different length scales and technologies is a significant challenge.

Scale and Miniaturization

Feature Size Mismatch: MEMS/NEMS structures can be significantly larger than CMOS transistors. Achieving a seamless integration of vastly different scales while maintaining performance is difficult.

Scaling Limitations: As devices continue to shrink, issues like stiction (micro-sticking) and wear become more prominent, affecting the reliability and lifespan of MEMS/NEMS structures.

Packaging and Encapsulation

Hermetic Sealing: MEMS/NEMS structures are often sensitive to the environment, requiring hermetic packaging to prevent contamination and degradation. Developing reliable, cost-effective packaging solutions is challenging.

Thermal Management: CMOS generates heat during operation, which can affect the performance and reliability of nearby MEMS/NEMS structures. Effective thermal management strategies are needed to prevent thermal-induced effects.

Signal Interference and Crosstalk

Electromagnetic Interference (EMI): CMOS circuits can produce electromagnetic fields that may interfere with the operation of MEMS/NEMS devices. Proper isolation and shielding are essential to mitigate EMI.

Mechanical Interference: Vibrations and mechanical stress from MEMS/NEMS components can impact the performance of nearby CMOS electronics, potentially leading to signal degradation or failure.

Power Dissipation and Energy Efficiency

Power Dissipation: The power consumption of CMOS electronics can generate heat, affecting the performance of MEMS/NEMS devices that are sensitive to temperature changes. Energy-efficient operation is crucial for avoiding these effects.

Power Management: Efficiently managing power distribution and consumption in hybrid systems requires specialized circuitry and strategies.

Reliability and Yield

Mechanical Reliability: MEMS/NEMS structures are susceptible to mechanical wear, fatigue, and stiction over time. Ensuring long-term mechanical reliability is challenging, especially in dynamic applications.

Yield Enhancement: Integrating multiple technologies increases the complexity of the fabrication process, impacting yield. Yield enhancement strategies are necessary to maintain cost-effectiveness.

Modeling and Simulation

Multi-Domain Modeling: Simulating the behavior of CMOS-MEMS/NEMS hybrid systems involves multiple physical domains (mechanics, electronics, electromagnetics). Developing accurate multi-physics models is complex and resource-intensive.

Process Variability: Integrating different technologies introduces variability that affects device performance. Modeling and mitigating process variability challenges are essential.

Overcoming these challenges requires collaboration between researchers and engineers from various disciplines, including material science, semiconductor processing, mechanical engineering, and electronics design. Additionally, advancements in fabrication techniques, materials development, simulation tools, and interdisciplinary knowledge exchange are critical for realizing the full potential of CMOS-MEMS/NEMS hybrid systems in various applications.

Integrating CMOS technology with MEMS/NEMS systems, also known as CMOS-MEMS/NEMS hybrid systems, presents both technical and engineering challenges. These challenges need to be overcome to fully realize the potential of these integrated systems. Here's an expanded look at some strategies to overcome these challenges:

(a) Process Compatibility and Integration

Challenge: CMOS and MEMS/NEMS fabrication processes often involve different materials and steps, making their integration complex.

Solution: Develop new fabrication techniques that allow for the seamless integration of CMOS circuits and MEMS/NEMS devices. This could involve modifying existing processes, creating new materials, or developing "middle-of-line" techniques that bridge the gap between the two technologies.

(b) Thermal Mismatch

Challenge: CMOS circuits and MEMS/NEMS devices can have different coefficients of thermal expansion, leading to mechanical stress and potential device failure.

Solution: Engineer the design of the devices to minimize thermal mismatches, possibly by incorporating stress-relieving structures or using materials with similar thermal properties. Thorough simulations and testing can help identify potential issues.

(c) Signal Interference

Challenge: CMOS components generate electrical noise that can interfere with sensitive MEMS/NEMS sensors, and the mechanical operation of MEMS/NEMS devices can induce noise in CMOS circuits.

Solution: Implement proper shielding, filtering, and grounding techniques to isolate sensitive components from unwanted noise. Design MEMS/NEMS devices and

CMOS circuits with appropriate spacing and isolation structures to minimize interference.

(d) Power Supply Compatibility

Challenge: CMOS circuits typically operate at low voltages, while MEMS/NEMS devices might require higher voltages to function optimally.

Solution: Develop voltage regulation circuits that can provide different voltage levels to different parts of the system. This could involve creating on-chip voltage converters or regulators that can adapt to the requirements of both CMOS and MEMS/NEMS components.

(e) Packaging and Assembly

Challenge: The packaging and assembly of CMOS-MEMS/NEMS hybrid systems require specialized techniques to ensure reliability and performance.

Solution: Invest in advanced packaging technologies that provide a hermetic seal, protect sensitive components from environmental factors, and maintain the necessary electrical connections. Wafer-level packaging techniques can streamline the assembly process and enhance yield.

(f) Scale and Cost

Challenge: Developing CMOS-MEMS/NEMS hybrid systems can be costly and time-consuming, particularly for small-scale production.

Solution: Collaborate with research institutions, industry partners, and semiconductor foundries to share resources and expertise. As the technology matures, economies of scale can help reduce production costs.

(g) Testing and Quality Assurance

Challenge: Ensuring the reliability and performance of CMOS-MEMS/NEMS hybrid systems requires specialized testing methods.

Solution: Develop testing protocols that account for the unique characteristics of both CMOS and MEMS/NEMS components. Implement advanced testing equipment to assess the functionality and longevity of integrated devices.

(h) Cross-Disciplinary Expertise

Challenge: Creating CMOS-MEMS/NEMS hybrid systems requires expertise in both electronics and micro/nano-mechanics, which are traditionally distinct fields.

Solution: Foster interdisciplinary collaboration and training programs that bring together engineers and researchers from various backgrounds. This can facilitate knowledge exchange and innovation.

Overcoming these challenges demands a combination of innovative engineering solutions, interdisciplinary collaboration, and investment in research and development. As the integration of CMOS technology with MEMS/NEMS devices continues to advance, the resulting hybrid systems have the potential to drive breakthroughs in diverse fields, from healthcare to communication to energy (Valle et al. 2021; Liu et al. 2018).

Conclusion

In conclusion, the integration of Micro/Nanoelectromechanical Systems (MEMS/NEMS) with Complementary Metal–Oxide–Semiconductor (CMOS) technology represents a transformative leap forward, offering numerous advantages such as high sensitivity, precision, and compact design across various applications. This fusion is a key enabling technology that will shape the future across industries like healthcare, telecommunications, transportation, and beyond, by transcending the limitations of individual disciplines and driving innovation to new heights. MEMS-based sensors, including accelerometers, gyroscopes, and pressure sensors, are now vital components in modern devices, with applications ranging from inkjet printer heads to biosensors and beyond.

As transistor size reaches its limit, new CMOS-compatible technologies such as CNTFET, Nanowire FET, Nanosheet FET, and TFET are emerging, while beyond-CMOS innovations like Spin FET, NCFET, Mott FET, and MTJ devices offer advanced current conduction techniques. These technologies will play a pivotal role in advancing the Internet of Things (IoT), providing the sensory and computational fabric needed for real-time data acquisition and smarter, more interconnected environments.

The integration of MEMS/NEMS with CMOS also promises to revolutionize healthcare with personalized diagnostics, remote monitoring, and minimally invasive treatments, while transportation is set for safer, more efficient mobility through real-time monitoring and autonomous decision-making systems. However, challenges remain, such as managing power consumption, thermal effects, and nanoscale fabrication, all of which demand cross-disciplinary collaboration.

Ultimately, this convergence between MEMS/NEMS and CMOS invites us to redefine technological boundaries, empowering a future world that is interconnected, efficient, and brimming with potential. It stands as a beacon of progress, guiding us toward new possibilities and a brighter future for humanity.

Competing Interests *The Authors Have no Conflicts of Interest to Declare that Are Relevant to the Content of This Chapter.*

References

Baltes, Henry et al (2002) CMOS MEMS-present and future. In: Technical digest. MEMS 2002 IEEE international conference. fifteenth IEEE international conference on micro electro mechanical systems (Cat. No. 02CH37266). IEEE

Chung HC, Enomoto T, Shinozuka M, Chou P, Park C, Yokoi I, Morishita S (2004) Real-time visualization of structural response with wireless MEMS sensors. In: Proceedings of 13th world conference on earthquake engineering, 121, pp 1–10

Deleonibus, Simon, ed (2021) Convergence of more moore, more than moore and beyond moore: materials, devices, and nanosystems. CRC Press

Frey A, Schienle M, Seidel H (2009) CMOS based sensors for biochemical analysis. TRANSDUCERS 2009—15th Int. Conf. Solid-State Sensors, Actuators Microsystems, pp 1670–1673, https://doi.org/10.1109/SENSOR.2009.5285752

Fang W et al (2013) CMOS MEMS: A key technology towards the "More than Moore" era. In: 2013 Transducers & Eurosensors XXVII: The 17th international conference on solid-state sensors, actuators and microsystems (Transducers & Eurosensors XXVII). IEEE

Ginet, Patrick, et al (2011) CMOS-compatible fabrication of top-gated field-effect transistor silicon nanowire-based biosensors. J Micromechanics Microengineering 21.6, 065008

Gong C, Hogan T (2014) CMOS compatible fabrication processes for the digital micromirror device. IEEE J Electron Devices Soc 2(3):27–32

Glueck M, Oshinubi D, Schopp P, Manoli Y (2013) Real-time autocalibration of MEMS accelerometers. IEEE Trans Instrum Meas 63(1):96–105

Hagleitner C, Kirstein K-U (2005) Circuit and system integration. CMOS—MEMS, 513–577

Ho, Chih-Ming, Yu-Chong Tai (1996) MEMS and its applications for flow control. 437–447

Islam M (2020) Saiful. Emerging Devices, Circuits, and Systems. Case Western Reserve University, Reconfigurable RF and Wireless Architectures Using Ultra-Stable Micro-and Nano-Electromechanical Oscillators

Izhaky, Nahum, et al (2006) Development of CMOS-compatible integrated silicon photonics devices. IEEE J Sel Top Quantum Electron 12.6, 1688–1698

Jo HC, Choi WY (2018) Encapsulation of NEM memory switches for monolithic-three-dimensional (M3D) CMOS–NEM hybrid circuits. Micromachines 9(7):317

Lin, Yen-Nan, Ching-Liang Dai (2018) Micro magnetic field sensors manufactured using a standard 0.18-μm CMOS process. Micromachines 9.8, 393

Li, Xiangyu et al (2020) Research on high-resolution miniaturized MEMS accelerometer interface ASIC. Sensors 20.24 7280

Liu, Haotian, et al (2018) Microhotplates for metal oxide semiconductor gas sensor applications—Towards the CMOS-MEMS monolithic approach. *Micromachines* 9.11, 557

Maleville, Christophe (2015) Engineered substrates for Moore and more than Moore's law: Device scaling: Entering the substrate era. In: 2015 IEEE SOI-3D-Subthreshold Microelectronics Technology Unified Conference (S3S). IEEE

Menon, Harshitha, et al (2008) Approximate High-Performance computing: a fast and energy-efficient computing paradigm in the Post-Moore Era. IT Prof 25.2, 7–15

Nasrollahpour, Mehdi, et al (2020) A low noise MEMS based CMOS resonator using magnetoelectric sensor. In: 2020 international SoC design conference (ISOCC). IEEE

Qu H (2016) CMOS MEMS fabrication technologies and devices. Micromachines 7(1):14

Ray, Vishva, et al (2008) CMOS-compatible fabrication of room-temperature single-electron devices. Nat Nanotechnol 3.10, 603–608

Riverola, Martin et al (2016) Dynamic properties of three-terminal tungsten CMOS-NEM relays under nonlinear tapping mode. IEEE Sens J. 16.13, 5283–5291

Riverola Borreguero, M (2018) Micro and Nano-electro-mechanical devices in the CMOS back end and their applications

Roy, Kaushik, et al (2013) Integrated systems in the more-than-moore era: designing low-cost energy-efficient systems using heterogeneous components. IEEE Design & Test 33.3, 56–65

Session, Third Annual. Physics for World and Society (2005)
Schropfer, Gerold, et al (2004) Designing manufacturable MEMS in CMOS-compatible processes: Methodology and case studies. MEMS, MOEMS, and Micromachining. 5455. SPIE
Valle J, Fernández D, Gibrat O, Madrenas J (2021) Manufacturing issues of BEOL CMOS-MEMS devices. IEEE Access 9:83149–83162
Verd, Jaume, Jaume Segura (2019) Editorial for the special issue on development of CMOS-MEMS/NEMS devices. Micromachines 10.4, 273
Verd, Jaume, Jaume Segura (2019) Development of CMOS-MEMS/NEMS devices. MDPI-Multidiscip Digit Publ Inst
Wise, Kensall D (2010) Microelectronics in the "More than Moore" era. In: 68th device research conference. IEEE
Žídek, Kamil, Alexander Hošovský, Vladislav Maxim (2012) Real-time safety circuit based on combined MEMS sensor data for rehabilitation device. In Proceedings of the 13th international carpathian control conference (ICCC), pp 786–790. IEEE

Chapter 7
CMOS-MEMS Sensors with FET-Based Transduction: A Paradigm Shift

Pramod Martha, Naveen Kadayinti, and V. Seena

Abstract MEMS sensors and devices have numerous applications spanning across consumer gadgets, automotive electronics, industrial medical, defence, aerospace, etc. Despite the wide and increasing demands for MEMS-based devices such as sensors and actuators in diverse applications, unlike semiconductor technologies like CMOS/VLSI, the design and fabrication of MEMS/Microsystems are not fully standardized. Over the years, the demand for microscale sensors with read out electronic circuits has grown. Most of the commercially available MEMS sensors are based on passive transduction mechanisms such as capacitive, piezoresistive or piezoelectric with inherent performance limitations. These conventional sensors require external amplification, limiting their scalability, design flexibility, fabrication, and on-chip Integrated Circuit (IC) compatibility. These challenges can be overcome by adopting integrated CMOS-MEMS sensor architectures paving the way for efficient, scalable solutions. FET-based active transduction schemes have the potential to overcome these limitations with ease in CMOS-MEMS integration as an additional merit. One of the FET-based architectures, "Suspended gate field effect transistor (SGFET)" is a suitable option for integrating sensors with built-in amplification. Its innovative design eliminates the need for specific interface conversion circuits, offering a streamlined and efficient solution compared to traditional capacitive transduction methods. Certainly, overcoming specific challenges associated with the SGFET-based active transduction technique is crucial to developing SGFET-based CMOS-MEMS sensors for practical applications. This chapter will address challenges and different aspects and overview of our attempts towards indigenous development of MEMS sensors

P. Martha · V. Seena (✉)
Department of Avionics, Indian Institute of Space Science and Technology (IIST), Thiruvananthapuram, India
e-mail: seena.v@iist.ac.in

P. Martha
Electronics and Communication Engineering, Manipal Institute of Technology, Manipal Academy of Higher Education (MAHE), Manipal, India

N. Kadayinti
Electrical Engineering, Indian Institute of Technology Dharwad, Dharwad, India

V. Pachkawade and K. Guha (eds.), *Micro- and Nano-Systems in 21st-Century*, Springer Tracts in Electrical and Electronics Engineering,
https://doi.org/10.1007/978-981-96-4659-3_7

with FET-based electromechanical transduction with scope for CMOS-MEMS integration. Design, simulation, and fabrication process integration of CMOS-MEMS Accelerometers SGFET-based electromechanical transduction with different device architectures for overcoming the inherent challenges of SGFET are discussed in this chapter.

Keywords CMOS-MEMS · Suspended gate field effect transistor · Active transduction · MEMS accelerometer · Closed Loop in plane movable SGFET · U-channel SGFET

Introduction

Sensors are devices that detect and respond to changes in its environment. They play a pivotal role in day-to-day life by detecting and measuring environmental signals in their surroundings or vital signals, offering valuable insights into the world we live in and the state of our own body. Since the Internet of Things (IoT) has been revolutionizing the way we live, sensor industry has been enjoying tremendous growth and endless demand. Manufacturing these miniature sensors and actuators, typically with dimensions from micro to nanoscale became possible due to advancements in the microelectronics industry. The microfabricated micromechanical and electrical components are integrated to form Microsystems or popularly known as MEMS (Micro-Electro-Mechanical Systems) (Liu 2012). The widespread adoption of MEMS can be attributed to their affordability, energy efficiency, durability, possibility for mass production, portability, and superior performance (Bogue 2013; Ciuti et al. 2015; Tanaka 2007).

MEMS technology allows the integration of mechanical, electrical, optical, fluidic, and thermal components into a single miniature system (Stepanovsky 2019; Ashraf et al. 2011; Ho and Tai 1996). Many complex micromechanical structures have been fabricated on silicon wafers to develop sensors and actuators such as accelerometers (Varanis et al. 2018), gyroscopes (Zhanshe et al. 2015), vibrometers, micromirrors (Pengwang et al. 2016), pressure and force sensors), microgrippers, switches, etc. MEMS devices find applications in aerospace (Flechtner et al. 2014), navigation (Zeng et al. 2018), bio-medical (Cornacchia et al. 2017), internet of things (IoT) (Bisio et al. 2016), unmanned aerial vehicle (UAV) (Youn et al. 2020), healthcare, consumer electronics (Zawawi et al. 2020), and so on. Despite the wide and increasing demands for MEMS-based devices such as sensors and actuators in diverse applications, unlike semiconductor technologies like CMOS/VLSI, the design and fabrication of MEMS/Microsystems are not fully standardized yet.

MEMS sensors are classified based on input signal, transduction mechanism, materials of fabrication, etc. The performance of MEMS sensors and actuators lies in the transduction technique, i.e. converting non-electrical signals into usable electrical signals. Based on the transduction principle, MEMS sensors are broadly categorized as optical, thermal, piezoelectric, piezoresistive, and capacitive or electrostatic

MEMS sensors. Among these, piezoelectric, piezoresistive, and capacitive sensors are very popular and widely used due to their relatively easy fabrication and simple working mechanism.

In the case of piezoresistive MEMS sensors, the resistivity of the piezoresistive material integrated with the MEMS structure varies with the stress due to external signals like force, pressure, and acceleration as depicted in Fig. 7.1a. Materials with a high gauge factor (ratio of fractional change in resistance to applied mechanical strain), such as doped silicon, indium titanium oxide (ITO), polyvinylidene fluoride (PVDF), metal alloys, and polysilicon, are preferred to be integrated in piezoresistive MEMS sensors.

In a piezoelectric MEMS accelerometer, a piezoelectric material is placed in the high stress region of the MEMS structure due to the external stimulus similar to piezoresistive type as depicted in Fig. 7.1b. The charge stored in piezoelectric material varies with stress and the amount of variation depends on the charge and voltage coefficients of piezoelectric materials such as zinc oxide (ZnO), aluminum nitride (AlN), Lead zirconate titanate (PZT), quartz, PVDF, $LiNbO_3$, $LiTaO_3$, and Tourmaline. The piezoresistive and piezoelectric sensing requires functional materials. Additionally, fluctuations in humidity and temperature pose challenges, impacting the piezoelectric behavior of materials and reducing sensor reliability. Hence, among these

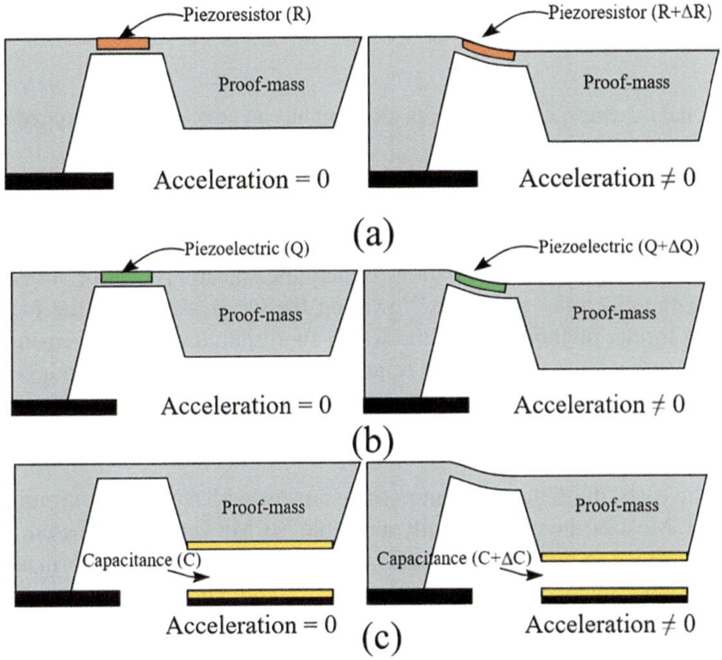

Fig. 7.1 MEMS sensors based on, **a** piezoresistive, **b** piezoelectric, **c** capacitive

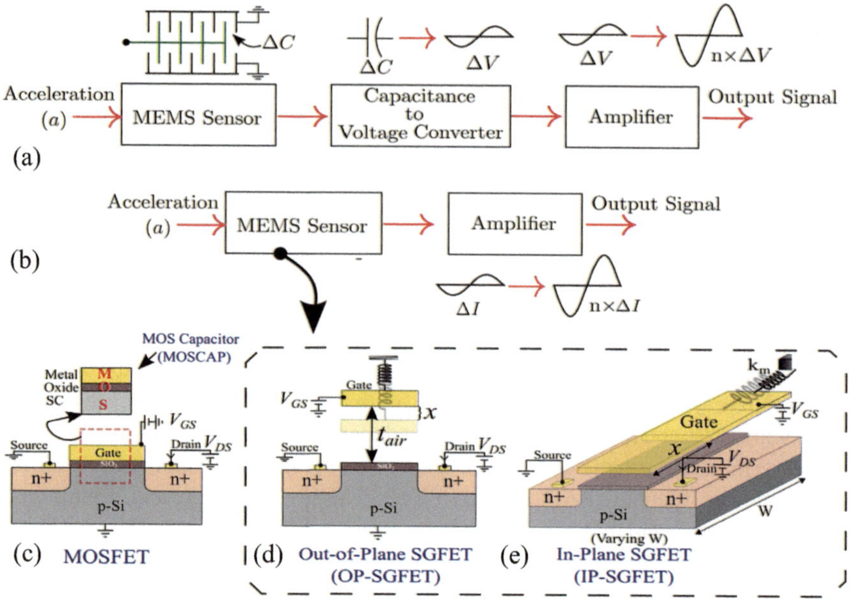

Fig. 7.2 **a** Block diagram of MEMS accelerometer based on capacitive transduction method, **b** block diagram of FET-based MEMS accelerometer, **c** out-of-plane Suspended gate FET (OP-SGFET), **d** in-plane SGFET (IP-SGFET)

conventional schemes, capacitive transduction has an edge over these piezoresistive or piezoelectric schemes.

In capacitive sensors, the capacitance between two electrodes varies with the input signal as illustrated in Fig. 7.1c. A system block diagram of capacitive MEMS accelerometer is shown in Fig. 7.2a. The sensitivity is the capacitance change corresponding to the input signal variation. Hence, the sensitivity can be improved by increasing the electrode area or by lowering the electrode gap to the nanometer range. The former method extends the sensor footprint, however, decreasing the air gap to the nanometer range brings in reliability issues such as the Casimir effect and Van der Waals forces. To read the capacitance changes, which are in a few femto farads to the atto-farad range, special capacitance to voltage conversion circuit is required.

Over the years, the demand for microscale sensors with read out electronic circuits has grown. Most of the commercially available MEMS sensors are based on these passive transduction mechanisms discussed above with inherent performance limitations. Readout integrated circuits (ROICs) are required to provide amplification, storage, temperature compensation, communication, etc. for the sensors and are realized separately using CMOS technology. These conventional sensors require external amplification, limiting their scalability, design flexibility, fabrication, and on-chip Integrated Circuit (IC) compatibility. These challenges can be overcome by adopting integrated CMOSMEMS sensor architectures paving the way for efficient, scalable solutions. Hence, CMOS and MEMS technologies can be integrated to offer,

- Miniaturization of sensor system
- Low-cost batch production
- Reduction of parasitic effects due to the use of on-chip interconnects
- Improvement of device sensitivity and signal-to-noise ratio due to the close proximity of the read-out circuits to the MEMS device.

Field effect transistor (FET) based active transduction schemes have the potential to overcome the limitations of capacitive MEMS sensors with ease in CMOS-MEMS integration as an additional merit. A MEMS accelerometer with FET-based active transduction technique is shown in Fig. 7.2b. A modified Metal Oxide Semiconductor Field Effect Transistor (MOSFET) (Fig. 7.2c), which inherently incorporates a capacitor (MOS capacitor) retains the benefits of capacitive transduction. One of the FET-based architectures, "Suspended gate field effect transistor (SGFET)" as depicted in Fig. 7.2d, e is a suitable option for integrating sensors with built-in amplification. Its innovative design eliminates the need for specific interface conversion circuits, offering a streamlined and efficient solution compared to traditional capacitive transduction methods.

The concept of SGFET was first coined in 1967 by Nathanson et al. (1967) when they demonstrated a resonator called resonant gate transistor (RGT). The beam vibration varied the air gap causing the change in the gate capacitance and hence, the drain current. Therefore, beam vibration signal was obtained from the current signal. In SGFETs, the output drain current could be modulated by various parameters. External stimulus can induce mainly two types of changes to the SGFET: (i) chemical changes which can cause changes in work function of the gate, gate dielectric properties, channel, and interface properties, and (ii) physical changes, which lead to the movement of the suspended gate of the SGFET with respect to the channel. The gate movement could either be out-of-plane in the case of out-of-plane SGFET (OP-SGFET) (Fig. 7.1d) or in-plane with the channel, thereby altering the width of SGFET in the case of in-plane SGFET (IP-SGFET) (Fig. 7.1e). SGFETs undergoing chemical changes find application as bio/chemical sensors suitable for environmental sensing. On the other hand, the physical sensors such as accelerometers force, and pressure could be implemented with SGFETs working with external physical stimulus.

In most of the SGFET-based MEMS gas sensors, a specific gas-sensitive layer deposited on the gate upon exposure to the target gas induces a change in the work function of the gate due to the diffusion of gas through the receptive layer. The reported specific adsorption layers deposited for different gases are palladium for hydrogen (Cassidy et al. 1986), polypyrrole for alcohol (Josowicz and Janata 1986), copper phthalocyanine (CuPc) for nitrogen oxide (NO_2) (Oprea et al. 2006), polyaniline (PANI) for gaseous cyanide (Li et al. 1993), and polyacrylic acid (PAA) for ammonia (NH_3) (Oprea et al. 2005). An SGFET pH sensor was reported with fluidic channel created between the gate and gate oxide in which the drain current (I_D) changes due to the presence of electrolyte in the channel (Silva Rodrigues et al. 2013). In 2012, Ankit Jain et al. proposed a biosensor based on SGFET named

Electro-Mechanical Nano-biosensor or Flexure-FET as shown in Jain et al. (2012a, b). On the adsorption of nano-molecules, the stiffness of the gate changes (from k to $k + \Delta k$), which changes the mechanical dynamics of the gate, which in turn is read out as a change in the drain current.

The first pressure sensor based on SGFET was reported by Lysko et al. (1995). The deflection of the gate due to pressure changes the gate capacitance, which changes threshold voltage and output drain current. Subsequently, SGFET-based MEMS pressure sensors were reported in Jachowicz and Azgin (2002), Dai et al. (2008), Gao et al. (2017, 2019). There are very few reports on force sensors based on SGFET. Gao et al. published a MEMS force sensor based on a laterally movable gate array field effect transistor (LMGAFET) in 2017 (Gao et al. 2017) and depletion type vertically movable gate array field effect transistor (VMGAFET) in Gao et al. (2019).

One of the MEMS physical sensors, MEMS accelerometer, dominates the MEMS sensor market due to the high demand for miniaturized sensor, primarily in automobile, navigation, and healthcare applications. In 2003, Plaza et al. reported the first FET-based accelerometer with a polysilicon MEMS structure that acted as the gate of FET (Plaza et al. 1997). A vertical movable gate FET was proposed in 2008 by Buschanowski et al. (2003). In 2008, Izumi et al. proposed a VMGFET-based force/acceleration sensor realized using an SOI wafer (Izumi et al. 2008). The gate was the heavily doped Si device layer with electroplated Nickel proof mass. Aoyagi et al. in 2011 proposed an SGFET-based accelerometer with a detection range of 0–1 g (Aoyagi et al. 2011). The accelerometer was fabricated using post-CMOS MEMS process flow on an SOI wafer. The sensitivity improvement was claimed based on the choice of flexible Parylene-C polymer as the beam material and high-density Nickel as proof mass. A dual VMGFET was fabricated using an SOI wafer and characterized by Song et al. (2014), Kang et al. (2015). The fabrication process was similar to the process in Aoyagi et al. (2011). However, the sensitivity improvement was achieved using a serpentine silicon MEMS structure.

Inherent amplification feature of SGFET has greatly enhanced sensing. However, there are some fundamental challenges with SGFETs. Overcoming these challenges associated with the SGFET-based active transduction technique is crucial to developing SGFET-based CMOS-MEMS sensors for practical applications. Designing and simulating SGFET-based accelerometer systems involves complex interactions across various domains such as mechanics, electrostatics, semiconductor physics, and circuits. Thorough analysis and optimization within each domain prior to fabrication is crucial. Present-day simulators are limited to specific domains will not be suitable for modeling and simulation of SGFET-based sensors. Pull-in instability poses a reliability challenge in capacitive parallel plate/comb MEMS devices. The OP-SGFET, being a modification of the capacitive parallel plate, faces similar concerns. The pull-in instability restricts gate movement, thereby limiting the detection range in OP-SGFET devices. In IP-SGFET, the gate moves horizontally, aligned parallel to the channel, causing the drain current to linearly change with the channel's width (IP-SGFETs are generally configured to have their gates movable along width direction). Consequently, the IP-SGFET-based sensor demonstrates exceptional linearity, broader dynamic range at the expense of reduced sensitivity. The presence of an air gap diminishes gate capacitance, limiting gate control over the channel, similar to short channel effects (SCE) in MOSFETs, where the drain electric field significantly impacts the channel. The SCEs remain a concern in SGFET despite the channel length being several μm. Hence, it is named pseudo-SCEs (P-SCE) and requires careful consideration at design time. The SGFET-based CMOS-MEMS accelerometer combines MEMS technology with CMOS process. The device can be realized with either bulk or surface micromaching technique by integrating CMOS and MEMS.

In order to overcome the challenges mentioned above and enhance the performance of SGFET-based CMOS-MEMS sensors, several strategies have been explored. These improvements focus on (i) employing a Look-up-table (LUT)-based technique for multidomain modeling and simulation, (ii) implementing a stepped suspended gate design to extend the stable driving range, thereby boosting both dynamic range and sensitivity, (iii) utilizing a differential approach with a fingered IP-SGFET implemented in a closed-loop configuration to enhance sensitivity, (iv) incorporating U-channel SGASOIFET to mitigate P-SCE and downscale the sensor, and (v) carefully planning the fabrication integration processes. Detailed discussions on these enhancements are presented in this chapter, illuminating the pathways to significantly enhance the overall performance of SGFET-based CMOS-MEMS sensors.

Suspended Gate Field Effect Transistor (SGFET)

The SGFET structure is very similar to that of a MOSFET, with only the gate dielectric stack modified. Therefore, the SGFET output current can be derived by using the MOSFET drain current expression, which is given as,

$$I_{\mathrm{DS}} = \begin{cases} 0, & \text{for } V_{\mathrm{GS}} < V_{\mathrm{TH}} \text{ (Cut-off region)} \\ \frac{\mu_n C_{\mathrm{eff}} W}{2L}\left(2(V_{\mathrm{GS}} - V_{\mathrm{TH}})V_{\mathrm{DS}} - V_{\mathrm{DS}}^2\right), & \text{for } V_{\mathrm{TH}} < V_{\mathrm{GS}} < V_{D\mathrm{sat}} \text{ (Linear region)} \\ \frac{\mu_n C_{\mathrm{eff}} W}{2L}(V_{\mathrm{GS}} - V_{\mathrm{TH}})^2, & \text{for } V_{D\mathrm{sat}} \le V_{\mathrm{GS}} \text{ (Saturation region)} \end{cases} \tag{7.1}$$

where μ_n is the electron mobility, W and L are the width and length of the MOSFET, respectively, C_{ox} is the oxide capacitance, the gate to source and drain to source bias voltages are V_{GS} and V_{DS}, respectively, and V_{TH} is the threshold voltage. The V_{Dsat} is the saturation drain voltage, defined as $V_{\mathrm{GS}} - V_{\mathrm{TH}}$. The threshold voltage of OP-SGFET is,

$$V_{\mathrm{TH}} = \phi_{\mathrm{MS}} + 2\psi_F - \frac{Q_D}{C_{\mathrm{eff}}} - \frac{Q_{\mathrm{eff}}}{C_{\mathrm{eff}}} \tag{7.2}$$

where ϕ_{MS} is the work function difference between gate metal (ϕ_M) and substrate (ϕ_S), $\psi_F = \frac{kT}{q}\ln\left(\frac{N_a}{n_i}\right)$ is the built-in potential of Si-substrate, Q_{eff} is effective charge density per unit area in the oxide and charges at the Si/SiO$_2$ interface, and $Q_D = -\sqrt{4\epsilon_{Si}qN_A\psi_F}$ is depletion region charge per unit area in substrate.

In MOSFET, most of the parameters such as gate capacitance, channel dimensions, threshold voltage, etc. remain constant post-fabrication except for the bias voltages. However, in the case of SGFET sensors, these parameters may vary based on the external stimulus such as force, pressure, acceleration etc. As mentioned in the previous section, the SGFET is categorized into two types: (i) OP-SGFET and (ii) IP-SGFET. A detailed analysis of each type will give insight into the operation of SGFET.

Out-of-Plane Movable Suspended Gate FET (OP-SGFET)

An *n*-type enhancement mode OP-SGFET device is shown in Fig. 7.3a. The gate is attached to a MEMS structure with spring constant k_m. The initial air gap is t'_{air} and the final air gap is $t_{air} = t'_{air} - x$, where x is the mechanical sensitivity (S_m) of the accelerometer given as,

$$x = \frac{m}{k_m}, \text{ m is the mass, } k_m \text{ stiffness of the MEMS structure.} \tag{7.3}$$

A capacitance model of OP-SGFET is shown in Fig. 7.3b. The gate capacitance per unit area is given as,

$$C_{eff} = \frac{C_{ox}C_{air}}{C_{ox} + C_{air}} = \frac{\epsilon_0}{t_{air} + \varsigma} \tag{7.4}$$

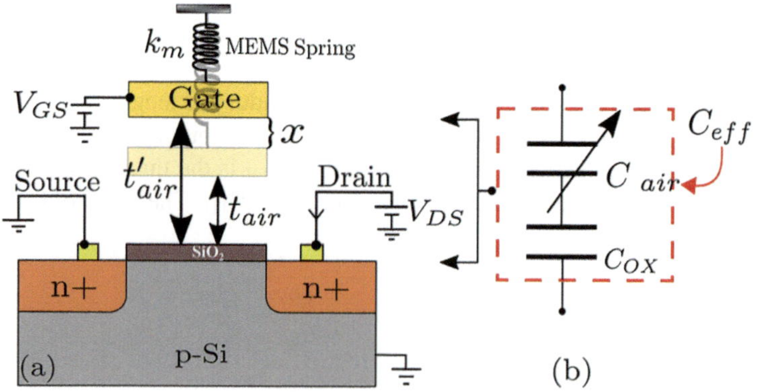

Fig. 7.3 **a** SGFET with gate is attached to a MEMS spring having stiffness k_m, **b** SGFET capacitance model with initial air gap t_{air}. The C_{air} varies with t_{air}

where $C_{ox} = \frac{\epsilon_{ox}}{t_{ox}}$ and $\varsigma = \frac{t_{ox}}{\epsilon_{rox}}$. Here, ϵ_{rox} (= 3.9) is the relative dielectric permittivity of gate oxide and ϵ_0 is the permittivity of free space.

The sensitivity of C_{eff} to air gap is given as,

$$\frac{dC_{eff}}{dt_{air}} = -\frac{\epsilon_0}{(t_{air} + \varsigma)^2} \tag{7.5}$$

The effective gate capacitance varies non-linearly with the air gap. The threshold voltage of OP-SGFET is same as given in Eq. 7.6,

$$V_{TH} = \phi_{MS} + 2\psi_F - \frac{Q_D}{C_{eff}} - \frac{Q_{eff}}{C_{eff}} \tag{7.6}$$

The work function of gate metal (ϕ_M) is material specific parameter, whereas work function of Si-substrate (ϕ_S) depends on substrate doping (N_A) as,

$$\phi_S = \chi + \frac{E_g}{2q} + \frac{kT}{q}\ln\left(\frac{N_a}{n_i}\right) \tag{7.7}$$

Here χ and E_g are the electron affinity and band gap (1.1 eV) of silicon respectively. $n_i = 1.45 \times 10^{10}$ cm^{-3} is the intrinsic carrier concentration. The threshold voltage (V_{TH}) varies as a function of air gap (t_{air}) and is obtained by rewriting the Eq. 7.6 as follows,

$$V_{TH}(t_{air}) = \beta + \delta(t_{air} + \varsigma)$$

where,

$$\beta = \phi_{MS} + 2\psi_F, \delta = -\frac{(Q_D + Q_{eff})}{\epsilon_0} \tag{7.8}$$

The air gap change has no effect on β and δ. Therefore, from Eq. 7.8, the threshold voltage (V_{TH}) varies linearly with air gap (t_{air}), and change in V_{TH} with air gap is,

$$\frac{dV_{TH}}{dt_{air}} = \delta = \text{constant} \tag{7.9}$$

The threshold voltage increases with the substrate doping, and to reduce V_{TH} in modern transistors a low V_{TH} implant is carried out in the substrate to lower the doping near the surface. However, in the case of SGFETs due to the presence of air gap, the threshold voltage would be inherently high if the typical values of substrate doping concentrations in CMOS are kept. Hence to avoid process complexity of counterdoping, it is better to maintain low substrate doping in the case of SGFETs.

The effective oxide charge in the oxide and Si/SiO_2 interface (Q_{eff}) is a process-dependent parameter and varies from fab to fab. However, the Q_{eff} is generally in the

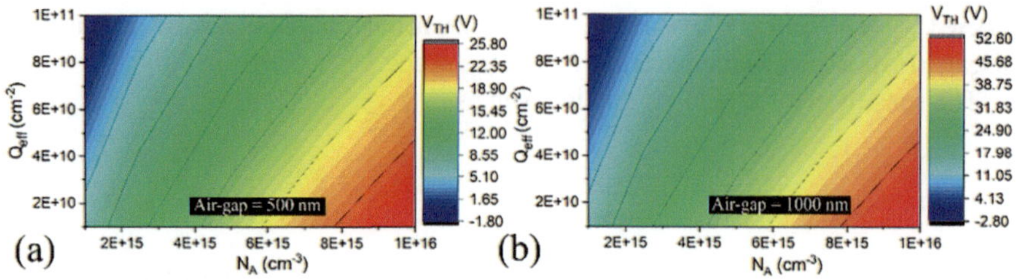

Fig. 7.4 Threshold voltage for different effective oxide interface charge (Q_{eff}) and different substrate doping (N_a) at an air gap of **a** 500 nm, **b** 1000 nm

10^{10}–10^{11} cm^{-2} range. This term has negligible or no effect on the V_{TH} in modern transistors as the gate capacitance is very high. In SGFET, this is considered an advantage because higher Q_{eff} can reduce the V_{TH}. The V_{TH} variation with Q_{eff} and N_A is illustrated in Fig. 7.4. The output drain current (I_{DS}) of OP-SGFET device in different regions of operation is given as,

$$I_{\text{DS}} = \begin{cases} 0, & \text{for } V_{\text{GS}} < V_{\text{TH}} \text{ (Cut-off region)} \\ \frac{\mu_n C_{\text{eff}} W}{2L}\left(2(V_{\text{GS}} - V_{\text{TH}})V_{\text{DS}} - V_{\text{DS}}^2\right), & \text{for } V_{\text{TH}} < V_{\text{GS}} < V_{D\text{Sat}} \text{ (Linear region)} \\ \frac{\mu_n C_{\text{eff}} W}{2L}(V_{\text{GS}} - V_{\text{TH}})^2, & \text{for } V_{D\text{Sat}} \leq V_{\text{GS}} \text{ (Saturation region)} \end{cases} \tag{7.10}$$

Substituting for C_{eff} and V_{TH} from Eqs. 7.4 and 7.8 respectively, I_{DS} is rewritten as a function of t_{air},

$$I_{\text{DS}}(t_{\text{air}}) = \begin{cases} 0, & \text{for } V_{\text{GS}} < V_{\text{TH}} \text{ (Cut-off region)} \\ \frac{\lambda}{t_{\text{air}}+\varsigma}\left(2(V_{\text{GS}} - \beta - \delta(t_{\text{air}} + \varsigma))V_{\text{DS}} - V_{\text{DS}}^2\right), & \text{for } V_{\text{TH}} < V_{\text{GS}} < V_{D\text{Sat}} \text{ (Linear region)} \\ \frac{\lambda}{t_{\text{air}}+\varsigma}(V_{\text{GS}} - \beta - \delta(t_{\text{air}} + \varsigma))^2, & \text{for } V_{D\text{Sat}} \leq V_{\text{GS}} \text{ (Saturation region)} \end{cases} \tag{7.11}$$

The parameter $\lambda = \frac{\mu_n \epsilon_0 W}{2L}$, which is constant for a given W and L. The C_{eff} and V_{TH} varies with the t_{air} and hence, modulate the drain current (I_{DS}). The sensitivity of the OP-SGFET device is derived from Eq. 7.11. The sensitivity in linear region ($V_{\text{TH}} \leq V_{\text{GS}} \leq V_{D\text{sat}}$) is,

$$\left(\frac{dI_{\text{DS}}}{dt_{\text{air}}}\right)_{\text{lin}} = \frac{\lambda}{(t_{\text{air}} + \varsigma)^2}\left[V_{\text{DS}}^2 - 2(V_{\text{GS}} - \beta)\right] \tag{7.12}$$

The sensitivity in saturation region ($V_{\text{GS}} \geq V_{D\text{sat}}$) is derived as follows,

$$\left(\frac{dI_{\text{DS}}}{dt_{\text{air}}}\right)_{\text{sat}} = p - \frac{q}{(t_{\text{air}} + \varsigma)^2} \tag{7.13}$$

where, $p = \lambda\delta^2$, and $q = \lambda(V_{GS} - \beta)^2$. The sensitivity varies with the air gap (t_{air}). The analysis shows that the OP-SGFET response is non-linear.

In-Plane Movable Suspended Gate FET (IP-SGFET)

An n-type enhancement mode IP-SGFET accelerometer is shown in Fig. 7.5a. The airgap (t_{air}) remains constant throughout the operation. The acceleration causes movement of the gate in either length or width direction. The gate movement in length direction ends up with an undesired underlap device. Therefore, mechanical design of the MEMS structure should ensure that the suspended gate movement is restricted to cause change only in width of the SGFET.

A part of the channel width that is covered with the gate is denoted as W_c, and it varies with input acceleration. The deviation in W_c with input acceleration is x, known as mechanical sensitivity (S_m). A capacitance model of IP-SGFET is shown in Fig. 7.5b, and it can be observed that the air gap capacitance per unit area C_{air} remains constant. The threshold voltage for IP-SGFET is the same as OP-SGFET given in Eq. 7.6. The width change does not affect any parameter present in Eq. 7.8 and therefore the V_{TH} remains constant. The output drain current (I_{DS}) of IP-SGFET device in different region of operation as a function of channel width is given as,

$$I_{DS}|W| = \begin{cases} 0, & \text{for } V_{GS} < V_{TH} \text{ (Cut-off region)} \\ \frac{\mu_n C_{eff} W}{2L}\left(2(V_{GS} - V_{TH})V_{DS} - V_{DS}^2\right), & \text{for } V_{TH} < V_{GS} < V_{DSat} \text{ (Linear region)} \\ \frac{\mu_n C_{eff} W}{2L}(V_{GS} - V_{TH})^2, & \text{for } V_{DSat} \le V_{GS} \text{ (Saturation region)} \end{cases} \tag{7.14}$$

At constant V_{GS} and V_{DS}, $I_{DS} \propto W$. The linear variation of I_{DS} varies linearly with the width, i.e., the output response of IP-SGFET is linear. The sensitivity of the

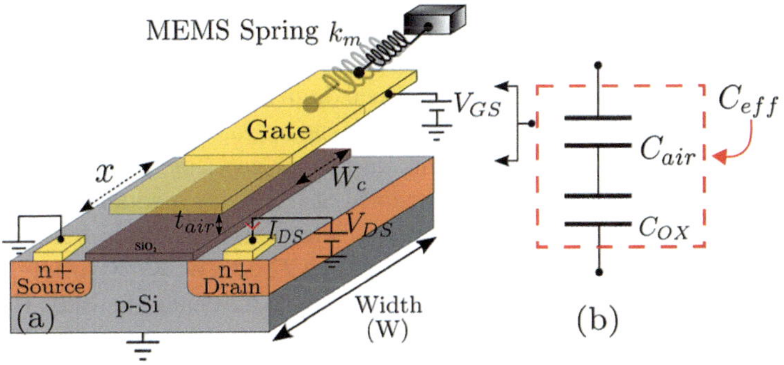

Fig. 7.5 **a** IP-SGFET with gate is attached to a MEMS spring having stiffness k_m. The airgap is t_{air}, and the initial channel width covered by the gate is W_c. The variation in W_c is x due to input acceleration, **b** IP-SGFET capacitance model, C_{Air} is constant as t_{air} remains constant

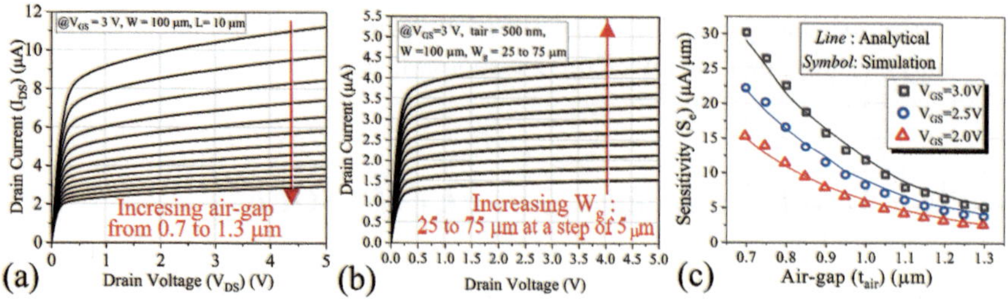

Fig. 7.6 Output characteristics ($I_{\mathrm{DS}} - V_{\mathrm{DS}}$) curve for **a** OP-SGFET with $N_A = 10^{15}$ cm^{-3}, $W = 100$ μm, $L = 10$ μm, $t_{\mathrm{ox}} = 3$ nm, $t_{\mathrm{air}} = 700$–1300 nm, $V_{\mathrm{GS}} = 3$ V, **b** IP-SGFET with $N_A = 10^{15}$ cm^{-3}, total width (W) = 100 μm, channel width covered by gate (W_c) = 25–75 μm, $L = 10$ μm, $t_{\mathrm{ox}} = 3$ nm, $t_{\mathrm{air}} = 500$ nm, $V_{\mathrm{GS}} = 3$ V, **c** sensitivity of OP-SGFET

IP-SGFET device is derived from Eq. 7.14. The sensitivity in linear region ($V_{\mathrm{TH}} \leq V_{\mathrm{GS}} \leq V_{D\mathrm{sat}}$) is,

$$\frac{dI_{\mathrm{DS}}}{dW} = \frac{\mu_n C_{\mathrm{eff}}}{2L}\left(2(V_{\mathrm{GS}} - V_{\mathrm{TH}})V_{\mathrm{DS}} - V_{\mathrm{DS}}^2\right) \tag{7.15}$$

The sensitivity in saturation region ($V_{GS} \geq V_{D\mathrm{sat}}$) is derived as follows,

$$\frac{dI_{\mathrm{DS}}}{dW} = \frac{\mu_n C_{\mathrm{eff}}}{2L}(V_{\mathrm{GS}} - V_{\mathrm{TH}})^2 \tag{7.16}$$

The sensitivity of the IP-SGFET remains constant, resulting in a linear response to input acceleration. However, compared to the OP-SGFET, the IP-SGFET exhibits lower sensitivity due to its linear dependence on width change. In Fig. 7.6, OP-SGFET and IP-SGFET output characteristics curves are depicted. The forthcoming section will delve into an extensive examination of process intricacies and device simulation methodologies. The primary goal here is to comprehend and investigate the behavior exhibited by both the OP-SGFET and IP-SGFET.

The threshold voltage (V_{TH}) proportionally increases with the air gap and is observed as 0.3 V at 700 nm, increasing to 1.8 V for 1.3 μm air gap. An increase in V_{TH} and a decrease in C_{eff} resulted in reduced drain current for higher air gaps. The saturation drain current ($I_{D\mathrm{sat}}$) was 11.22 μA at an air gap of 700 nm which reduced non-linearly with air gap to 2.94 μA at 1.3 μm air gap (Fig. 7.6a). In contrast to OP-SGFET, the V_{TH} of IP-SGFET remained at 0.2 V for a width (W_c) variation from 25 to 75 μm at an air gap of 500 nm. An increase in width made drain current to increase linearly from 3.29 μA ($W_c = 25$ μm) to 9.72 μA ($W_c = 75$ μm) (Fig. 7.6b).

The sensitivity of IP-SGFET remains constant with the change in width. However, the sensitivity of OP-SGFET varies with the air gap as depicted in Fig. 7.6c. The sensitivity improves with reducing air gap and increasing the V_{GS}. Yet, aggressive reduction in the air gap and escalation of V_{GS} trigger reliability concerns, including Casimir force, van der Waals force, and electrostatic pull-in instability, which will be discussed in the forthcoming section. At $V_{\mathrm{GS}} = 3$ V and $V_{\mathrm{DS}} =$

5 V, the sensitivity of OP-SGFET and IP-SGFET was 13.71 μA/μm and 0.13 μA/μm, respectively. Notably, IP-SGFET exhibited lower sensitivity than OP-SGFET, demonstrating minimal nonlinearity (compared to ~8% observed in OP-SGFET) at output.

The sensitivities of both OP-SGFET and IP-SGFET increase with V_{GS}. Is there any limitation on it? One aspect we are yet to discuss is the MEMS structure responsible for the change in air gap or width of the FET and is being covered in the next section.

Pull-in Voltage

The gate voltage (V_{GS}) must be higher than V_{TH} to turn on SGFET. However, there is another electromechanical parameter that decides the upper limit of V_{GS} known as Pull-in voltage (V_p). In Fig. 7.7, the SGFET with pull-in and without pull-in are shown. The V_{GS} creates an electrostatic force (F_e) (Liu 2012),

$$F_e = \frac{1}{2}\frac{C_t V_{GS}^2}{(t_{air} - x)^2} \tag{7.17}$$

where C_t is the total capacitance between the two electrodes. The F_e pulls the gate towards the channel and mechanical spring force (F_m) opposes the movement of gate.

The F_m is given by,

$$F_m = -k_m x \tag{7.18}$$

The net force on gate is,

$$F_{net} = F_e + F_m = 0 \tag{7.19}$$

The Eq. 7.19 is called the force balance equation. The system stabilizes as both forces reach equilibrium. While F_e decreases quadratically with air gap variation,

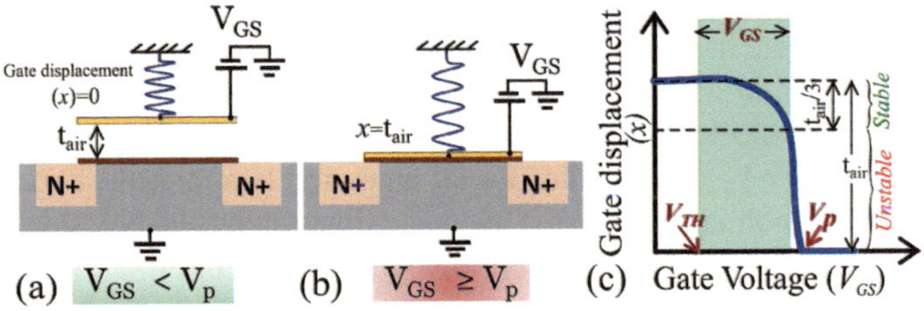

Fig. 7.7 **a** Without pull-in, $V_{GS} < V_p$, **b** With pull-in, $V_{GS} \geq V_p$, **c** gate displacement curve for different gate voltage, stable travel range is $t_{air}/3$ and shaded region is the range of V_{GS} for stable operation of SGFET-based accelerometer

F_m exhibits linear dependency on displacement. Eventually, the equilibrium breaks beyond a specific threshold, causing the gate to collapse onto the channel. This region of gate movement, ensuring a stable equilibrium, defines the stable travel range (x_{ps}) and is expressed as (Liu 2012):

$$x_{\text{ps}} = \frac{t_{\text{air}}}{3} \tag{7.20}$$

The system comes out of stable equilibrium at a certain V_{GS} known as pull-in voltage (V_p) and is given as,

$$V_p = \sqrt{\frac{8k_m t_{\text{air}}^3}{27\epsilon_0 A_e}} \tag{7.21}$$

where $A_e = W \times L$ is the area of gate electrode. The above instability phenomenon is called Pull-in instability and is found in capacitive sensors, actuators, etc. Due to the pull-in instability, the gate can collapse and rest permanently on the substrate leading to the failure of the accelerometer. Therefore, the gate voltage (V_{GS}) should be lower than the V_p throughout the operation to avoid any instability. The boundary conditions on V_{GS} as depicted in Fig. 7.7c is,

$$V_{\text{TH}} < V_{\text{GS}} < Vp \tag{7.22}$$

Here, k_m is the spring constant of the MEMS structure in z-direction. In OPSGFET, the gate moves out-of-plane and hence, lower stiffness (k_m) is needed to obtain higher sensitivity. However, in IP-SGFET, the gate movement in the z-axis is restricted. Therefore, the electrostatic force between the suspended gate and channel is constant, and pull-in instability does not occur.

The value of k_m depends on the dimensions, design, and material of the MEMS structure, highlighting the critical role of proper design optimization in realizing the MEMS accelerometer. Criteria for selecting a suitable MEMS structure encompass various factors like application area, sensitivity, frequency, feasibility, and reliability. Sensitivity, frequency, feasibility, and area take precedence as primary considerations.

Addressing both mechanical and electromechanical aspects, it's pivotal to design, analyze, and simulate the MEMS structure to cover detectable accelerations within the range. Utilizing different air gap values for SGFET design and simulation corresponding to varied accelerations allows checking and comparison of V_{TH} and V_p. If the condition $V_{\text{TH}} < V_{\text{GS}} < V_p$ holds for all input accelerations, the current design undergoes completion and analysis. Otherwise, fine-tuning the MEMS structure's dimensions is necessary until meeting the condition stated in Eq. 7.22.

The analysis showed,

- In OP-SGFET-based accelerometer, the V_{TH} varies with air gap. Hence, at each input acceleration, the condition $V_{\text{TH}} < V_{\text{GS}} < V_p$ needs to be verified, and according to the analysis, the dynamic range or detection range of acceleration needs to be fixed. In other words, the pull-in instability limits the dynamic range of the accelerometer. The response of OP-SGFET is non-linear. However, the sensitivity is high.

- In an IP-SGFET-based accelerometer, the air gap does not vary, and therefore, the parameters, V_{TH}, V_{GS}, and V_p need to be fixed in the beginning such that $V_{\mathrm{TH}} < V_{\mathrm{GS}} < V_p$, so as to avoid pull-in instability during the operation. Therefore, there is no limitation on dynamic range due to pull-in instability. The response of IP-SGFET is linear. However, the displacement sensitivity is typically inferior as compared to OP-SGFET.

Multi-domain Modeling and Simulation of SGFET Sensors

The development of SGFET-based accelerometers involves the design of MEMS, semiconductor transistor devices, and their corresponding read-out integrated circuits. An efficient simulation framework is crucial for enabling rapid design optimization of sensors, leading to reduced development costs.

There are several models and simulation methods. The widely used models are the lumped parameter model, EKV model, and FEA or multiphysics solvers (Tsamados et al. 2008; Reddy et al. 2012; Abele et al. 2005; Abele 2007; Ionescu et al. 2002). The SGFET serves as the core in SGFET-based MEMS sensors, and the significance of SGFET's fabrication process parameters is crucial for precise modeling. However, most of the earlier reported literature discusses SGFET models that focused solely on specific domains. For example, in the case of the lumped parameter model, the focus was on solely on MEMS modeling (Abele et al. 2005; Ionescu et al. 2002). The models developed for SGFET as switches or inverters for low-power applications disregard comprehending fabrication considerations. Conversely, the focus solely rested on presenting the fabrication process concerning SGFET's application as a sensor. Present simulators often confine themselves to limited analysis, failing to capture the intricate dynamics of such sensors. This gap has been addressed by a look-up-table approach adopted to model and simulate the SGFET-based CMOS-MEMS sensor, encompassing all domains and their interdependencies, including the interface circuit (Martha et al. 2021a).

Multiphysics Solver

Multiphysics solver solves complex problems that involve multiple physical domains. SGFET involves mechanics, electrostatics, and semiconductor domains. Therefore, a hybrid numerical simulation framework combining ANSYS multiphysics and ISEDESSIS modeled and simulated inverter with SGFET as shown in Fig. 7.8a (Tsamados et al. 2008). The COMSOL multiphysics solver simulated SGFET, and the

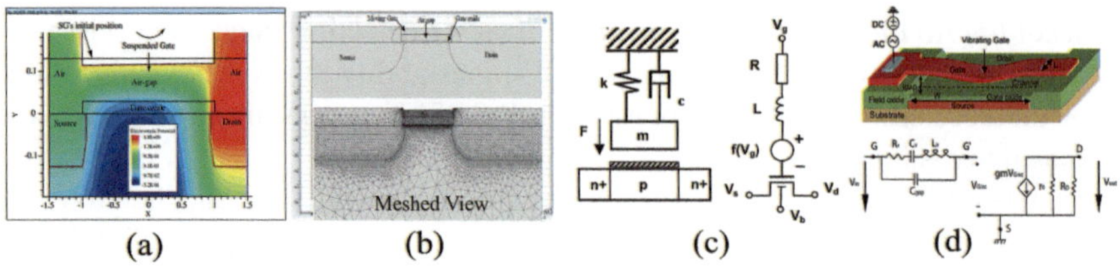

Fig. 7.8 **a** Hybrid simulation framework with ANSYS and ISE-DESSIS (Tsamados et al. 2008), **b** COMSOL multiphysics (Reddy et al. 2012), **c** equivalent models (Ionescu et al. 2002), **d** Abele et al. (2005)

response of SGFET for different air gaps was obtained (Fig. 7.8b) (Reddy et al. 2012). The MEMS structure was simplified and taken as a fixed–fixed beam, and process parameters were ignored. The multiphysics solvers are computationally intensive and often take longer simulation time and large memory. Therefore, many simplifications and approximations are taken to solve equations faster and more efficiently, which may lead to inaccurate results.

Equivalent Models

The spring-mass-damper system is often modeled as an *R* (resistor)-*L* (inductor)-*C* (capacitor) circuit. The air gap capacitance is replaced with a variable voltage source depending on the gate voltage (V_g) (Ionescu et al. 2002). Some gate voltage dropped across *R-L*, and the rest acted as a normal MOSFET with $f(V_g)$ as the gate voltage as illustrated in Fig. 7.8c. The small signal model of SGFET was derived and used as a resonator (Fig. 7.8d) (Abele et al. 2005; Abele 2007). The models oversimplified a complex problem. The fabrication process parametric variations and non-linear effects, such as the fringing field effect and pseudo short channel effects in SGFET, were not added. Hence, these models are not universal, and using those for SGFET-based MEMS sensors will not be justified.

Nearly all SGFET modeling concentrated on sub-100 nm air gaps for switching purposes. However, in sensor applications, the air gap in SGFETs extends to a few hundred nanometers, possibly causing increased non-linear effects. The MEMS component's impact on sensitivity, operational frequency, and gate voltage constraints in SGFETs was overly simplified during modeling. Accurately addressing these intricacies is vital for precise representation in sensor applications.

Look-Up-Table (LUT)-Based Technique for Modeling and Simulation of SGFET-Based Accelerometer

The SGFET-based accelerometer with an interface circuit is modeled and simulated using the look-up tables (LUT). The complete procedure is outlined in Fig. 7.9. The fundamental approach is to optimize the accelerometer in each domain separately, considering the boundary conditions of the same domain as we as the other domains. The MEMS design and simulations need to be carried out by a FEM simulator like CoventorMP or COMSOL. The air gap for different input accelerations is calculated from the mechanical sensitivity. The air gap and gate dimensions from MEMS mechanical and electromechanical simulation were used for the TCAD process and device simulation of SGFET. The MEMS results and the response of SGFET for different air gap and bias conditions were kept in a table format known as Look-up-table (LUT). Using Verilog-A code, the LUT models of MEMS and SGFET were invoked into Cadence Virtuoso for a complete SGFET-based accelerometer with an interface circuit. An OPSGFET-based accelerometer modelling and simulation using the LUT approach and discussed here.

OP-SGFET-Based MEMS Accelerometer

There are two parts in OP-SGFET-based MEMS accelerometer, (i) the MEMS structure with the gate of the FET attached to the proof-mass, (ii) the silicon substrate with $n+$ source, drain, and gate dielectric. The MEMS structure shown in Fig. 7.10a, is an 8-flexure MEMS structure. However, the MEMS spring can be different for different applications while retaining the advantages of the FET-based transduction technique.

(A) ***MEMS Design and Simulation***: The FEM and electromechanical simulations are performed using the CoventorMP tool. The spring constant (k_m) and resonant frequency (f_r) of the MEMS structure in z-axis are given as,

$$k_m = \frac{8EW_bT_b^3}{L_b^3}, f_r = \frac{1}{2\pi}\sqrt{\frac{k_m}{m}} \tag{7.23}$$

where E is the Young's modulus of the beam material, m is the effective mass of the MEMS structure, W_b, L_b, and t_b are the width, length, and height of the beam, respectively. The S_m is the mechanical sensitivity defined as the proof-mass displacement with the gate under 1 g acceleration. In Fig. 7.10b displacement of gate for acceleration in different axis is shown. The resonant frequency is 2.21 kHz. From Eq. 7.23, it can be observed that the spring constant and resonant frequency are inversely proportional, leading to a phenomenon in MEMS called sensitivity-bandwidth trade-off. Therefore, the MEMS structure

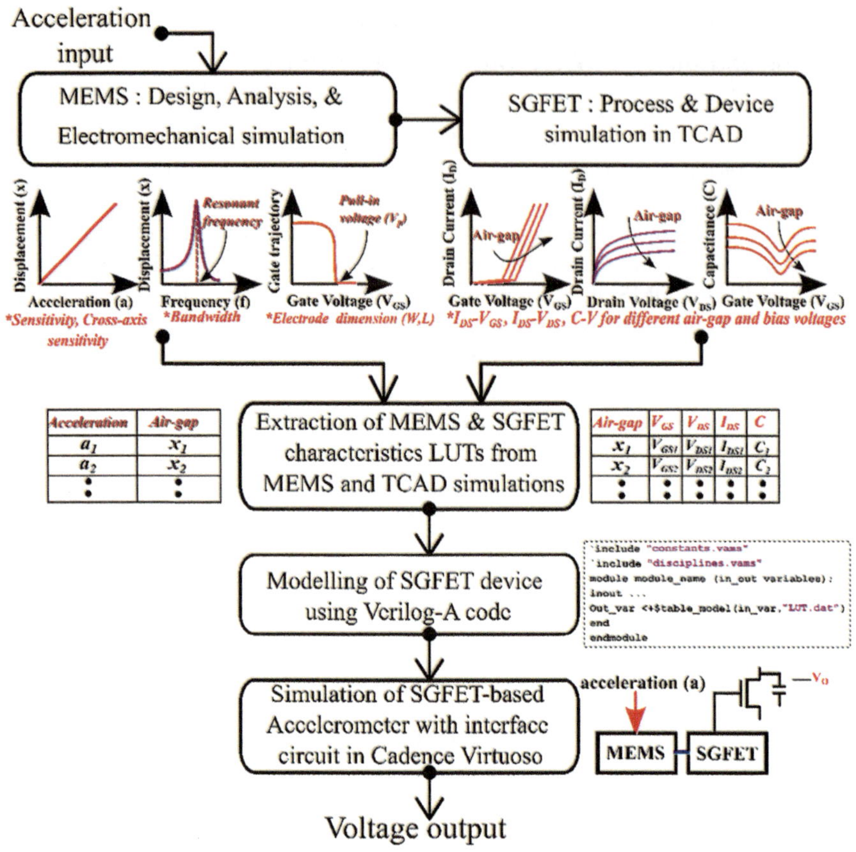

Fig. 7.9 Multi-domain simulation flow for SGFET-based accelerometer

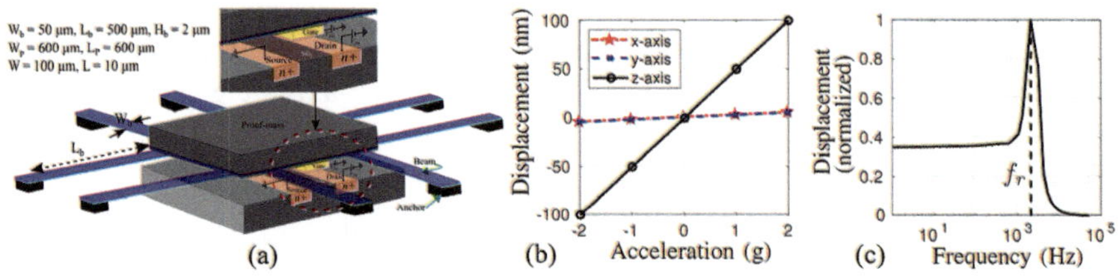

Fig. 7.10 **a** A 8-flexure proof-mass MEMS structure with gate attached to the proof-mass, **b** proof-mass displacement for acceleration in different axis, **c** resonant frequency = 2.21 kHz

dimensions must be chosen per the required specifications. The initial air gap is 500 nm, and as per pull-in analysis as discussed in the previous section, the stable driving range is 166.66 nm. Hence, the dynamic range is fixed at ±2 g to avoid pull-in instability. The MEMS simulation results are taken into the

TCAD for design and simulation of SGFET to obtain output drain current and capacitances at different bias voltage corresponding to input accelerations.

(B) ***OP-SGFET design and simulation in TCAD***: The OP-SGFET was designed for different air gap in the Sentaurus SProcess in line with the standard 180 nm self-aligned CMOS process.

The starting substrate is a <100> p-type (Boron-doped) silicon. The analytical calculations indicated that the V_{TH} is very high for higher N_A and will not be suitable for low-supply voltage applications. Therefore, substrate doping (N_A) of 1×10^{15} cm^{-3} is considered. Further, the gate dielectric is grown by thermal oxidation (Fig. 7.11a) followed by deposition and patterning of 200 nm thick polysilicon (Fig. 7.11b). The lightly doped source and drain formed by ion implantation of Arsenic (As) of dose 2.7×10^{14} cm^{-2} at 10 keV energy (Fig. 7.11c). Nitride spacer (Si_3N_4) is isotropically deposited and etched anisotropically to define the spacer (Fig. 7.11d) followed by deep $n+$ source and drain formation by ion implantation of Arsenic (As) of dose 5.14×10^{15} cm^{-2} at 65 keV energy (Fig. 7.11e). Further, spacer nitride and polysilicon are stripped from the substrate (Fig. 7.11f).

A user-defined dielectric material is created and named "air" in TCAD. All properties of air gap are emulated into the new material air with dielectric constant 1. An air gap of thickness t_{air} is deposited and patterned (Fig. 7.11g). The gate, source, and drain contacts are formed by deposition and patterning of 100 nm Aluminum (Al) metal (Fig. 7.11h). The OP-SGFET designed in SProcess is shown in Fig. 7.12a. The designed OP-SGEFT was simulated in Sentaurus SDevice. For carrier transport, the default drift–diffusion model was utilized along with the Caughey-Thomas model (Caughey and Thomas 1967) and Arora et al.'s doping-dependent mobility model (Arora et al. 1982) to address carrier mobility.

Additionally, the Slotboom model (Slotboom and Graaff 1976) was applied for band gap narrowing, while the temperature-independent model was utilized to describe the density of states. Furthermore, the simulation incorporated the Shockley-Read Hall (SRH) recombination model (Shockley and Read 1952) to encompass carrier generation and recombination mechanisms. The drain

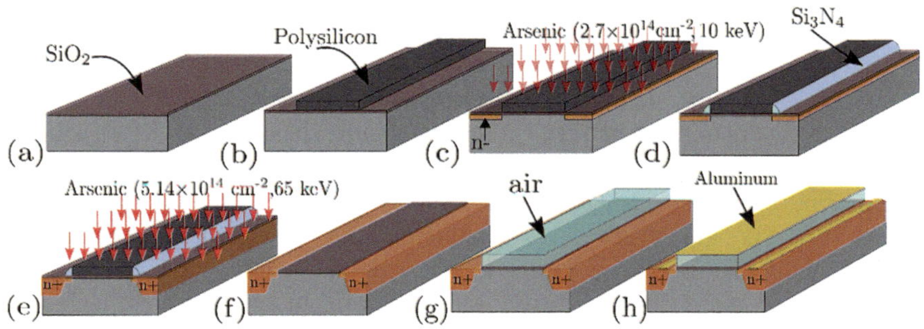

Fig. 7.11 TCAD process flow for OP-SGFET

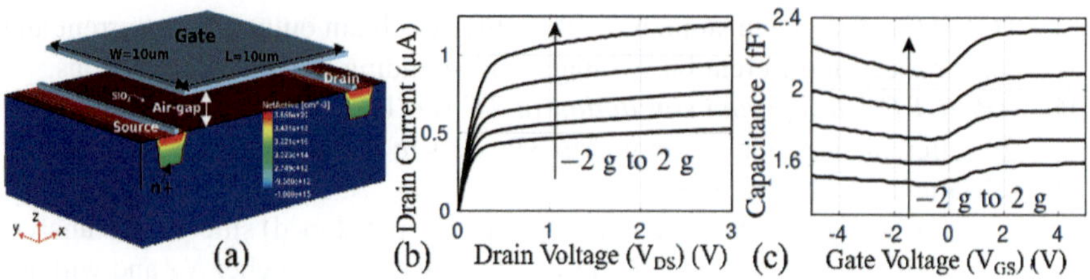

Fig. 7.12 **a** OP-SGFET designed in SProcess following a modified 180 nm self-aligned process flow, **b** $I_{\mathrm{DS}} - V_{\mathrm{DS}}$ for different acceleration, **c** $C-V$ curve for different acceleration

current (I_{DS}) for different gate and drain voltages was obtained from the DC simulations. The threshold voltage at 0 g acceleration was found to be 0.27 V. The current sensitivity (S_e) of 0.22 μA/g was calculated from the output characteristic plot as illustrated in Fig. 7.12b. The AC simulation was carried out at a frequency below $f_r/4$ to obtain the capacitance–voltage (C–V) curve. The gate capacitance and the parasitic capacitance between the nodes were found and plotted in Fig. 7.12c.

(C) *Look-up-tables*: The LUT approach approximates data using the pre-computed data stored in table format. Based on the simulation results, the following tables are prepared,

- MEMS table: *Input*: Acceleration, *Output*: Air gap.
- SGFET table: *Input*: Air gap, Gate voltage (V_{GS}), Drain voltage (V_{DS}), *Output*: Drain current (I_D).
- Capacitance tables: *Input*: Air gap, *Output*: Capacitance.

The LUTs are invoked into the Cadence Virtuoso tool by using the Verilog-A code. These modules are then interconnected to represent an SGFET-based accelerometer.

(D) *SGFET-based accelerometer with Interface circuit*: The SGFET output is connected to an interface circuit to obtain voltage output. A common source amplifier configuration of SGFET with a load resistor is used to obtain the voltage signal. Although other circuit elements could provide improved outcomes, resistors were initially selected for simplicity. Subsequent sections will explore SGFET configurations combined with different circuit components. The SGFET common source amplifier was simulated in Cadence Virtuoso. The acceleration is a mechanical signal given as input to the circuit as an electrical AC signal. The complete circuit schematic with LUTs is illustrated in Fig. 7.13a, and the symbol of SGFET with resistive load is depicted in Fig. 7.13b. The change in output voltage for the ±2 g acceleration range was 34 mV, and the ΔV_O was found to be 8.5 mV/g. The transient response of the SGFET-based accelerometer system is shown in Fig. 7.13c. The LUT-based technique comprehensively accounts for multiple domains, including

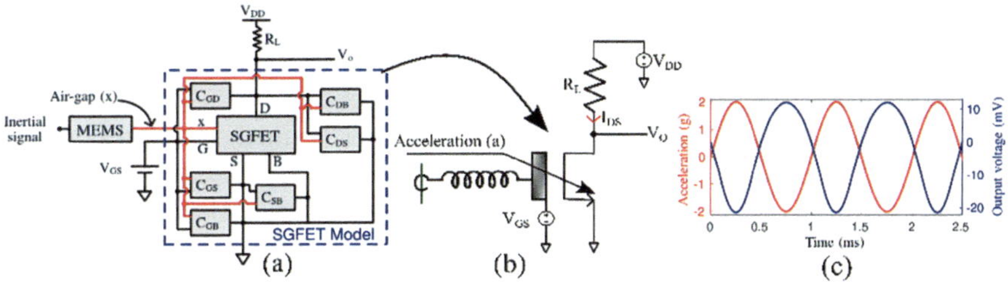

Fig. 7.13 **a** A common source amplifier with SGFET sensor. Inertial signal and air gap are modeled as voltage signals. The elements in shaded boxes are LUT models, **b** SGFET symbol, **c** output for a sinusoidal acceleration of ±2 g

semiconductor physics, microfabrication process variability, mechanics, electrostatics, and circuitry, evaluating their interdependencies while simulating the SGFET-based CMOS-MEMS accelerometer in detail.

Extending Stable Travel Range of Gate with Stepped Suspended Gate FET (SSGFET)

Electrostatic pull-in instability detrimentally affects performance and reliability in parallel plate capacitive MEMS sensors and actuators. As the electrostatically coupled gate electrode over the air gap of the variable metal oxide semiconductor (MOS) capacitor is the most important element in FET-based transduction, pull-in instability severely limits dynamic range, sensitivity, and overall reliability. Addressing pull-in instability in capacitive MEMS could offer a solution to mitigate this issue in SGFET-based MEMS accelerometers. There are three ways to do this, and almost all methods developed for extending the stable driving range and reducing the possibility of pull-in instability fall into these three categories: (i) Modifying the mechanical spring restoring force on the top electrode, (ii) Modifying the electrostatic force between two electrodes, (iii) Adding mechanical stoppers.

In the leveraged bending method as shown in Fig. 7.14a, the electrostatic force is applied only to some portion of the structure while the rest of the structure acts as a Lever and can move a larger range beyond the pull-in limit. However, the electrostatic actuated portion deflects within the pull-in range, and the rest can move the entire gap (Senturia 2001; Hung and Senturia 1999).

Electrostatic levitation or repulsive electrode configuration involves three fixed electrodes at the bottom, and the top electrode is suspended over the middle electrode at a certain air gap as shown in Fig. 7.14b. The extra two electrodes effectively produce the repulsive force on the top electrode. The suspended and fixed middle electrodes are kept at the same voltage or grounded. The side electrodes produce an electrostatic fringe field, which attracts the top of the suspended beam rather than the bottom side, resulting in a net upward force. The force is attractive, acting in the opposite direction to the substrate, and termed repulsive to avoid confusion with parallel plate attractive force. As the fringe field is a weak electrostatic force, the side electrode bias voltage needs to be sufficiently high (Daeichin et al. 2021; Pallay and Towfighian 2018; Allameh and Shafai 2021).

A two-beam method extends the x_{ps} to 50%. However, the method is quite complex and will increase the fabrication complexity during integration with SGFET. The closed-loop feedback technique closely monitors the position of E1 during operation and dynamically modifies the voltage V_G to avoid pull-in and achieve the desired position of E1 (Qiao et al. 2007). However, a complex circuit is required to accurately estimate the position of E1. Extra circuitry requirements of precise control of charge and parasitic capacitances are the bottleneck in charge-based control techniques. Due to process variations, the external adaptive control circuits are optimized for a specific MEMS device and are unlikely to work for others (Qiao et al. 2007).

Several other methods have been reported to extend the pull-in range in capacitive sensors, such as curved microbeams (Hu et al. 2010) and the external electrode approach (Rosa et al. 2004). However, these techniques come with challenges such as fabrication complexity, large area requirements, and the complex read-out circuit. Furthermore, realizing the structures, as mentioned earlier, as the gate of an SGFET is challenging and likely to be incompatible for CMOS-MEMS integration (Martha et al. 2021b).

The pull-in instability restricts the maximum travel range of the gate to 33% of the initial air gap. Hence, increasing the mechanical sensitivity will limit the detection range of accelerometer. Therefore, expanding the stable travel range beyond this 33% threshold offers a potential solution to enhance the detection range without compromising sensitivity. The dynamic range of existing out-of-plane SGFET-based accelerometers, as reported in Kang et al. (2015), Izumi et al. (2008), Aoyagi et al.

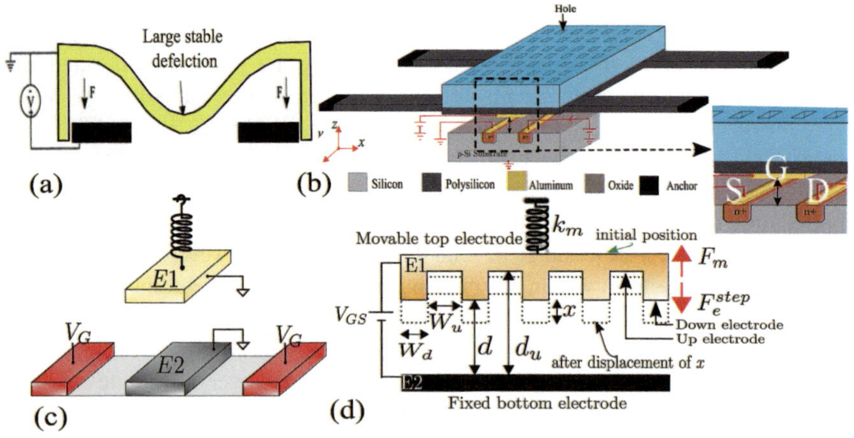

Fig. 7.14 **a** Leveraged bending method (Hung and Senturia 1999), **b** repulsive electrode configuration (Daeichin et al. 2021; Pallay and Towfighian 2018). **c** SGFET-based accelerometer with **d** stepped gate electrode (Martha et al. 2021b)

(2011), Song et al. (2014), typically remains within ±1 g due to the limitations imposed by pull-in instability.

Stepped Electrode MEMS Structure

The pull-in voltage (V_p) and stable driving range (x_{ps}) for a conventional two plate electrode separated by an air gap (t_{air}), is given as,

$$x_{ps} = \frac{t_{air}}{3}, \text{ and } V_{ps} = \sqrt{\frac{8k_m t_{air}^3}{27\epsilon_0 A_e}} \tag{7.24}$$

The stepped electrode configuration aims to alter the electrostatic force, reducing the risk of pull-in instability and widening the stable travel range of the gate. Segmentation of the top planar gate into n stepped divisions, as illustrated in Fig. 7.14d, is one way to reduce the electrostatic force. The arrangement is seen as an array of variable capacitors C_{up} and C_{dn} operating in parallel, where C_{up} and C_{dn} are capacitances due to the upper and lower electrodes of E1, respectively.

The net electrostatic force $\left(F_e^{step}\right)$ is the summation of the electrostatic forces due to upper $\left(F_e^{up}\right)$ and lower $\left(F_e^{dn}\right)$ part of the top electrode was derived as,

$$F_e^{step}(x) = \epsilon_0 L \frac{V^2}{2}\left[\frac{nW_d}{(d-x)^2} + \frac{(n-1)W_u}{(d_u - x)^2}\right] \Rightarrow F_e^{step}(x) = F_e^{up}(x) + F_e^{dn}(x) \tag{7.25}$$

The F_e^{step} is balanced by the mechanical spring force, $F_m = -k_m x$ acting opposite to it before pull-in happens.

Variations in the air gap ratio $\left(\frac{d_u}{d}\right)$ and width ratio $\left(\frac{W_u}{W_d}\right)$ directly influence the forces F_e^{up} and F_e^{dn}, consequently impacting F_e^{step} as described in Eq. 7.25. Consequently, the pull-in parameters are dependent on the ratio between F_e^{up} and F_e^{dn}. The ratio between F_e^{up} and F_e^{dn} can be expressed as:

$$\frac{F_e^{up}}{F_e^{dn}} = \left(\frac{n-1}{n}\right)\left(\frac{W_u}{W_d}\right)\left(\frac{d-x}{d_u - x}\right)^2 \tag{7.26}$$

According to Eq. 7.26, the ratio $\frac{F_e^{up}}{F_e^{dn}}$ is influenced by factors such as the number of steps n, the width ratio $\left(\frac{W_u}{W_d}\right)$, and the air gap ratio $\left(\frac{d_u}{d}\right)$. Given that the influence of the number of steps (n) is insignificant, the stable drive range (x_{ps}) and pull-in voltage (V_{ps}) were predominantly determined by the ratios $\left(\frac{d_u}{d}\right)$ and $\left(\frac{W_u}{W_d}\right)$. When $F_e^{up} > F_e^{dn}$ (indicating dominance of the upper electrode), there was a tendency for

x_{ps} to approach (1/3)rd of the upper air gap, denoted as $d_u/3$. Therefore, the condition for extending x_{ps} was deduced as follows:

$$F_e^{up} - F_e^{down} > 0 \tag{7.27}$$

$$\Rightarrow \left(\frac{W_u}{W_d}\right) > \left(\frac{n}{n-1}\right)\left(\frac{d_u - x}{d - x}\right)^2 \tag{7.28}$$

Similarly, when $F_e^{up} < F_e^{dn}$, the stable drive range x_{ps} converges toward (1/3)rd of the down air gap, denoted as $d/3$. Thus, given a specific $\left(\frac{d_u}{d}\right)$ and $\left(\frac{W_u}{W_d}\right)$, there exists a critical value of x where $F_e^{up} = F_e^{dn}$; let's denote this as x_c for the given $\left(\frac{d_u}{d}\right)$ and $\left(\frac{W_u}{W_d}\right)$. Under a constant lower air gap (d) for a given $\left(\frac{W_u}{W_d}\right)$, the electrostatic force due to the lower electrodes remains fixed. However, an increase in the upper air gap (d_u) results in an increment in $\left(\frac{d_u}{d}\right)$, leading to a reduction in F_e^{up} following (7.25), consequently decreasing x_c. Once $\left(\frac{d_u}{d}\right)$ surpasses a certain threshold, F_e^{up} significantly diminishes compared to F_e^{dn}, causing $F_e^{step} \simeq F_e^{dn}$, allowing dominance of the lower electrode in the system. Consequently, x_{ps} reaches its maximum value and subsequently decreases.

Based on this analysis, the condition ensuring that F_e^{up} remains a substantial contributor to the net electrostatic force $\left(F_e^{step}\right)$ is expressed as:

- For a given $\left(\frac{W_u}{W_d}\right)$, the value of x_c determined for a particular $\left(\frac{d_u}{d}\right)$ should surpass the previously determined x_{ps} value for the prior $\left(\frac{d_u}{d}\right)$, denoted as $x_c\left[\left(\frac{d_u}{d}\right)\right]i > x_{ps}\left[\left(\frac{d_u}{d}\right)\right]_j$, where, $i > j$, and $i, j \in 1, 2, 3 \ldots$

Once the stipulated criterion is no longer met, the influence of F_e^{up} on F_e^{step} diminishes, allowing F_e^{dn} to take control of the system. As a result, the value of x_{ps} decreases, as illustrated in Fig. 7.15a. In the context of the stepped electrode configuration, the magnitude of F_e^{step} consistently remains lower than the electrostatic force observed in the planar top electrode structure (F_e). Hence, the value of x_{ps} consistently surpasses $\frac{d}{3}$.

The expanded stable driving range (x_{ps}) and pull-in voltage (V_p) showed an increase concerning d_u/d, as demonstrated in Fig. 7.15a. However, the subsequent reduction in x_{ps} and V_p can be attributed to the earlier explanation. The electrostatic force due to the upper electrode (F_e^{up}) increases with the width of the upper electrode (W_u), resulting in concurrent increases in both X_{ps} and V_{ps} with W_u/W_d, as observed in Fig. 7.15b. Figure 7.15c presents a comparative analysis of pull-in parameters between devices employing different air gaps with planar and stepped gate configurations. Notably, the pull-in voltage (V_{ps}) exhibited almost a twofold increase compared to conventional capacitive types. At the same time, the stable driving range nearly reached 50% of the air gap, as compared to the 33% typically achieved in classic parallel plate MEMS. For instance, for $W_u/W_d = 30$ and $d_u/d =$

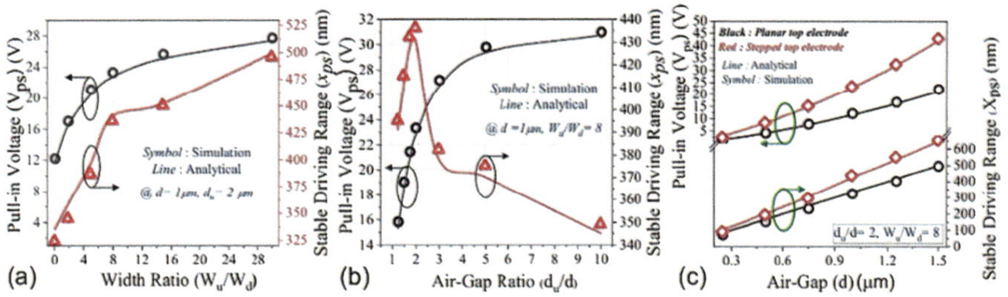

Fig. 7.15 Stable driving range (X_{ps}) and pull-in voltage (V_{ps}) variation, **a** with air gap ratio (d_u/d), **b** with width ratio (W_u/W_d), **c** comparison of planar two plate and stepped electrode two plate MEMS structure (Martha et al. 2021b)

Table 7.1 Pull-in parameter comparison of Stepped and planar top electrode two plate MEMS structure ($d = 1$ μm, $d_u = 2$ μm, $W_u = 21.5$ μm, $W_d = 2.7$ μm)

		Planar gate electrode	Stepped gate electrode
Pull-in voltage (V_{ps})	Analytical (V)	11.57	22.7
	Simulation (V)	11.91	23.2
	% Error	2.78	2.16
Stable driving range (x_{ps})	Analytical (nm)	333.33	425
	Simulation (nm)	326.35	432.65
	% Error	2.10	2.11
Dynamic range (g)		4	6

2, the top electrode could move up to 50% (500 nm) of the initial air gap (1 μm) (Table 7.1).

Stepped Suspended Gate FET (SSGFET)

Electrical Model

The segregation of gate electrodes in SGFET can be accomplished in three manners: (i) along the channel, (ii) across the width, and (iii) throughout the entire gate area plane. Maintaining a robust electrostatic coupling of the gate along the channel's entire length is crucial for optimal gate control. Moreover, due to carrier scattering issues, segregating the gate as a stepped structure along the channel's length can negatively impact the transistor's performance. Similarly, segregating gate electrodes across the entire gate area plane can aggravate carrier scattering and complicate modeling and simulation efforts. Therefore, a scheme with the segregation of gate electrodes across the width of the channel is preferred. This approach strikes a balance, ensuring effective gate control while minimizing adverse effects on carrier behavior and simplifying modeling and simulation.

A parallel configuration of n SGFETs with an air gap of d, width W_d, and carrying current I_D and $(n-1)$ SGFETs with an air gap of d_u, width W_u, and carrying current I_{Du} is used to model the SSGFET device. According to the SSGFET model, the net drain current is expressed as,

$$I_D^{\mathrm{net}} = nI_D + (n-1)I_{\mathrm{Du}} \tag{7.29}$$

The threshold voltages are V_{TH}^d and V_{TH}^u for n SGFET's with air gap d and $n-1$ SGFET's with air gap d_u respectively, as given below,

$$V_{\mathrm{TH}}^d = \phi_{\mathrm{MS}} + 2\psi_F - \frac{(Q_D + Q_{\mathrm{fix}})}{C_{\mathrm{eff}d}} \tag{7.30}$$

$$V_{\mathrm{TH}}^u = \phi_{\mathrm{MS}} + 2\psi_F - \frac{(Q_D + Q_{\mathrm{fix}})}{C_{\mathrm{eff}u}} \tag{7.31}$$

The transistor elements with air gap (d) operate in the inversion region when the applied gate voltage $(V_{\mathrm{GS}}) > V_{\mathrm{TH}}^d$, while the remaining SSGFET elements remain in the sub-threshold region. However, the entire SSGFET channel inversion occurs when $V_{\mathrm{GS}} > V_{\mathrm{TH}}^u$.

The sensitivity of SSGFET is given by using (7.29) and given as,

$$S_e = n\left[p_d - \frac{q_d}{(d+\varsigma)^2}\right] + (n-1)\left[p_u - \frac{q_u}{(d_u+\varsigma)^2}\right] \tag{7.32}$$

where $p_d = \lambda_d\delta^2, q_d = \lambda_d(V_{\mathrm{GS}} - \beta)^2, \lambda_d = \frac{\mu_{n_0} W_d}{2L}, p_u = \lambda_u\delta^2, q_u = \lambda_u(V_{\mathrm{GS}} - \beta)^2$, and $\lambda_u = \frac{\mu_{n_0} W_u}{2L}$. The SSGFET's sensitivity rises as the air gap narrows. The transistor's width (W) can be increased, or more SSGFET can be added to an array connected in parallel to increase the SSGFET's effective width.

The SSGFET is designed in SPROCESS with a modified 180 nm self-aligned CMOS process flow, as shown in Fig. 7.16a. The SDEVICE tool was used to do the SSGFET device simulations.

The agreement between the simulated and analytically derived values of Q_D and V_{TH} was remarkably close. Specifically, for an air gap of 1 μm, the analytical solution yielded $Q_D = 1.39 \times 10^{-4}$ C and $V_{\mathrm{TH}} = 0.96$ V, while simulation results showed $Q_D = 1.43 \times 10^{-4}$ C and $V_{\mathrm{TH}} = 1.16$ V. Enhancing the sensitivity of the SSGFET involves reducing the air gap between the gate and substrate, albeit at the expense of linearity. However, factors like pull-in, Casimir effect, and van der Waals force tend to induce instability in SGFETs with smaller air gaps (Akarvardar et al. 2008). Hence, to ensure operational stability, the SSGFET is operated with an air gap above 500 nm.

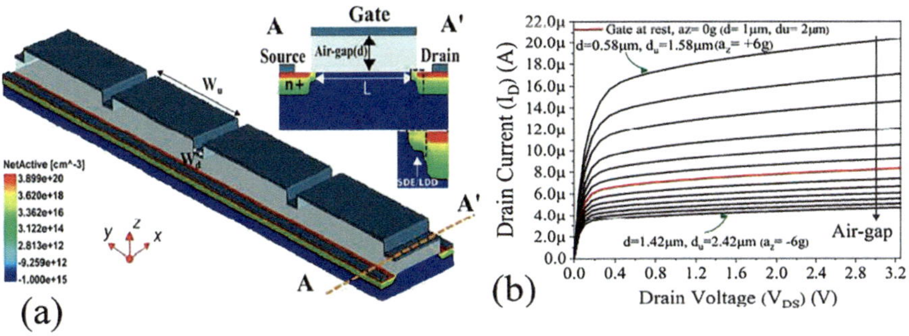

Fig. 7.16 **a** Designed SSGFET in SPROCESS and the cross-sectional view, **b** Simulated $I_D - V_D$ plot for different air gap (d and d_u) corresponding to accelerations (a_z) ranging from −6 to +6 g at a step of +1 g at $V_{GS} = 3$ V, $W_u/W_d = 8$, and initial $d_u/d = 2$ (Martha et al. 2021b)

One of the potential applications for SSGFET is the MEMS accelerometer and the improvement in the accelerometer performance with the implementation of stepped suspended gate architecture is worth discussing. The design methodology followed is LUT-based (Look-Up Table) approach.

SSGFET-Based Accelerometer

The four-flexure polysilicon MEMS structure discussed in Sect. "Stepped Electrode MEMS Structure" acts as an accelerometer. The finite element analysis has already been discussed. Although the pull-in analysis was performed already and found that the dynamic range is improved with a stepped structure, its potential to increase the electrical sensitivity still needs to be discussed. Note that the increase in the stable driving range of the gate helped improve the accelerometer's performance in two ways. With fixed mechanical sensitivity, the dynamic range was improved. In another way, the mechanical sensitivity can be enhanced by keeping a fixed dynamic range. The improvement in pull-in voltage helps improve electrical sensitivity. The gate is displaced by 420 nm at +6 g input acceleration, and the gate electrode dimensions were chosen in such a way that stable driving range (x_p) > 420 nm, ensuring stable operation. The gate width (W) was selected as 100 μm in Sect. "Stepped Electrode MEMS Structure", and the pull-in voltage was 11.64 V under +6 g acceleration. The operating voltage of SSGFET was 3 V, with a nearly 9 V difference. Therefore, W can be increased to increase the sensitivity until the $V_p > 3$ V. The optimized width was 800 μm with the $V_p = 4.10$ V. Accommodation gate with a width 800 μm needs a large proof-mass area, which is undesired. Hence, arraying SSGFET is preferred, which also reduces the contact resistance. The first natural frequency was obtained at 1.91 kHz. The cross-axis sensitivities were 3.2% (x-axis) and 9.7% (y-axis).

The SSGFET output response for the range of input acceleration is obtained from the TCAD simulations for different bias conditions as depicted in Fig. 7.16b, and the

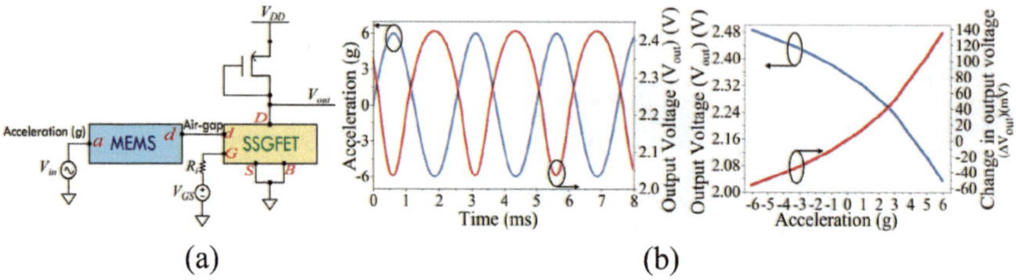

Fig. 7.17 **a** SSGFET-based common source (CS) amplifier with diode connected PMOS load. The elements in the shaded boxes are LUT models, **b** response of the proposed SSGFET-based accelerometer at $V_{GS} = 3$ V, $V_{DD} = 3.3$ V for an acceleration ranging from −6 to +6 g (Martha et al. 2021b)

current sensitivity was 1.41 μA/g. The resistive load is not preferred in the common source amplifier design and is replaced with a diode-connected active PMOS load, which acts as a resistor. The complete schematic is shown in Fig. 7.17a. Using the LUT method discussed in the previous section, the complete accelerometer was simulated, and the transient and DC response is shown in Fig. 7.17b, c. The SSGFET and the PMOS load transistor were in saturation throughout the operation. The non-linearity at the output was 5.3% due to the non-linear response of SSGFET (Martha et al. 2021a). The final sensitivity of the accelerometer was 38.33 mV/g over a dynamic range of ±6 g. The bias voltages were kept below 3.3 V.

SSGFET-based CMOS-MEMS accelerometers have a wider dynamic range and higher sensitivity as compared to other SGFET-based MEMS accelerometers and this is attributed to the stepped gate electrode geometry.

Closed-Loop IP-SGFET (CLIP-SGFET) Accelerometer

In OP-SGFET, pull-in instability is a significant challenge, as discussed in the previous section, which restricts the gate movement in an out-of-plane direction. The air gap varies with an input signal, thereby varying air gap and channel current density, which resulted in non-linear drain current variation. In-plane SGFET (IP-SGFET) offers better performance than OP-SGFET in terms of linearity at output, and larger gate movement without any instability, which allows for a wider range of signal detection. The sensitivity of IP-SGFET is given as,

$$S \propto \frac{\delta I_{DS}}{\delta W'} = \frac{1}{2}\mu_n C_{eff}\frac{1}{L}(V_{GS} - V_{TH})^2 \tag{7.33}$$

The sensitivity of the device (when biased in saturation region), can be derived as,

$$S_a = S_m \times S_e \times r_o \tag{7.34}$$

The larger gate movement increases mechanical sensitivity (S_m). However, electrical sensitivity (S_e) is poor compared to OP-SGFET. The state-of-the-art accelerometers should have higher sensitivity, linearity, wider detection range, stability, and reduced sensitivity to parametric variations. These are achievable only with a negative feedback control. A fingering technique was proposed in Thomas et al. (2017) to improve the sensitivity of IP-SGFET. A suitable MEMS sensor with an actuator is needed to implement negative feedback control. An IP-SGFET-based accelerometer with high linearity and high open-loop sensitivity was reported in Ciuti et al. (2015). In addition to the fingering technique, IPSGFETs were arranged in a differential configuration to suppress common mode noise and cross-axis sensitivity (Song and Ajmera 2004). The negative feedback control brings the back gate to its original position, achievable with a MEMS actuator. A suitable MEMS sensor with an actuator is essential for the closed-loop IP-SGFET (CLIP-SGFET) accelerometer.

Differential Fingering IP-SGFET

Differential Configuration

Initially, in differential IP-SGFET as shown in Fig. 7.18a, the drain current in both pairs of IP-SGFET is identical and has zero output differential current as the gate covers half of the channel width. However, with the input signal, the current change in IP-SGFET is ΔI_D due to ΔW width change, and the differential output current becomes $2\Delta I_D$. Differential amplifier configuration nullifies the common mode noise input, cross-axis sensitivity (as same current change in both IP-SGFET), and doubles the signal swing (Song and Ajmera 2004; Martha et al. 2022).

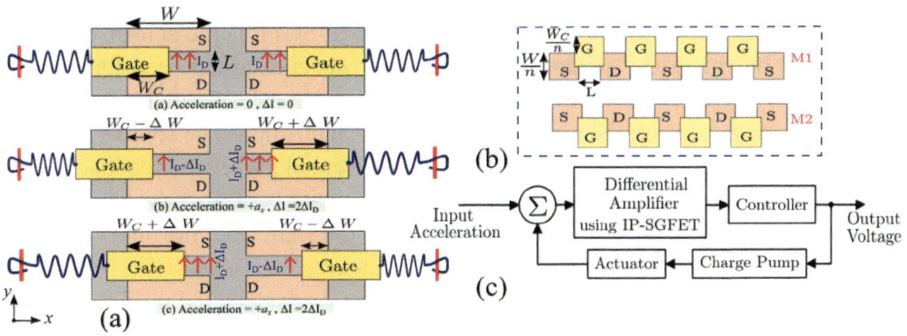

Fig. 7.18 **a** Differential arrangement and operation of IP-SGFET, **b** fingered IP-SGFET, **c** block diagram of CLIP-SGFET (Martha et al. 2022)

Fingering Configuration

In CMOS technology, the fingering technique helps mitigate gate resistance and parasitic capacitances by replacing MOSFETs with a large width with smaller-width MOSFETs, keeping effective width unchanged. The fingering technique helps in improving the sensitivity of IP-SGFET (Thomas et al. 2017). The sensitivity of fingered IP-SGFET is,

$$S \propto \frac{\delta I_{\mathrm{DS}}}{\delta W_T} = \frac{\delta I_{\mathrm{DS}}}{\delta \frac{W}{n}} = n \times \frac{\delta I_{\mathrm{DS}}}{\delta W} \tag{7.35}$$

where W is the width of single IP-SGFET, n is the number of IP-SGFET with width $W_T = (W/n)$, connected in parallel as illustrated in Fig. 7.18b. The block diagram of closed-loop IP-SGFET (CLIP-SGFET) accelerometer is shown in Fig. 7.18c, where an actuator is also required for negative feedback control along with sensor.

Closed-Loop IP-SGEFT (CLIP) Accelerometer

MEMS Sensor with Comb-Drive Actuator: An actuator is an essential part of the negative feedback path, and the actuator design and analysis are the first steps before moving into closed-loop implementation. A MEMS structure with lower stiffness in the lateral direction is essential for integration with IP-SGFET. Flexible levers and compliant mechanisms have been implemented to improve the sensitivity of the MEMS capacitive accelerometer in the in-plane axis. The comb drive type is used for both in-plane acceleration detection and measurement. The comb drive structure with a large electrode height (Tang et al. 2007), or lowering the gap between, can be realized with a bulk micromachining process. However, aspect-ratio limitations in DRIE pose challenges (Jansen et al. 2009). In IP-SGFET, the sensor output is represented by the drain current rather than the capacitance between the combs. Nevertheless, the comb drive serves as an actuator and is incorporated into negative feedback control to return the back gate to its nominal position. The comb actuator is a suitable structure for push–pull driving and an essential part of the feedback loop. The comb actuator is responsible for bringing the gate back to zero acceleration position by generating electrostatic force between the static and movable combs. A folded beam featuring a comb-drive structure is integrated with IP-SGFET to reduce cross axis coupling and facilitate closed-loop operation as depicted in Fig. 7.19a–c.

The spring constant of the folded beam MEMS structure is given by,

$$k_x = \frac{2EW_b^3 H_b}{L_b^3} \tag{7.36}$$

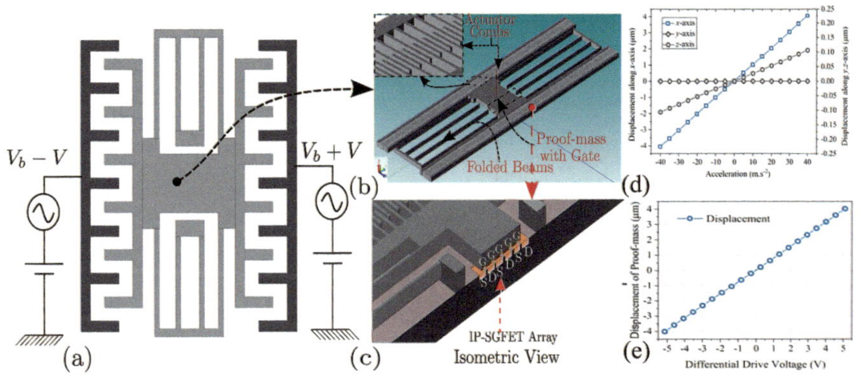

Fig. 7.19 **a** Top view of folded beam MEMS structure with push–pull actuator (Martha et al. 2022; Bao 2005), **b** isometric view of IP-SGFET transistor embedded in a folded beam accelerometer (Martha et al. 2022), **c** the gate is on the bottom side of the proof mass, **d** sensor response to acceleration along the three axes, **d** actuator response (Martha et al. 2022)

The device consists of 2 fingered SGFETs arranged to have a differential output. The structural material is heavily doped p-type Silicon (10^{20} cm^{-3}). The spring constant of the folded beam structures is 0.44 N/m and resonant frequency of the structure is 505.18 Hz. Hence, the operating bandwidth of the sensor is restricted to about 100 Hz.

The sensor's response to acceleration in the three axes is shown in Fig. 7.19c. The structure has a linear response with a displacement sensitivity (S_M) of 1 μm/g. The mechanical cross axis sensitivity is 13.2 pm/g along y-axis and 26.5 nm/g along z-axis. This implies that the air gap is almost constant; hence, device characteristics do not change. In general, the actuator response is non-linear due to the square law dependence on the actuation voltage. The linearity is improved substantially by using a push–pull (differential) drive (Toshiyoshi 2008). In this technique, movable electrode is grounded and a differential voltage with a DC bias is applied to the two stator combs as shown in Fig. 7.19a. For a DC bias voltage (V_b) of 10.5 V, the displacement vs differential drive voltage is shown in Fig. 7.19e. It shows that a maximum differential voltage of ±5.12 V is required to counterbalance the proof mass against a ±4 g acceleration. The sensitivity of the actuator (S_a) is 0.78 μm/V. The width of the gate electrode (W) attached to the proof mass varied from 6 to 14 μm for input acceleration varying from −4 to 4 g and fed to Sentaurus TCAD for design and simulation of IP-SGFET.

IP-SGFET Design and Simulation in TCAD: The IP-SGFET device is designed with 5 fingered SGFETs with channel length of the IP-SGFET chosen as 10 μm. The gate width of one IP-SGFET varied from 6 to 14 μm and 4 to 6 μm for other IP-SGFET for acceleration ranging from −4 to +4 g. The IP-SGFET with different width was designed and simulated in Sentaurus TCAD. The designed IP-SGFET is shown in Fig. 7.20a. The threshold voltage of IP-SGFET was obtained as 0.27 V from the input characteristics (Fig. 7.20b) for fingered IP-SGFET at zero acceleration

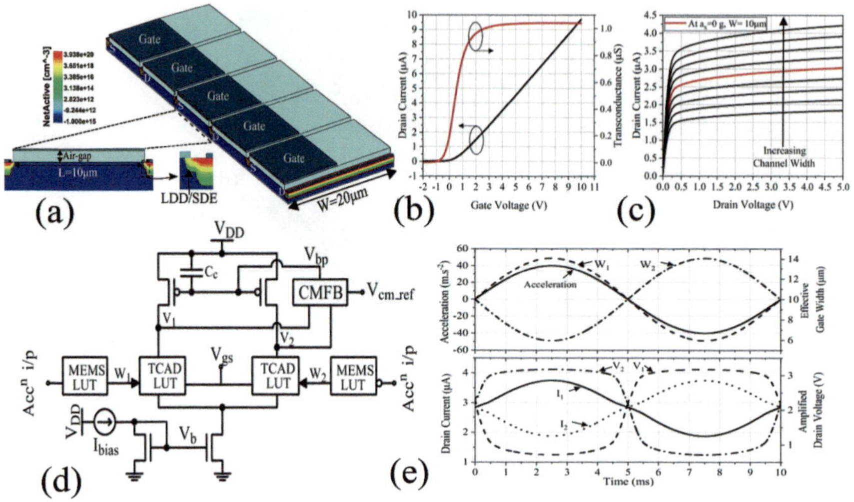

Fig. 7.20 **a** Designed IP-SGFET with cross-sectional views and doping concentration, **b** input characteristics for zero acceleration and $V_{DS} = 0.1$ V, **b** Output characteristics of the SGFET for device width corresponding to different acceleration values from −4 to +4 g, simulated at V_{GS} = 2 V, **d** Schematic of the fully differential amplifier comprising of the IPSGFET, **e** Transient simulation of the IP-SGFET accelerometer in open loop for acceleration of ±4 g at 100 Hz (Martha et al. 2022)

at V_{DS} of 0.1 V. Such a low V_{TH} at air gap of 500 nm was observed due to low substrate doping concentration of 10^{15} cm^{-3}. As per the application, IP-SGFET can be operated either in saturation or subthreshold regimes. The output response of IP-SGFET (Fig. 7.20c) was recorded for different width and bias conditions and formed the LUTs. The drain current sensitivity of the IP-SGFET device was found to be 0.38 μA/g.

Open-Loop IP-SGFET (OLIP-SGFET): The circuit is a standard tail current-biased differential amplifier with current source loads as shown in Fig. 7.20d. The gate voltage (V_{GS}) and air gap remain constant. Therefore, the device threshold voltage (V_{TH}) and the sensitivity remain constant. The input acceleration converted to the width of IP-SGFET through MEMS LUT and the output response of IP-SGFET for different bias conditions and input accelerations obtained from TCAD LUT. These LUT models are used to simulate complete differential amplifiers with current source loads that are biased with common mode feedback (CMFB) circuits. Complementary source follower buffers serve as an effective common mode detector (Daliri and Maymandi-Nejad 2008), offering a superior dynamic range in common mode detection. Figure 7.20e shows the simulated response of this circuit for an input sinusoidal acceleration at 100 Hz ranging from −4 to +4 g. In proportional to acceleration, output voltage and current varied. However, output voltage saturates due to high gain, and the small signal gain is 927 mV/g. The high gain is essential to design the closed-loop negative feedback circuit for the sensor.

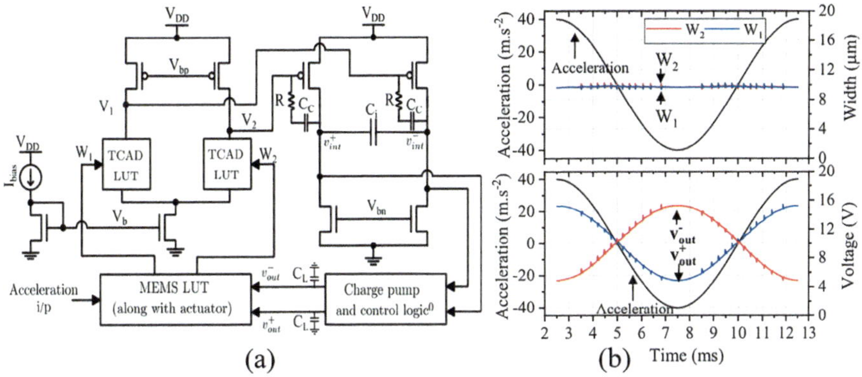

Fig. 7.21 **a** Complete schematic of the CLIP-SGFET accelerometer, **b** transient simulation of the CLIP-SGFET accelerometer in closed loop for sinusoidal acceleration input of 4 g at 100 Hz (Martha et al. 2022)

Closed Loop IP-SGFET (CLIP) Accelerometer: As depicted in Fig. 7.18c, a closed-loop accelerometer features an IP-SGFET and an actuator within feedback mechanism. Input acceleration causes a displacement (x) from the initial state ($W/2$) to $(W/2) + x$, where x can be positive or negative based on the acceleration direction. The gate returns to its nominal position ($W/2$) through feedback, maintaining it within an error margin determined by the loop gain. The complete closed-loop system schematic is illustrated in Fig. 7.21a.

The front end retains similarity to the open-loop circuit, with the MEMS Look-Up Table (LUT) now encompassing the actuator behavior alongside acceleration and feedback voltage characteristics. This LUT encompasses acceleration, feedback voltage, and the corresponding IP-SGFET width. Figure 7.21a depicts capacitor C_L as the effective input capacitance of the actuator, followed by a Gm C integrator as the second stage. This integrator output is directed to the charge pump and its control electronics to generate a substantial bias for the actuator drive. The necessary actuator voltage range spans ±5.12 V over a 10.5 V DC bias, discussed in Sect. "Closed-Loop IP-SGEFT (CLIP) Accelerometer". Depending on the immediate acceleration, the charge pump produced different v_{out} values, varying from 5.38 to 15.62 V (10.5 ± 5.12 V).

Figure 7.21b plot illustrates the CLIP-SGFET accelerometer's response to a ±4 g sinusoidal acceleration input at 100 Hz. Notably, a steady-state error of 10.7×10^{-3} g is observed. The apparent ripple in the response stems from the 50 MHz charge pump, aligned with the charge pump clock frequency. This ripple noise is effectively subdued due to the system's resonant frequency of 505.14 Hz and the 4.6 μs integrator time constant. It's noteworthy that the spikes, with a width of merely 30 μs, corresponds to less than 0.3% of the time period of the highest frequency of expected acceleration (100 Hz).

The negative feedback system precisely controls the charge pump output voltages and accurately tracks the acceleration input, maintaining the gate at its intended nominal position. Utilizing the fingered IP-SGFET and differential sensing technique enhances both sensitivity and stability while minimizing non-linearity and cross-axis acceleration at the output.

Pseudo Short Channel Effect (P-SCE) in SGFET and Its Mitigation

Miniaturization of integrated circuits through CMOS scaling has been instrumental in improving the circuit speed, density and lowering the power consumption. However, short-channel effects (SCEs) pose challenges to the scaling of conventional MOSFET designs. In order to combat SCEs, CMOS device architectures with HALO implants, Strained-Si MOSFETs, DG-MOSFETs, FD-SOIFETs, GAAFETs, and FinFETs were introduced. SCEs arise when the lengths of the channel and source/drain depletion regions are comparable, reducing the gate's control over carriers in the channel and thereby allowing the drain to govern carrier flow in the channel (Taur and Ning 2013).

The scaling of SGFET has several challenges. In the case of SGFET-based switch, With the scaling of the air gap down to 10 nm, a 1 nm of roughness in SiO_2 dielectric results in a drop in the on-current by three orders and a lowering of the pull-in voltage, raising reliability issues in the MEMS switch (Bardon et al. 2010). However, in the case of SGFET as a sensor, the air gap is typically a few 100 nm, and hence the roughness might not play a significant role. In SGFET as an accelerometer, the channel length is typically more than tens of micrometers (Aoyagi et al. 2011; Song et al. 2014; Kang et al. 2015; Abele et al. 2005) due to which SCE has never been discussed as a concern in the earlier reported literature on SGFET. However, the gate control over the channel is poor due to the air gap and the drain electric field having more effect on the channel than the gate field, similar to short channel effects (SCE) in nanoscale transistors. The SCEs are present in SGFET even though the channel length is several μm. Hence, it can be named pseudo-SCEs (P-SCE) and needs to be taken care of while scaling the SGFET channel length. (Martha et al. 2021c). The performance of SGFET can be increased by lowering the channel length. Sensitivity increases with the scaling of the gate electrode in the following ways,

- Reduction in channel length (L) increases the output drain current and the sensitivity. $S_e \propto I_{DS} \propto L^{-1}$.
- As pull-in voltage increases, there is a scope for increasing the operating gate voltage (V_{GS}) or adding more SGFET in an array, thereby increasing the sensitivity (Liu 2012).

In SGFETs, the air gap in the dielectric stack weakens the gate capacitance despite the source and drain being distanced by several micrometers. This allows the lateral field from the drain to interfere with the perpendicular field generated by the gate voltage, resulting in a two-dimensional field within the channel. Consequently, as observed in the electrostatic potential contour, the gate loses sole control over the channel, illustrating a two-dimensional electric potential distribution (Fig. 7.22a). The electrostatic potential contour profiles of SGFET for different channel lengths at an air gap of 1 μm obtained from TCAD simulations are illustrated in Fig. 7.22a. The potential contours are parallel to the silicon-oxide interface and are one-dimensional for most of the parts of the channel in SGFET with channel lengths of 15 and 10 μm. However, the potential contours are curvilinear and 2-dimensional for channel lengths 5 and 3 μm. This resemblance to the short channel effects seen in sub-micrometer MOSFETs and is attributed to the pseudo short channel effects (P-SCE) in long channel SGFETs. These effects are reflected in the output characteristic of SGFET. A parameter α (V^{-1}) was defined to quantify the P-SCE. The α (V^{-1}) is the slope of the normalized $I_D - V_D$ plot (Fig. 7.22) and is given as,

$$\text{Slope}(\alpha) = \frac{dI_{\text{Dn}}}{dV_{\text{DS}}} \tag{7.37}$$

where I_{Dn} is the normalized drain current. Ideally, $\alpha = 0$, and devices with minimum α have better resistance to the P-SCE. The normalized drain current (I_{Dn}) versus the drain voltage (V_{DS}) for different channel lengths are plotted in Fig. 7.22b. The slope of the current plot gradually increases with the reduction in channel length. Drain-induced barrier lowering (DIBL) is another crucial short-channel effect observed in submicron MOSFETs. Therefore, the DIBL and α are obtained from the simulations of SGFET with different channel lengths to analyze the severity of P-SCEs and the values are depicted in Fig. 7.22c. The α is 0.24 for 3 μm SGFET and reduced to 0.028 for 12 μm channel length. The large area requirement of SGFET to avoid P-SCE limits the performance of the sensor and a U-channel SG-SOIFET was presented to mitigate the P-SCE in lower channel length SGFET (Martha et al. 2021c).

U-Channel SG-SOIFET (USG-SOIFET) to Mitigate P-SCE

The UTBB FDSOI FET, featuring a U-channel Planar ultra-thin body and box design, aims to shield the drain electric field, notably enhancing the DIBL and subthreshold slope of the transistor, particularly for sizes below 5 nm (Muralidhar et al. 2018). This structure seamlessly integrates into SGFETs, effectively alleviating P-SCE while enhancing DIBL and α. Moreover, its fabrication requires fewer steps, reducing complexity and costs in realizing the sensor. The schematic of the USG-SOIFET-based sensor is depicted in Fig. 7.23a.

The effective channel length of the USG-SOIFET is given as,

$$L_{\text{eff}} = L_F \text{ (flat region)} + 2L_C \text{ (curved region)} \tag{7.38}$$

The flat region of the channel (L_F) stays constant and relies on the isotropic etching parameters, while the curved areas (L_C) are contingent on the junction depth

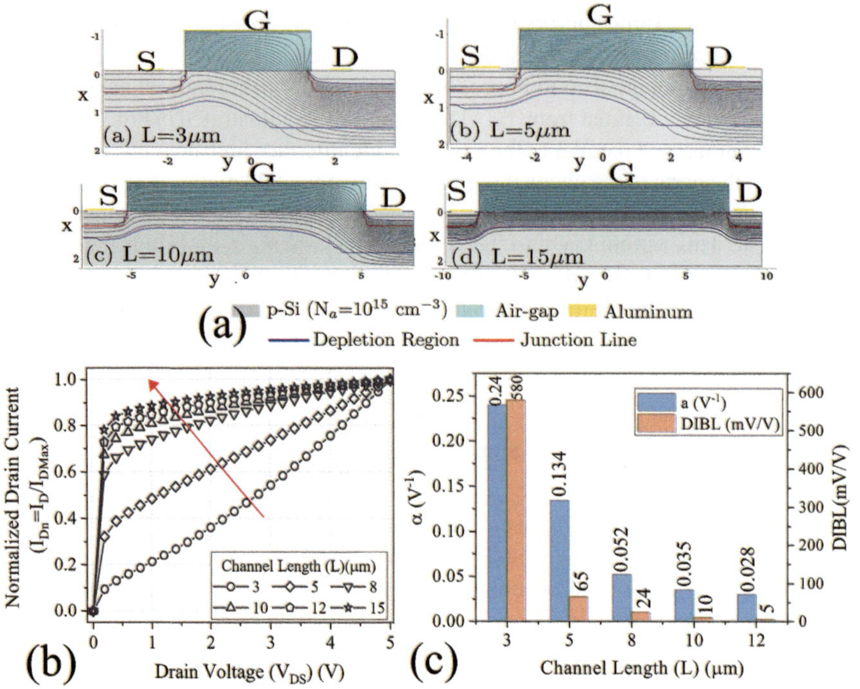

Fig. 7.22 **a** Simulated constant potential contour profiles at $V_{GS} = 2$ V, $V_{DS} = 3.3$ V, **b** normalized $I_{DS} - V_{DS}$ curves for SGFET with different channel length (Martha et al. 2021c), **c** the α and DIBL values for SGFET with different length are compared

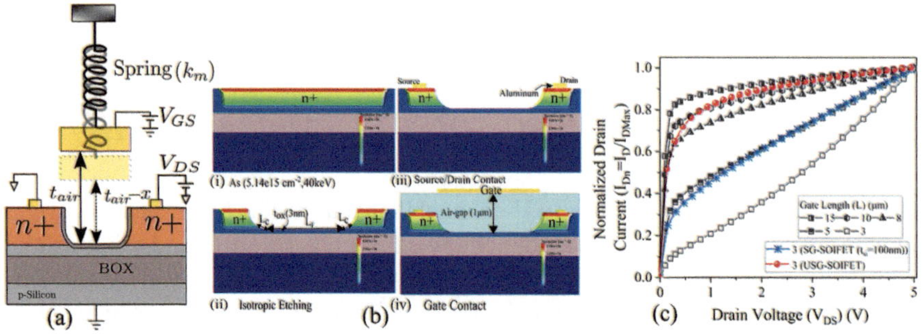

Fig. 7.23 **a** USG-SOIFET schematic, **b** process flow of USG-SOIFET, **c** normalized $I_D V_D$ plot for different devices at $V_{GS} = 3$ V (Martha et al. 2021c)

(x_j) of the source and drain, largely determined by ion-implantation energy. Higher ion-implantation energy increases x_j, reducing L_C and aligning L_{eff} closer to L_F. However, this can amplify P-SCE effects as the channel length decreases despite a fixed gate length. Hence, in design, careful selection of ion-implantation parameters is essential to match the channel length with the gate length.

The standard fabrication processes were followed to design n-USG-SOIFET, followed by the simulation using appropriate models in TCAD. The first process is to grow a thin oxide (a few nm) by thermal oxidation on an RCA-cleaned SOI wafer whose device layer thickness was 600 nm and p-type doped (1×10^{15} cm^{-3}). The next step was arsenic (As) ion implantation at an energy of 40 keV (Fig. 7.23b(i)), followed by the stripping of thin oxide and isotropic etching of 500 nm silicon, leaving 100 nm for carrier flow (Fig. 7.23b(ii)). The isotropic etching ensures the physical separation of the source and drain. Further, thermal oxidation was carried out to grow SiO_2 of 3 nm followed by deposition and patterning of aluminum (Al) for source and drain contacts (Fig. 7.23b(iii)). The air gap is created by depositing the user-defined material 'Air' followed by gate metal deposition and patterning (Fig. 7.23b(iv)).

The USG-SOIFET response to parametric variations is studied through TCAD simulations employing suitable models as discussed in Sect. "OP-SGFET design and simulation in TCAD". Notably, as gate length increases, α decreases due to minimized P-SCE impact. An increase in ion-implantation energy extends x_j, shortening the channel length and amplifying drain electric field penetration near the source, leading to an observed α increment. A thinner channel depth enhances gate control, whereas deeper depths amplify P-SCE. The α value was 0.013 at a 50 nm depth, escalating to 0.7 at 500 nm, indicating susceptibility to pseudo-short channel effects. Process variations affecting substrate depth cause around $\pm 10\%$ fluctuations in channel thickness, altering α by approximately $\pm 11\%$. For instance, at 90 nm thickness, α stands at 0.024, while at 110 nm, it registers at 0.03, minimally impacting the transistor's output.

The normalized $I_D - V_D$ plot for 1 μm air gap SGFET for different channel lengths compared with the 3 μm-USGFET with the substrate depth of 100 nm as depicted in Fig. 7.23c. Also, an SG-SOIFET with 100 nm substrate depth is designed and simulated to compare with USG-SOIFET. The performance of SGFET with a 10 μm channel length is similar to the USG-SOIFET with a gate length of merely 3 μm. The α is calculated from the output characteristic curves. The sensitivity of the USGFET was found to be 26.66 μA/μm, higher than any reported SGFET-based sensors (Aoyagi et al. 2011; Kang et al. 2015). The α and DIBL observed in the 3 μm USG-SOIFET are close to those of a 12 μm SGFET. Despite reducing the gate area by four times, the performance of the SGFET remains uncompromised. Lowering the gate length also helps increase the drain current, leading to an increase in sensitivity.

Further, a MEMS structure can be explored to realize USG-SOIFET-based accelerometer with a lesser footprint than previous works.

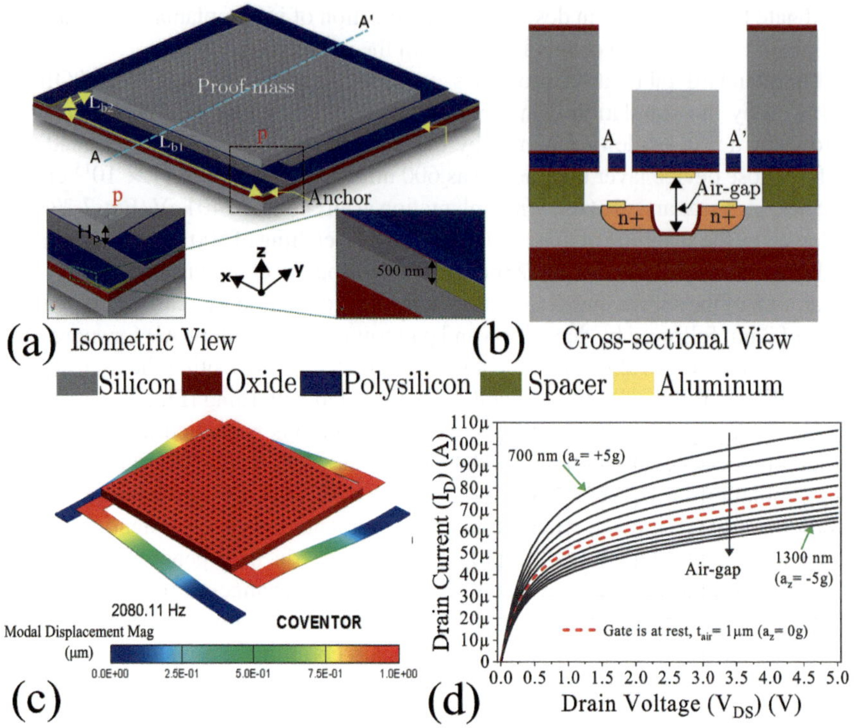

Fig. 7.24 **a** Crab leg four flexure SGFET-based accelerometer, **b** cross-sectional view of the accelerometer, **c** mode-1 from MEMS FEA Simulations, **d** At $V_{GS} = 3$ V, simulated $I_D - V_D$ plot for different air gap (t_{air}) corresponding to *z*-axis accelerations (a_z) ranging from −5 to +5 g at a step of 1 g (Martha et al. 2021c)

USG-SOIFET-Based CMOS-MEMS z-Axis Accelerometer

The SGFET is downscaled to maintain performance while reducing area, yet the MEMS structure remains relatively large, necessitating scaling for smaller sensor footprints. The four-flexure crab leg MEMS structure provides a low spring constant with a small footprint. The beam material is polysilicon with Young's modulus of 160 GPa, and the proof mass is silicon, as shown in Fig. 7.24a, b. The proof mass displaced by 60 nm/g is the gate of USG-SOIFET and the resonant frequency was 2.07 kHz (Fig. 7.24c). The pull-in voltage (V_p) for the gate electrode of width 100 μm and length 3 μm was 13.45 V and 8.23 V at 0 g and +5 g acceleration in vertical direction, respectively. The USG-SOIFET with optimized gate dimensions is designed and simulated in TCAD.

The threshold voltage (V_{TH}) is varied from 0.68 to 0.77 V in ±5 g acceleration. The output response of USG-SOIFET over the dynamic range of the accelerometer at V_{GS} = 3 V is shown in Fig. 7.25a. In saturation, the asymmetrical change in I_{DS} is observed because of the non-linear behavior of USG-SOIFET. However, by lowering the

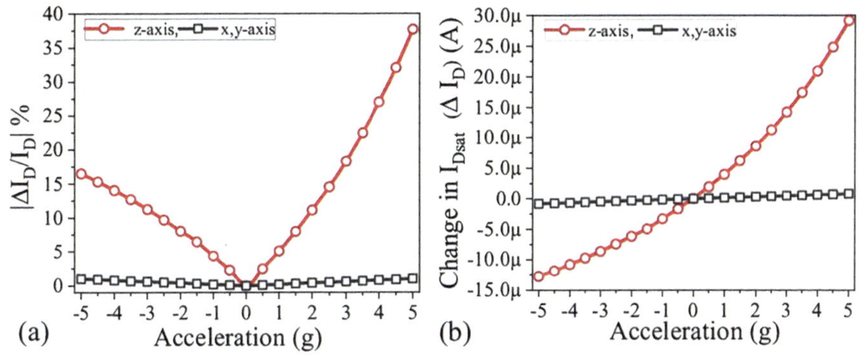

Fig. 7.25 **a** Percentage change in drain current with respect to initial current (when device is at rest, $x = 0$), **b** sensitivity of proposed accelerometer (Martha et al. 2021c)

dynamic range, non-linearity at output will be reduced. The accelerometer sensitivity is given as,

$$S = S_m \times S_e = \frac{\Delta I_{\mathrm{DS}}}{g} \tag{7.39}$$

The percentage change in the output current from its initial value $\left(\frac{\Delta I_D}{I_D}\%\right)$ for different input accelerations along the x, y, and z-axis are plotted. Figure 7.25a depicts the percentage change in the output current from its zero-acceleration value $\left(\frac{\Delta I_D}{I_D}\%\right)$.

The sensitivity (S) plot of the accelerometer is illustrated in Fig. 7.24a. The sensitivity was 4.18 μA/g with a cross-axis sensitivity of 0.16 μA/g for acceleration in the x and y-axis, which is 3.82%, and the device non-linearity was 4.96% for a DR of ±5 g and 0.3% for ±1 g. The USG-SOIFET-based accelerometer gives improved sensitivity, even with the reduced effective area, compared to other SGFET-based MEMS accelerometers.

The USG-SOIFET-based CMOS-MEMS accelerometer addressed the scaling issues in SGFET and reduced the accelerometer footprint with improved sensitivity. Tailored CMOS-MEMS process technology has the potential to realize this novel device architecture (Martha et al. 2021b, c, 2022).

Challenges and Opportunities

The challenges and research opportunities in integrating SGEFT MEMS sensors as CMOS-MEMS platform have been multi-fold. The strategies to overcome some of the specific challenges associated with the SGFET-based transduction technique were discussed. There are other significant challenges yet to be solved for the practical implementation of SGFET based CMOS-MEMS senor platform. One of the major concerns is related to the modification of existing CMOS process integration to implement SGFET MEMS sensor. Heterogeneous integration would always remain as an intermediate step only towards the fully integrated CMOS-MEMS sensor system as it is essential to integrate the MEMS sensor elements and signal conditioning circuit elements in close proximity to maintain good signal-to-noise ratio. The noise modelling of SGFET MEMS sensors needs to be fully understood and developed for developing a comprehensive device and system level models for these sensors and thereby predict the limit of detection and dynamic range for such sensors for various applications.

Competing Interests All the authors declare no conflicts of interest regarding this chapter of the book.

References

Abele N (2007) Design and fabrication of suspended-gate MOSFETs for MEMS resonator, switch and memory applications. Technical Report EPFL

Abele N, Fritschi R, Boucart K, Casset F, Ancey P, Ionescu AM (2005) Suspended-gate MOSFET: bringing new MEMS functionality into solid-state MOS transistor. In: IEDM technical digest, pp 479–481

Akarvardar K, Eggimann C, Tsamados D, Chauhan YS, Wan GC, Ionescu AM, Howe RT, Wong H-SP (2008) Analytical modeling of the suspendedgate FET and design insights for low-power logic, 55(1):48–59

Allameh M, Shafai C (2021) Tri-electrode mems electrostatic actuator with lower control voltage and higher stroke for actuator array implementations. J Electrostat 114:103635

Aoyagi S, Suzuki M, Kogure J, Kong T, Taguchi R, Takahashi T, Yokoyama S, Tokunaga H (2011) Accelerometer using MOSFET with movable gate electrode: Electroplating thick nickel proof mass on flexible parylene beam for enhancing sensitivity. In: Proceedings of the 16th international solid-state sensors, actuators and microsystems conference. IEEE, pp 2030–2033

Arora ND, Hauser JR, Roulston DJ (1982) Electron and hole mobilities in silicon as a function of concentration and temperature. IEEE Trans Electron Dev 29(2):292–295

Ashraf MW, Tayyaba S, Afzulpurkar N (2011) Micro electromechanical systems (MEMS) based microfluidic devices for biomedical applications. Int J Mol Sci 12(6):3648–3704

Bao M (2005) Chapter 3—Air damping. In: Bao M (ed) Analysis and design principles of MEMS devices. Elsevier Science, Amsterdam, pp 115–174

Bardon MG, Neves HP, Puers R, Van Hoof C (2010) Scaling the suspendedgate FET: impact of dielectric charging and roughness. IEEE Trans Electron Dev 57(4):804–813

Bisio I, Delfino A, Lavagetto F, Sciarrone A (2016) Enabling IoT for in-home rehabilitation: accelerometer signals classification methods for activity and movement recognition 4(1):135–146

Bogue R (2013) Recent developments in MEMS sensors: a review of applications, markets and technologies. Sens Rev 33:09

Buschnakowski S, Bertz A, Brauer W, Heinz S, Schuberth R, Ebest G, Gessner T (2003) Development and characterisation of a high aspect ratio vertical FET sensor for motion detection. In: TRANSDUCERS '03. 12th international conference on solid-state sensors, actuators and microsystems. Digest of Technical Papers (Cat. No.03TH8664), vol 2, pp 1391–1394

Cassidy J, Pons S, Janata J (1986) Hydrogen response of palladium coated suspended gate field effect transistor. Anal Chem 58(8):1757–1761

Caughey DM, Thomas R (1967) Carrier mobilities in silicon empirically related to doping and field. Proc IEEE 55(12):2192–2193

Ciuti G, Ricotti L, Menciassi A, Dario P (2015) MEMS sensor technologies for human centred applications in healthcare, physical activities, safety and environmental sensing: a review on research activities in Italy. Sensors 15(3):6441–6468

Cornacchia M, Ozcan K, Zheng Y, Velipasalar S (2017) A survey on activity detection and classification using wearable sensors 17(2):386–403

Daeichin M, Miles RN, Towfighian S (2021) Large-stroke capacitive MEMS accelerometer without pull-in. 21(3):3097–3106

Dai C-L, Kao P-H, Tai Y-W, Wu C-C (2008) Micro FET pressure sensor manufactured using CMOS-MEMS technique. Microelectron J 39(5):744–749

Daliri M, Maymandi-Nejad M (2008) A 0.8 V 420 nW CMOS switched-opamp switched-capacitor pacemaker front-end with a new continuous-time CMFB. In: 2008 15th IEEE international conference on electronics, circuits and systems, pp 758–761

da Silva Rodrigues B, Siarkowski A, De Sagazan O, Crand S, Mohammed Brahim T, Morimoto N (2013) Sensors based in suspended gate field effect transistors for pH measurements. J Phys: Conf Ser 421:012006 (IOP Publishing)

Flechtner F, Morton P, Watkins M, Webb F (2014) Status of the GRACE follow on mission. In: Gravity, geoid and height systems. Springer, pp 117–121

Gao W, Zhao L, Jiang Z, Xia Y, Guo X, Zhao Z, Zhao Y, Sun D (2017) A novel mems force sensor based on laterally movable gate array field effect transistor (LMGAFET). In: 2017 IEEE 12th international conference on nano/micro engineered and molecular systems (NEMS). IEEE, pp 723–727

Gao W, Jia C, Jiang Z, Zhou X, Zhao L, Sun D (2019) The design and analysis of a novel micro force sensor based on depletion type movable gate field effect transistor 28(2):298–310

Ho C-M, Tai Y-C (1996) MEMS and its applications for flow control

Hu Y, Yang J, Kitipornchai S (2010) Pull-in analysis of electrostatically actuated curved micro-beams with large deformation. Smart Mater Struct 19:065030

Hung E, Senturia S (1999) Extending the travel range of analog-tuned electrostatic actuators. J Microelectromech Syst 8(4):497–505

Ionescu AM, Pott V, Fritschi R, Banerjee K, Declercq MJ, Renaud P, Hibert C, Fluckiger P, Racine GA (2002) Modeling and design of a low-voltage SOI suspended-gate MOSFET (SG-MOSFET) with a metal-over-gate architecture. In: Proceedings of the international symposium on quality electronic design, pp 496–501

Izumi H, Matsumoto Y, Aoyagi S, Harada Y, Shingubara S, Sasaki M, Hane K, Tokunaga H (2008) Development of MEMS capacitive sensor using a MOSFET structure. IEEJ Trans Sens Micromach 128(3):102–107

Jachowicz RS, Azgin ZM (2002) FET pressure sensor and iterative method for˙ modelling of the device. Sens Actuators A: Phys 97–98:369–378 (Selected papers from Eurosenors XV)

Jain A, Nair PR, Alam MA (2012a) Flexure-FET biosensor to break the fundamental sensitivity limits of nanobiosensors using nonlinear electromechanical coupling. Proc Natl Acad Sci 109(24):9304–9308

Jain A, Islam ÀE, Alam MA (2012b) On the electro-mechanical reliability of NEMFET as an analog/digital switch. In: 2012 IEEE international reliability physics symposium (IRPS). IEEE, pp ME–1

Jansen HV, de Boer MJ, Unnikrishnan S, Louwerse M, Elwenspoek MC (2009) Black silicon method X: a review on high speed and selective plasma etching of silicon with profile control: an in-depth comparison between Bosch and cryostat DRIE processes as a roadmap to next generation equipment. J Micromech Microeng 19(3):033001

Josowicz M, Janata J (1986) Suspended gate field effect transistors modified with polypyrrole as alcohol sensor. Anal Chem 58(3):514–517

Kang HS, Lee K-H, Yang D-Y, You BH, Song I-H (2015) Microaccelerometer based on vertically movable gate field effect transistor. Nanomicro Lett 7(3):282–290

Li J, Petelenz D, Janata J (1993) Suspended gate field-effect transistor sensitive to gaseous hydrogen cyanide. Electroanalysis 5(9–10):791–794

Liu C (2012) Foundations of MEMS. Pearson Education India

Lysko JM, Jachowicz RS, Krzycki MA (1995) Semiconductor pressure sensor based on FET structure. IEEE Trans Instrum Meas 44(3):787–790

Martha P, Ganga KM, Sebastian A, Seena V, Kadayinti N (2022) A closed loop in-plane movable suspended gate FET (CLIP-SGFET) sensor with a dynamically reconfigurable charge pump. IEEE Sens J 22(22):21550–21560

Martha P, Sebastian A, Seena V, Kadayinti N (2021a) A technique for modeling and simulating transistor based MEMS sensors. In: 2021 IEEE international symposium on inertial sensors and systems (INERTIAL), pp 1–2

Martha P, Kadayinti N, Seena V (2021b) CMOS-MEMS accelerometer with stepped suspended gate FET array: design & analysis. 68(10):5133–5141

Martha P, Kadayinti N, Seena V (2021c) A CMOS-MEMS accelerometer with U-channel suspended gate SOIFET 21(9):10465–10472

Muralidhar R, Dennard RH, Ando T, Lauer I, Hook T (2018) Advanced FDSOI device design: the U-channel device for 7 nm node and beyond 6:551–556

Nathanson H, Newell W, Wickstrom R, Davis J (1967) The resonant gate transistor. IEEE Trans Electron Dev 14(3):117–133

Oprea A, Simon E, Fleischer M, Frerichs H-P, Wilbertz C, Lehmann M, Weimar U (2005) Flip-chip suspended gate field effect transistors for ammonia detection. Sens Actuators B: Chem 111–112:582–586 (Eurosensors XVIII 2004)

Oprea A, Weimar U, Simon E, Fleischer M, Frerichs H-P, Wilbertz C, Lehmann M (2006) Copper phthalocyanine suspended gate field effect transistors for NO_2 detection. Sens Actuators B: Chem 118(1–2):249–254

Pallay M, Towfighian S (2018) A reliable MEMS switch using electrostatic levitation. Appl Phys Lett 113:213102

Pengwang E, Rabenorosoa K, Rakotondrabe M, Andreff N (2016) Scanning micromirror platform based on MEMS technology for medical application. Micromachines 7(2):24

Plaza J, Benitez M, Esteve J, Lora-Tamayo E (1997) New FET accelerometer based on surface micromachining. Sens Actuators A: Phys 61(1):342–345 (Proceedings of EUROSENSORS X)

Qiao D-Y, Yuan W-Z, Li X-Y (2007) A two-beam method for extending the working range of electrostatic parallel-plate micro-actuators. J Electrostat 65(4):256–262

Reddy J, Malhi C, Pratap R, Bhat N (2012) Coupled numerical analysis of suspended gate field effect transistor (sgfet), pp 141–144

Rosa MA, Bruyker DD, Volkel AR, Peeters E, Dunec J (2004) A novel external electrode configuration for the electrostatic actuation of MEMS based devices. J Micromech Microeng 14:446–451

Senturia SD (2001) Energy-conserving transducers. Microsyst Des 125–147

Shockley W, Read W Jr (1952) Statistics of the recombinations of holes and electrons. Phys Rev 87(5):835

Slotboom J, De Graaff H (1976) Measurements of bandgap narrowing in Si bipolar transistors. Solid-State Electron 19(10):857–862

Song I-H, Ajmera PK (2004) Differential laterally movable gate FETs (LMGFETs) as a position sensor. In: Smart structures and materials 2004: smart electronics, MEMS, BioMEMS, and nanotechnology, vol 5389, pp 267–274 (International Society for Optics and Photonics, 2004)

Song I-H, Forfang WBD, Cole B, You BH (2014) Characterization of a vertically movable gate field effect transistor using a silicon-on-insulator wafer. J Micromech Microeng 24(10):105002

Stepanovsky M (2019) A comparative review of MEMS-based optical cross-connects for all-optical networks from the past to the present day. IEEE Commun Surv Tutor 21(3):2928–2946

Tanaka M (2007) An industrial and applied review of new mems devices features. Microelectron Eng 84(5–8):1341–1344

Tang M, Liu A, Agarwal A, Habib M (2007) A single-mask substrate transfer technique for the fabrication of high-aspect-ratio micromachined structures. J Micromech Microeng 17(8):1575

Taur Y, Ning T (2013) Fundamentals of modern VLSI devices. Cambridge University Press

Thomas TC, Reddy VS, Seena V (2017) Novel CMOS-MEMS accelerometer with fingered differential LMGFET architecture with linear response. In: 2017 IEEE 12th nanotechnology materials and devices conference (NMDC). IEEE, pp 125–126

Toshiyoshi H (2008) 2.01—Electrostatic actuation. In: Comprehensive microsystems. Elsevier, Oxford, pp 1–38

Tsamados D, Chauhan YS, Eggimann C, Akarvardar K, Wong H-SP, Ionescu AM (2008) Finite element analysis and analytical simulations of suspended gate-FET for ultra-low power inverters. Solid-State Electron 52(9):1374–1381 (Papers Selected from the 37th European Solid-State Device Research Conference—ESSDERC'07)

Varanis M, Silva A, Mereles A, Pederiva R (2018) MEMS accelerometers for mechanical vibrations analysis: a comprehensive review with applications. J Braz Soc Mech Sci Eng 40:1–18

Youn W, Choi H, Am Cho SK, Rhudy MB (2020) Accelerometer fault tolerant model-aided state estimation for high-altitude long endurance UAV 69(10):8539–8553

Zawawi SA, Hamzah AA, Majlis BY, Mohd-Yasin F (2020) A review of MEMS capacitive microphones. Micromachines 11(5)

Zeng Q, Wang J, Meng Q, Zhang X, Zeng S (2018) Seamless pedestrian navigation methodology optimized for indoor/outdoor detection 18(1):363–374

Zhanshe G, Fucheng C, Boyu L, Le C, Chao L, Ke S (2015) Research development of silicon MEMS gyroscopes: a review. Microsyst Technol 21:2053–2066

Part V
Materials and Packaging

Chapter 8
New and Emerging Materials for MEMS/NEMS: Properties and Characterization

Olga Jakšić and Zoran Jakšić

Abstract Emerging materials with novel properties are of utmost interest for the very existence of MEMS and NEMS systems. The devices in which only standard planar technology materials are used (silicon, silica, noble metals, polymers—photoresists) can only achieve a fraction of multifunctionalities attainable with innovative materials. Further, the new materials must be able to retain their functions at nanoscale, a requirement that a vast majority of conventional functional materials do not fulfill. Here we consider several classes of structured materials, some of which were only recently discovered, while most of them are under intensive research at the moment of writing this text. We consider a possible classification of such materials, to further focus our attention on some of the most important ones. These include nanoscale composites with engineered properties, architectured nanomaterials, different metamaterials and metasurfaces, novel nano-biomaterials, quantum materials. Among the functionalities we consider are nanoscale ferroelectricity, nanomagnetism, nano-piezoelectricity, thermal management for micro and nanosystems, nanoscale phase change materials, and nanoscale shape memory alloys. Some attractive applications of multifunctionalities offered by these materials are considered, together with their integration with trending technologies like Internet of Things, Artificial Intelligence, etc. The chapter is concluded with a short summary of the presented insights.

Keywords Functional materials · MEMS · NEMS · Nanomaterials · Nanocomposites · Metamaterials

---

O. Jakšić (✉) · Z. Jakšić
Department of Microelectronic Technologies, Institute of Chemistry, Technology and Metallurgy – National Institute of the Republic of Serbia, University of Belgrade, Belgrade, Serbia
e-mail: olga.jaksic@ihtm.bg.ac.rs

Z. Jakšić
e-mail: jaksa@nanosys.ihtm.bg.ac.rs

V. Pachkawade and K. Guha (eds.), *Micro- and Nano-Systems in 21st-Century*, Springer Tracts in Electrical and Electronics Engineering,
https://doi.org/10.1007/978-981-96-4659-3_8

Introduction

It was expected that the twenty-first century will be an era of peace, sustainable development, and growth if mankind succeeds in achieving the seventeen sustainable development goals (SDG) defined by the Department of Economic and Social Affairs of United Nations (https://sdgs.un.org/2030agenda). However, in spite of all serious commitments, actions, and plans for people, planet, and prosperity, in the third decade of the twenty-first century, we still have hunger, unnecessary health and nature disasters, economic turbulences, and wars in many parts of the globe, our 2030 Agenda is in risk of failing (Pradhan 2030). Our post-2030 global Agenda, which is yet to be announced, is envisaged to address the issues in a more scientific way and set the desired targets rather quantitatively than in an aspirational way (Cormier and Elliott 2017). Both ways, achieving those goals is not possible without the multidisciplinary technological innovations and transnational research and development in high tech products based on novel materials.

Materials science, reinforced first by the advancements of nanotechnologies, and then by the advancements in computer science through digitalization and technologies assisted by Artificial Intelligence (AI), is seen as a strong pillar of the fourth industrial revolution (Alviz-Meza et al. 2017), and a provider of one of the key technologies for a prosperous Society 5.0 where the digitalized world, enriched by IoT (internet of things) and smart cyber physical systems, is human-centered (personalized healthcare based on human-interface technologies, biotechnologies, smart hospitality, advanced infectious disease surveillance, participatory-style assisting system, next-generation urban transportation systems...) (Coelho et al. 2023). In order to enable and support achieving these anticipated technologies (4.0, 5.0...) the current research needs to comply with new, responsible approach, which assumes investigations conform to the 'Do not Significant Harm' (DNSH) principle. This is valid with respect to climate change mitigation and adaptation, protection of water and marine resources, transition to a circular economy, pollution prevention and control, protection and restoration of biodiversity and ecosystems... Awareness of the depletion of Critical Raw Materials (CRM) and the consequences of high emissions of gases made Responsible Research a necessity and that affected the fields of Micro/Nanotechnology and Materials a lot.

In a Micro-Electro-Mechanical System (MEMS) or a Nano-Electro-Mechanical System (NEMS) there is an integration of mechanical elements, sensors, actuators, and electronics on a common substrate. The production of MEMS/NEMS and microsystems is based on the fabrication technologies used for microelectronics, many materials used in MEMS/NEMS microsystems are materials technologically mature, dimensionally stable, well established in microelectronics, like semiconductors, silicon (Si), quartz, Pyrex or gallium arsenide (GaAs)... However, in spite of the qualities of these materials, there is a demand for using other materials such as polymers (Pergal et al. 2023), plastics, ceramics, composites, etc. (Mohapatra 2020). The only challenge for the use of a variety of novel materials in MEMS/NEMS-based devices is their compatibility with the mature and well-established fabrication and packaging techniques.

The use of novel materials can be application-driven, based on their performances, like ever since the advent of the awareness of first materials (bronze replaced iron in Iron Age weapons, aluminum replaced wood in airframes of early flying vehicles...) but with time, the criteria for the decision on the use of materials are more and more complex, today we have restrictive legislation with strong concern for sustainability and the environment (Ashby 2011). Nowadays, material properties are not just something we discover only, they became rather something an application demands and we design (Estrin et al. 2021). The review on architecturing materials at mesoscale portrays architectured lattice building block materials along with a large number of various engineered unit ceils and assemblies of their interlocked planar or non-planar blocks emphasizing the benefits of knowing the differences between biological and engineered materials and the benefits of mimicking biological features (Estrin et al. 2021). Inspired by the biological growth where a small number of light elements after self-assembly can form different hierarchical structures at all size levels and exhibit environmentally-caused adaptations and self-healing, the development of new engineered materials resulted in materials whose production or operation is based on new concepts. They comprise bioinspired engineered tissues, shape-morphing materials, bioinspired optical materials, polymers with novel functionalities, composites of hierarchically ordered building blocks where the composite properties are controlled by phenomena at the interfaces between their building blocks or lithomimetic materials (Estrin et al. 2021; Beygelzimer et al. 2021) whose architectured layered metallic structure at mesoscale resemble patterns of lithosphere rocks which, in spite of the apparent difference between metals and rocks (and the length scales of the patterns they develop), equally obey the thermodynamic principle of the maximum rate of mechanical energy dissipation that governs the formation of meso structures during plastic deformation and can be described by similar rheological models.

The integration of novel bioinspired engineered materials into MEMS/NEMS devices resulted in more than just mimicking. Synthetic counterparts to the natural originals allow for additional functionalities, due to the possibilities offered by the contemporary nanotechnologies. In that context, one example is a bio-inspired pressure sensor, bio-e-skin, made of gelatin nanofiber, which is a stand-alone component that is wearable and additionally, self-powered (Ghosh et al. 2017). Similarly, available functionalities of synthetic nanomembranes sometimes surpass the ones of their biological counterparts (Jakšić and Jakšić 2020; Jakšić et al. 2022a). It is not valid just for nanomembranes, in general, the variety of constituents of artificial structures is greater, contrary to the biological ones they can operate in a wider span of temperature and humidity conditions and the range of additional functionalities is ever growing, it is feasible and common to equip MEMS/NEMS devices with additional piezoelectric, plasmonic, magnetic, electric and optical enhancements (Jakšić et al. 2022b).

Recent developments in computational materials science and AI (artificial intelligence) assisted design empower the discovery of new materials and their employment in novel MEMS/NEMS devices in applications relevant to almost every aspect of contemporary life, from health, transport, food, and feed, earth, and space communications. Then again, for the discovery, fabrication, and characterization of novel materials for MEMS/NEMS devices, nature can be a role model. Such is the case with metaheuristic optimization algorithms, perfected through the evolution of various species: swarm intelligence—the collective behavior of large groups of social animals (insects, fish, birds, mammals), tracing paths between nests of ant colonies or hunting routes of humpback whales, to name a few (Jakšić et al. 2023).

The field of materials science is broad and constantly developing through many diverse and multidisciplinary research areas, so in the current literature it is not easy to come across a compendium of different materials, their properties, fabrication and testing techniques, and applications. This chapter does not aim to be exhaustive but it aims to give a concise yet useful summary of information relevant to the use of innovative materials in current and foreseen applications in MEMS and NEMS-based devices.

The chapter is organized in the following manner. After the Introduction, Sect. "The Matter that Matters in MEMS/NEMS" gives the classification of the emerging materials of interest and shortly considers their main properties. Section "Historical Roots: Materials for Integrated Microelectronic and MEMS Devices" describes the materials base used in conventional microelectronics and MEMS, which is a starting point for our journey to novel materials and functionalities. Section "Some Emerging Materials" considers some emerging materials with an accent on nanomaterials and their classification. The next Section analyzes selected functional nanomaterials. Section "Characterization and Testing of Novel Materials" shortly considers methods for characterization and testing of the novel materials, mentioning AI and machine learning. Finally, conclusion provides a brief summary and outlook.

The Matter that Matters in MEMS/NEMS

The matter, in a material sense, is crucial for operation of any system not just MEMS or NEMS. The way we perceive matter, a material, is based on its behavior in certain circumstances, as said by Einstein "... nothing happens until something moves". Based on the material behavior in an electric field, a magnetic field, under mechanical stress or after a temperature change we can speak of its electrical, magnetic, mechanical or thermal properties of a material and classify them. For instance, based on a measured electric resistivity ρ we classify a given material as an insulator ($\rho > 10^8\ \Omega$ cm), a semiconductor ($10^{-3} < \rho < 10^8\ \Omega$ cm) or a conductor ($\rho < 10^{-3}\ \Omega$ cm). Material properties are fundamentally related to the material structure, processing, and material performance in applications. The structure, properties, processing, and performances are four interconnected pillars of materials science.

This section starts with a classification of materials, delving into their essential properties. Subsequent sections digest other aspects of novel materials for MEMS/NEMS devices, examining traditional materials like silicon, polymers, and metals alongside emerging materials such as nanomaterials, functional and biomaterials, proceeding with their fabrication, characterization, and applications.

Classifications

The categorization of engineering materials is a fundamental aspect of materials science and engineering, providing a systematic framework for understanding and organizing materials based on their properties and applications.

One of the oldest categorizations of materials followed a fundamental classification system, based on their properties, chemical composition, and atomic structure: metals, ceramics, and polymers, although obviously many materials fall into intermediate composite groups while some of them deserve special classes due to their important functionalities like semiconductors (González-Viñas and Mancini 2004).

Nowadays, besides the materials that have appeared since, there are new categories of emerging materials, important because of their integration within cyber-physical systems (smart materials), for their integration within BioMEMS devices (biomaterials), and for their exceptional architecture (nanomaterials).

With time, various criteria for material classification have gained importance, depending on the context. They may be economical (cost-effective versus high performance materials), environmental (recyclable, biodegradable materials), resource-oriented (natural, synthetic) or application-based (structural, functional, multi-functional, medical)... But, a selection based on physical properties is easiest to automate. For industrial purposes, automated material identification and material selection can be also performed by using machine vision and machine learning, based on pre-trained datasets with pictures of material surfaces (Penumuru 2020).

Regarding the novel nanomaterials, a study on classification, reference material issues, and strategies for physical-chemical characterization of nanomaterials showed that the properties of interest are building blocks size distribution, their agglomeration state/dispersion, crystal structure, chemical composition, surface area/porosity, surface chemistry, surface charge, shape and morphology, dissolution/solubility (Penumuru 2020). Additionally, it has shown that when it comes to the characterization of novel materials, the existence of the reference/test material may become an issue.

What follows is a brief digest of selected material properties, categorized into logical groups.

Physical Properties

In terms of engineering, a physical property is a characteristic of a substance that can be observed or measured without changing the substance's chemical or atomic structure.

Examples of physical properties include density, color, melting point, boiling point, electrical and thermal conductivity, etc. Physical properties are often categorized as intensive or extensive, where intensive properties are independent of the size or volume of a system, as well as of the quantity of the matter in the object, while extensive properties exhibit an additive relationship.

In the context of micro and nanomaterials, this difference becomes pivotal. On the micro and nano scale, intensive properties assume great importance as they do not vary with the size of the system or the number of particles. These properties constitute the fundamental characteristics of materials at these small scales, including melting points, electrical conductivity, and magnetic properties.

Extensive properties, on the other hand, may demonstrate an additive relationship, meaning that the behavior of the material on a macroscopic level can be predicted by summing the properties of its microscopic components. This is particularly significant in composite materials, where the combination of micro and nanoscales often dictates the ultimate macroscopic properties.

Physical properties of MEMS/NEMS materials, although being extensively studied are not present in the literature per se, but rather as a part of more complex studies and characterizations that incorporate other properties particularly relevant for a specific application, for instance alongside mechanical properties which reveal the materials response in a force field (Stanimirović and Stanimirović 2009; Namazu 2023).

Chemical Properties

A chemical property manifests during or after a chemical reaction, representing a quality that can only be observed by changing the chemical identity of a substance. Chemical property unfolds when the substance's internal structure undergoes significant transformation. Considering the interplay of all involved substances, the catalytic behavior is also a chemical property.

In the context of MEMS/NEMS devices, catalytic properties play a great role in microfluidic reactors where a catalyst, material capable of increasing the rate of chemical reactions, gains higher catalytic efficiency and catalytic power due to a larger surface to volume ratio i.e. larger contact area (Rogolino and Savio 2021). It is also correct to say the other way around, MEMS microfluidic reactors with in situ and in operando spectroscopy play a great role in the research on the innate behavior of catalytic sites. With materials transparent to electromagnetic radiation it is possible to fabricate microfluidic MEMS devices capable of performing Raman, UV-Vis, or IR spectroscopy directly at the point of the reaction (Rizkin et al. 2019).

Novel materials for MEMS/NEMS devices can also have tailored properties, including a nuanced response to changes in temperature, pressure, or the presence of specific gases, which is instrumental in the development of sensors and actuators with finely tuned functionalities. Surface chemistry plays a pivotal role in the performance of many micro and nanoscale devices. Novel materials are designed with tailored surface functionalities that enhance adhesion, modify wettability, or facilitate specific chemical interactions. These engineered surface characteristics contribute to the overall efficiency and reliability of MEMS/NEMS devices.

A hallmark of novel materials is their dynamic tunability. Tailoring surface functionalities of novel materials also enables enhanced behavior with respect to other chemical properties like corrosion resistance, hygroscopy, pH, reactivity, specific internal surface area, surface energy, and surface tension.

Mechanical Properties

Mechanical property is a characteristic of a substance that describes how it behaves when it is deformed or subjected to a force. As the essential characteristic of all mechanical properties is their ability to describe the material's capacity to resist deformation, they determine the scope and limits of a material's functionality, as well as establish the expected service life or performance, which is of great importance for many industries.

Examples of mechanical properties include elasticity, stiffness, toughness, strength, ductility, hardness, and impact resistance (Namazu 2023), etc. Among them, the most examined material feature is hardness. Although the main unit is Pa, there is a number of relevant tests and related units. Brinell Hardness Number (HB), Vickers Hardness Number (HV), Rockwell Hardness Number, and Leeb Hardness Value (HRA, HRB, HRC, HLE, etc.) are a few of the regularly used units for hardness measurement. Investigating hardness or rebound hardness by performing impressions under a steady load by indenters, e.g. sphere, pyramid is a class of many filed patents at WIPO (World Intellectual Property Organization), where G01N3 denotes the class (e.g., Li and Herrmann (2006)). Here, MEMS devices can successfully serve as measurement devices: a MEMS-nanoindenter chip for performing nanoindentation on a specimen is patented in Europe (Beyeler et al. 2018).

Mechanical properties of materials are not constant, they continuously change when exposed to various conditions, such as heat or loading rate. Additionally, being related to the extensive physical properties, mechanical properties can change with scaling. Micro and nanomaterials can exhibit significantly enhanced mechanical characteristics compared to their macroscopic counterparts. Strengthening materials at these small scales often results in increased strength, wear resistance, and durability. The development of novel materials exerts the development of appropriate methods for testing and characterization (Stanimirović and Stanimirović 2009).

Acoustical Properties

Since sound is a mechanical wave that propagates through a medium (air, liquids, solids), acoustic energy can be harnessed (Yuan et al. 2019), sensed (Kabir et al.

Table 8.1 Relevant properties of piezoelectric materials: PVDF—Polyvinylidene difluoride, $BaTiO_3$—Barium titanate, PZT—Lead zirconate titanate

Property	Units	PVDF	$BaTiO_3$	PZT
Relative dielectric constant	–	12	600–1200	1000–4000
Piezo charge constant δ_{31}	pC/N	20	-60 to -30	-600 to -100
Electromechanical coupling factor	%	11	21	30–75
Young's modulus	10^{10} N/m^2	0.3	11–12	6–9
Density	kg/m^3	1780	5300–5700	7500–7700

2018; Ozevin 2020; Voiculescu and Nurashikin Nordin 2012) or generated (Wang et al. 2021a) by MEMS devices.

Transduction mechanisms responsible for interaction of MEMS devices with the acoustic waves are capacitive, electrostatic, resonance-based, piezoresistive, or piezoelectric (Ozevin 2020; Kumar et al. 2024). The mechanisms of acoustic energy harvesting (AEH) include the Helmholtz resonator approach, the quarter-wavelength resonator approach, and the acoustic metamaterial approach (Yuan et al. 2019). Table 8.1 shows relevant properties of piezoelectric materials used in acoustic MEMS/NEMS devices (Yuan et al. 2019) (remark: δ_{31} means that the strain is in x_3 direction, when electrode is on $x_2 - x_3$ face, similarly δ_{32} would mean that the strain is in x_3 direction when electrode is on $x_1 - x_3$ face. Piezoelectric charge constant δ is an anisotropic physical quantity).

Beside acoustic energy sensing, generating, and harvesting, MEMS devices are capable of acoustic insulation in a way not possible otherwise. Acoustic barriers are hard to achieve without blocking the medium too, sound finds its way through the air. An artificial sparse metasurface structure can be used for both sound isolation and air ventilation (Cheng et al. 2015). With the advancements in design and fabrication of multifunctional structures based on metamaterials, tailoring acoustic properties also becomes possible (Yuan et al. 2021).

Electrical Properties

Understanding electrical properties is fundamental for the development and optimization of MEMS/NEMS devices within a diverse range of applications. It is also fundamental for the understanding of other related properties (for instance the optical ones, due to the linear relation between the dielectric permittivity and the refractive index).

Scaling down to the micro and nanoscale in MEMS/NEMS devices introduces unique challenges and opportunities, particularly concerning electrical properties. Capacitance is influenced by the size and separation of electrodes. Reducing dimensions can lead to higher capacitance and improved energy storage capabilities (Patil and Lee 2019). Ultimately, the downscaling of electrical systems gives rise to quantum effects, due to spatial confinement imposed on electrons at the nanoscale and the appearance of collective effects like e.g. entanglement.

Electrical resistivity and conductivity are fundamental for the interconnects and the overall performance of electronic components. Capacitance, piezoelectricity, and dielectric properties play vital roles in the functionality of sensors and actuators, while pyroelectric and thermoelectric properties are crucial for converting mechanical and thermal energy into electrical energy. However, transduction can be highly complex.

Contemporary research explores harvesting energy of thermoelastic bending due to the interplay between thermal expansion and flexoelectricity, the coupling between electric polarization and strain gradient—an effect where strong strain gradients can appear in nano-sized and nano-structured systems and lead to electric polarization, as in analogy to piezoelectricity but without restrictions regarding the material symmetry (Yurkov et al. 2019).

The intense focus of the recent research is also placed on ferroelectric materials, which can be defined as dielectric materials in which permanent polarization exists even after removing the applied electric field (Cao et al. 2023) because they exhibit flexoelectric coefficients three to five orders higher than typical dielectrics (Patel 2023). Such high permittivity ferroelectrics are BTO ($BaTiO_3$), $(BaSr)TiO_3$, $Ba(TiSn)O_3$, $Pb(ZrTi)O_3$, $(KNa)NbO_3$ (Patel 2023), perovskites, hafnium and wurtzite (aluminum scandium nitride) in Mikolajick et al. (2021) named next generation ferroelectric materials for semiconductor process integration. Along with providing theoretical models developed in the field of flexoelectricity, a review on flexoelectric energy harvesting (Tripathy et al. 2021) points out the vast potential applications in sensors, actuators, transducers also.

Recent research on the use and engineering of thermoelectric materials in MEMS/NEMS devices has resulted in solutions for thermoelectric generators for harvesting energy at near room temperature (Bhatnagar and Vashaee 2022). Thermoelectric materials, as substances that can directly convert heat energy into electrical energy and vice versa through the Seebeck effect, due to temperature gradient across a material, are utilized in thermoelectric devices for applications such as power generation and temperature sensing (Bhatnagar and Vashaee 2022).

Magnetic Properties

Integration of magnetic materials into MEMS and NEMS devices provides responsiveness to both electric and magnetic fields and, due to magnetoelectric coupling, enables plethora of additional functionalities, at the same time providing downsizing of elements and rise in the scale of integration (for instance in antenna applications) (Will-Cole et al. 2022). Magnetoelectric coupling can be manifested in magnetoelectric MEMS/NEMS devices through direct or converse magnetoelectric effect (direct magnetoelectric effect being related to the control of electric polarization by changing the magnetic field, converse magnetoelectric effect being related to the control of magnetization oscillation by changing the electrical field).

Besides the permeability and susceptibility, one of the main properties of single phase magnetoelectric materials, both multiferroics and composite magnetoelectric materials, is Curie temperature, the critical temperature above which intrinsic magnetic behavior is lost and can only be induced. Other features important for the synthesis and characterization of materials for magnetic MEMS devices are the hysteresis loop, magnetic saturation (measured in emu), coercivity (Oe), and squareness (calculated as the ratio of the Retentivity and Magnetic saturation) (Poongodi and Thangaraj 2023). In case of composite magnetoelectric materials where magnetic nanoparticles are embedded into a matrix, the exact phase composition and presence of magnetic/paramagnetic ions is also a detrimental factor for their magnetic behavior (Stiharu et al. 2023).

The change in the magnetization intensity with temperature is known and used for determining Curie points of materials (Chynoweth 1958), but there are materials, like gadolinium, that additionally exhibit a magnetocaloric effect (they heat up inside the magnetic field and lose thermal energy to the environment, after exiting the magnetic field they become cooler than while being inside the field). The research on magnetic properties is scientifically attractive on a conceptual level (Petelin et al. 2023), and for specific materials like $DyAl_2$ single crystals (de Ribeiro et al. 2023) or antiperovskites (Mehmood et al. 2023), however magnetic cooling is still an unconquered field at least for our everyday needs. On the other hand, many other magnetic field-dependent properties of materials for MEMS applications have proven versatile and beneficial due to effects like piezomagnetism (Kumar and Kaur 2023), magnetoresistance (de Ribeiro et al. 2023), or magnetostriction (Minegishi 1963).

Optical Properties

For any material property, there is a way to put it into use in some application. In that context, the research in the optical domain that followed the scaling from macro to micro to nano resulted in the most astonishing effects. Many of new phenomena observed about a decade ago, are still under thorough theoretical and experimental investigation (an example is the extraordinary optical transmission, i.e. passage of light through optically opaque films perforated by subwavelength apertures (Liu and Lalanne 2008)).

Considering the possibilities for the light-matter interaction and the possibilities for micro and nanofabrication of metastructures, it is clear that there are many ways to achieve tunability of optical properties through magneto-optical, thermo-optical, electro-optical, ionic-optical, mechano-optical, piezo-optical, photo-elastic or phase change effects, as given in the overview (Ling and Yoo 2023). With a proper design, properties like transmittance, absorbance, reflectance, hence the refractive index (be it independent of light polarization, or showing birefringence or trirefringence) (Meng et al. 2022), also beam shape, polarization, phase can all be tailored. Tools for that are still being scientifically fine-tuned themselves. In the theoretical domain, known models are still being used and improved (like the use of Mie theory to derive absorption, extinction efficiencies, the scattering phase function, and the asymmetry factor for a particle embedded in an absorbing medium (Fu and Sun 2001)). In the domain of computational electromagnetics new software solutions are being developed for the use of Transfer Matrix method, Finite Difference Frequency Domain method (FDFD), Finite Difference Time Domain method (FDTD), Plane Wave Expansion Method (PWEM), Rigorous Coupled-Wave Analysis (RCWA), Method of Moments, etc. (Sumithra and Thiripurasundari 2017; Rumpf 2022).

The interplay of theoretical, numerical, and experimental approach is important in many ways, some of them being optimized production and responsible research. More than that, based on transmission coefficients calculated from simulations, for a nanoengineered and nanostructured materials, it is possible to determine effective far field parameters like permittivity (ε) and permeability (μ) as if it were a homogeneous material (Smith et al. 2002). It can be done analytically, for some ordered nanoengineered structures (Tanasković et al. 2016), or numerically for complex nanocomposites (Kari et al. 2007a, b, 2008).

Historical Roots: Materials for Integrated Microelectronic and MEMS Devices

Materials used for integrated microelectronic devices (integrated circuits, IC) before the era of material properties engineering and before prioritizing responsible research with regard to spending too much on hazardous chemicals and critically deficient raw materials, have not lost their importance.

Materials used for passive or active substrates (aimed to mechanically support active elements like sensors, actuators, waveguides and transducers or general microelectronic devices,) like the ones with high crystallographic purity and mechanical robustness (Si, Si compounds, Ge, GaAs), or transparent ones like glass or quartz or different types of glass, need to be dimensionally stable and insensitive to environmental conditions. Besides mechanical stability, other desirable properties of these materials include high Young modulus, high melting point, low thermal expansion coefficient, no mechanical hysteresis, and easy formation of flat and adherent coatings and thin films. The equipment and the top-down fabrication processes—planar technologies (methodologies of doping, oxidizing, etching, metallizing, bonding, etc.) are well developed, optimized, and mature. The compatibility with this developed technology is one of the necessary prerequisites for the integration with building blocks made of novel, engineered, and emerging materials.

The approach to material selection for the mechanical design developed by Ashby for macro systems and products (who also implemented it in a software CES EduPack (Ashby 2011)) has been also successfully used for the selection of materials in MEMS and NEMS devices, however an important *caveat* should be placed at the change of properties with the scaling (Mehmood et al. 2018).

Among often utilized silicon compounds, silicon dioxide is used as a thermal and electric insulator and as a sacrificial layer in surface micromachining. Silicon carbide is used as a passivation (protective) layer because it has dimensional and chemical stability at high temperatures, it can resist common silicon etchants such as potassium hydroxide (KOH), ethylenediamine pyrocatechol (EDP), tetramethylammonium hydroxide (TMAH), hydrofluoric acid (HF) and nitric acid (HNO_3). Silicon nitride is used as encapsulant, to prevent diffusion of unwanted fluids into the substrate and as a high-strength electric insulator. Polycrystalline silicon (polysilicon) has microcrystals in random sizes and orientations, thus it can be treated as isotropic. In ICs it can serve for Ohmic contacts, resistors, field effect transistor gates, thin film transistors... Quartz, being abundant, thermally stable, and transparent to UV light is used for sensors, piezoelectric devices, resonators.

Piezoelectric materials, whose ability is to change their shape due to applied voltage or to generate voltage due to mechanical stress between their faces are used for a plethora of resonators: those based on surface acoustic waves (SAWs), bulk acoustic waves (BAWs), Lamb waves (LWs). They can exhibit various vibration modes: flexural, bulk, shear, torsional, etc. (Will-Cole et al. 2022). These materials are employed for fast devices with high precision in applications such as catalysis, energy harvesting, sensors, biomedical engineering, and additive manufacturing (Sekhar et al. 2023).

The list of the conventional materials would not be complete if we skipped polymers (elastomers, thermoplastics, and thermosetting polymers) (Pergal et al. 2023)—materials offering features like easy and cost-effective manufacturing, high flexibility, and customizable characteristics (biocompatibility, conductivity) are polymers (elastomers, thermoplastics, and thermosetting polymers) (Pergal et al. 2023). Their versatility allows for a very wide range of applications, from photolithographic masks, molds, and substrates to devices and systems like lab-on-a-chip systems, point-of-care diagnostic devices or wearable electronic devices. They are a crucial constitutive part of many irreplaceable emerging materials in contemporary MEMS and NEMS systems.

Some Emerging Materials

The variety of materials for contemporary MEMS and NEMS building blocks now vastly surpasses those that have been used at the earliest stages of the MEMS technologies development. As already mentioned above, at the very beginnings, the modest available material set was more or less identical to what has already been developed for planar technologies and used for the production of integrated circuits and generally microelectronic devices—semiconductors (silicon), silicon dioxide (silica) metals (usually gold) and polymers (notably photoresists).

A plethora of novel materials have emerged since as a consequence of the advent of nanotechnologies. However, devices downscaling alone, i.e. their size reduction, is far from being the sole reason for the employment of emerging materials. Besides offering an increased productivity and cost competitiveness stemming from their batch fabrication, there is a long list of added values and multifunctionalities that justify not only the employment of emerging materials but also the huge investment in their further research. Nanomaterials and the structures that include them like e.g. nanocomposites, metamaterials, plasmonic materials, biomaterials, biomimetic materials, etc. all exhibit their own specific advantages and disadvantages, both those known and those yet undiscovered.

Nanomaterials Definitions

Nanomaterials can be somewhat arbitrarily defined as solid particles of various shapes that have at least one of their three dimensions not larger than 100 nm and not smaller than 1 nm (Sannino 2021). Both top and bottom limits are a matter of agreement among the users since no material parameters change exactly at 100 or 1 nm. Actually, while the top limit remains a matter of consensus, some documents define the bottom limit of the constituent dimension as a single atom, i.e. 0.2 nm (Nanoscience and nanotechnologies: opportunities and uncertainties | Royal Society 2004) (Chap. 2, paragraph 6, p. 5). Such definitions are sufficiently wide that they can even include homogeneous chemical compounds or chemical elements if they are adequately processed (e.g. pulverized) that their dimensions satisfy the said geometrical requirement—for instance, large amounts of nanoparticles of pure compounds or element allotropes belong to nanomaterials if they satisfy the dimension requirement. A number of various definitions have been collected in a JRC report from 2010 (Loevestam, et al. 2010). Finally and probably the most importantly, currently the most recent definition of nanomaterials as recommended by the EU legislature in Commission Recommendation of 10 June 2022 (Text with EEA relevance) 2022/C 229/01, published in the Official Journal of the European Union is

> 'Nanomaterial' means a natural, incidental or manufactured material consisting of solid particles that are present, either on their own or as identifiable constituent particles in aggregates or agglomerates, and where 50% or more of these particles in the number-based size distribution fulfil at least one of the following conditions: (a) one or more external dimensions of the particle are in the size range 1 nm to 100 nm; (b) the particle has an elongated shape, such as a rod, fibre or tube, where two external dimensions are smaller than 1 nm and the other dimension is larger than 100 nm; (c) the particle has a plate-like shape, where one external dimension is smaller than 1 nm and the other dimensions are larger than 100 nm. In the determination of the particle number-based size distribution, particles with at least two orthogonal external dimensions larger than 100 μm need not be considered. However, a material with a specific surface area by volume of $<6\ m^2/cm^3$ shall not be considered a nanomaterial.

It is implied that the properties of nanomaterials are defined by their nanoscale. In other words, their physical, chemical, mechanical or biological behavior is inherently determined by their minuscule size. Thus nanotechnological approach to these materials directly tailors their targeted properties, often enabling unusual and sometimes even counter-intuitive elementary traits.

Classification of Nanomaterials Geometries

This subsection classifies some of the possible building blocks of nanomaterials according solely to their geometry. Their required dimensions being defined in the previous subsection, they can be grouped by the number of Cartesian axes that do not belong to the nanometer domain as the following three groups.

0D (or zero-dimensional): their all three dimensions are in the range below 100 nm, i.e. from the macroscopic point of view none of their dimensions is above the nanoscale. Examples include various nanoparticles regardless of their shape, material or production method. The structures that belong to this set are, e.g., quantum dots and quantum wells, various metal, dielectric or semiconductor nanoparticles, including core-shell nanoparticles (single and multiple shells), atomic clusters, fullerenes, nanodiamonds, biomimetic nanoparticles, organic molecules, biological nanoparticles, etc. The shapes of nanoparticles may strongly vary. The exact terminology and classification of different nanoparticles still remain unclear and undefined. The uncountable, unregulated and basically infinite number of different nanoparticles forms has sometimes been jovially denoted as the "nanoparticle zoo." (Perkin and Gubala 2017). Figure 8.1 shows some examples of 0D nanomaterials among the many existing.

1D: Only a single dimension out of three is below 100 nm. This encompasses nanofibers, nanotubes (including the well-known carbon nanotubes), nanorods, nanowires, nanopillars, and generally any elongated fiber-like structures. However, one can encounter contradictory definitions of 1D structures in literature, since some authors denote ultrathin plate-like structures as one-dimensional, in spite of their two dimensions not being in nano-domain. Examples include 1D photonic crystals (Lee and Yao 2002) which actually represent sheet-like structures with only their strata thickness in the mesoscopic domain and 1D metamaterials (Goncharenko et al. 2014) which are actually sheets with nanometer-scale thickness. Figure 8.2 shows some examples of 1D nanomaterials.

2D: Here two out of three dimensions are below 100 nm. Such nanomaterials include various nanosheets, nanoflakes, nanoplates, nanopores (Byakodi et al. 2022), freestanding nanomembranes (Jakšić and Jakšić 2020), metasurfaces, etc. Concrete materials include chemical element allotropes like the well-known graphene, various other freestanding monatomic sheets like silicene, germanene, 2D bismuth, borophene, phosphorene (black phosphorus), blue phosphorus, stanene, plumbene, inorganic compound sheets among which maybe the most well-known are MXenes (transition metal carbides, nitrides or carbonitrides), molybdenum disulfide, tungsten diselenide, etc. Other inorganic 2D freestanding structures include metallic ones, metal-nanocomposite and alloy-based, various diamond, diamond-like, and diamondoid nanomembranes, etc. Organic nanomembranes, flakes, and sheets may be single-compound (pure), polymer-composite (copolymer)-based. These materials also include artificial replicas of the living cellular membranes, synthetic lipid bilayer nanomembranes, also known as black lipid membranes, and painted bilayers. Figure 8.3 shows some examples of 2D nanomaterials, including the lipid bilayers.

Fig. 8.1 Some examples of 0D nanomaterials

3D: All three dimensions of an object may or may not exceed 100 nm, but the dimensions of sub-building blocks of such complex hierarchical structures are in the nanoscale. Some examples of 3D nanomaterials include nanostars, nanoprisms, nanoflowers, nanotrees, nanodendrites, nanoleaves, metal oxide frameworks (MoF), calixarenes, generally nanocages, nanocubes, dendrimers, liposomes, micelles, ferritin, etc. Figure 8.4 shows some examples of 3D nanomaterials, including liposomes and micelles.

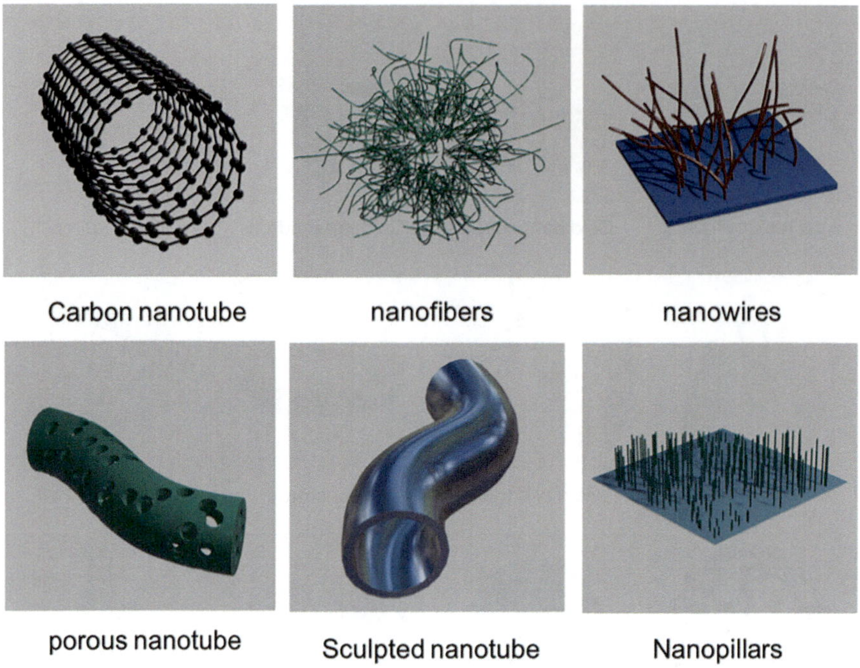

Fig. 8.2 Some examples of 1D nanomaterials

Nanocomposites

Nanocomposites could be defined as one or more of 0D, 1D, 2D or 3D nanomaterials distributed as nanofillers in a matrix (host) of a homogeneous material. Thus a nanocomposite itself will be a macroscopic structure with nanomaterial constituents as its building blocks. Typically such compositing will result in vastly enhanced performance and will often impart additional functionalities to the nanocomposite that were not present in its separate constituents.

Usually, three main groups of matrix materials are cited, namely metals, ceramics, and polymers (Rajamehala et al. 2022). However, besides these most often mentioned hosts, there is a number of alternative matrix materials. They include various types of glass or glass-ceramics, metal-organic frameworks with their porous crystalline structure, different kinds of biomimetic matrices, for instance biomineralized ones, liquid crystal matrices, etc.

The nanoscale dimensions of the fillers mean a higher surface to volume aspect ratio. This typically results in higher chemical reactivity, changes of electrical transport properties due to nano-confinement of charge carriers, the appearance of quantum effects, and modification of overall physical traits (for instance, enhanced mechanical properties.) Nanocomposites can be made with increased hardness and abrasion resistance. The possibility to make arbitrary combinations of chosen

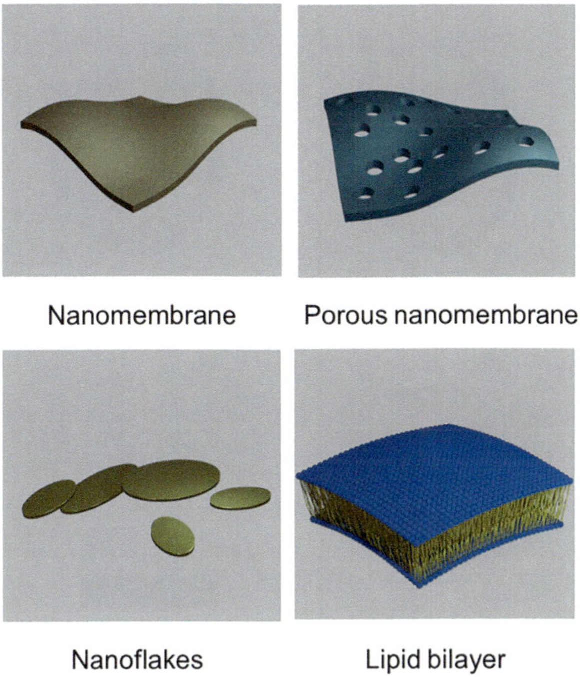

Fig. 8.3 Some examples of 2D nanomaterials

nanomaterials with chosen matrices enables the manufacturer to choose from an extremely wide toolbox and many parameters can be tailored to desired values, often simultaneously.

The application fields of nanocomposites are stupendously extensive and cover a vast number of areas. A typical example represents mechanical enhancements exceeding those of the original constituents by orders of magnitude and the usability in industrial machines, but also in transportation, be it automotive industry (lightweight and sturdy building blocks with lower abrasion and with lower engine power demands), shipbuilding (marine applications) or aircraft design. Diverse applications include environment-friendly energy production (e.g. the fabrication of proton exchange membrane-based fuel cells or biofuel generation, hydrogen production by solar-based photocatalysis, etc.), environmental protection through pollutant removal by adsorption-based filtering. In agriculture, nanocomposites can be used for environment-friendly production of essential fertilizers, for the enhancement of plant germination, for producing targeted herbicides and pesticides, etc.

There are numerous applications of nanocomposites in healthcare and biomedicine. They can be used as tissue regeneration scaffolds in tissue engineering, itself a part of regenerative medicine, bio-implants, wound healing, contrasting agents for magnetic resonance and CAT scanners, MEMS and NEMS drug delivery systems, etc.

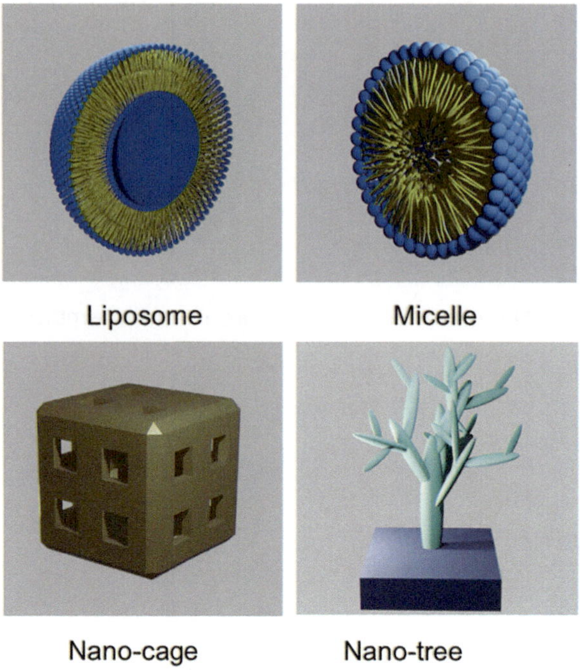

Fig. 8.4 Some examples of 3D nanomaterials

In optics and photonics, the ordered arrays of mesoscopic or subwavelength fillers ensure enhanced performance of photonic crystals, metamaterials/metasurfaces, etc. This ensures an extended range of applications of such structures.

Some other fields include pharmaceutical drugs production, food industry additives, food packaging, waste management, and environmental protection. Nanocomposites also serve as building blocks for electronic and optoelectronic devices (due to their tailorable conductivity, thermal management abilities, and adjustable optical traits), different kinds of MEMS/NEMS sensors (the fillers can be chosen to react with a specific measurand or analyte) and many more.

Architectured Nanomaterials

The development of newly synthesized engineering materials is diverse, it is growing in many different ways and it is based on the employment of many different approaches. Developing new materials by designing their inner architecture rather than manipulating atomic or phase composition is an emerging trend in materials engineering referred to as architecturing materials where the term is not restricted solely to MEMS or NEMS devices but can also refer to the design of architectured materials at mesoscale. The intrinsic mechanisms of architectured materials and their adjustable properties ensure their prominence in the future within a variety of applications.

In search for materials with novel properties, the fact that the material properties are rooted in its structure inspired the process of architecturing materials, where the micro and nanoscale elements are interconnected as the basic components of some volume element with representative properties at the mesoscale. The representative properties of such volume elements are qualitatively superior to properties of its building blocks. For instance, segmented ceramic structures exhibit improved bendability and increased fracture resistance, metallic building blocks enhance toughness and strength, polymers contribute to the ease of assembly and disassembly... Research shows, one way for assembly is topological interlocking, appropriate for building blocks of mutually suitable nonplanar shapes (Estrin et al. 2021). In such a topological assembly, the nanoscale building blocks complement each other geometrically and fit into larger ordered formations, i.e. they form nanocomposites without the matrix material.

The same research (Estrin et al. 2021) also presented the benefits of machine learning and methods based on artificial intelligence for the design of architectured materials, especially the ones presentable by stick and balls model, lattice materials. While AI can find the relations not easy to spot by the naked eye or grasp by one's mind, the design often relies on the opposite challenge, that of reproducing a phenomenon observed in nature or in a living organism.

Natural beings (chitin), structures (lithosphere), and phenomena (search for food and habitat procedures), evolved through eons, and have properties that are easily observed. Biomimicking brings benefits to momentous applications based on self-healing, adaptivity to applications: regenerative medicine, soft robotics, microrobots, flexible electronics, neuromorphic electronic materials... Heterogeneous zones in architectured nanomaterials with mechanical and physical properties radically different from each other between the zones, whose properties vary by more than 100% (heterogeneous lamellae, laminate structured, gradient structured, harmonic (core-shell), dual/multi-phase structured, and multi-modal materials) exhibit significant synergy that goes beyond the predictions of the rule of mixtures. The idea behind this architecturing process is directed toward harnessing the potential to transfer size-dependent properties of individual clusters to bulk materials, enriching the resulting materials properties.

Architecturing materials can go beyond the sophisticated mixing of materials and their properties. An illustrative example for that is the design of metamaterials which, due to the effects of light propagation through a structure with subwavelength details, provokes extraordinary and previously unexpected effects in the interplay between light and matter.

Metamaterials and Metasurfaces

Metamaterials can be generally defined as artificial structures with properties not usually found in natural materials. They often exhibit peculiar and even anti-intuitive properties that far surpass those of the natural structures or strongly deviate from them. The firstly discovered metamaterials were structures with negative refractive index. They were theoretically proposed in 1967 as a scientific curiosity (Veselago 1968); a way to fabricate them was described much later, in 1999 by sir John Pendry and the first experimental structures were produced as late as in 2001 (Smith et al. 2000).

As an example, Fig. 8.5 shows two planar “meta-atoms”, the structures defining the unit cell of the artificial “crystal”. One type of these are split ring resonators (with circular or square structure) while the other are fishnet metamaterials (arrays of subwavelength holes in a metal-dielectric-metal sandwich). Many other metamaterial geometries exist besides these given as illustrations.

Many unusual properties and applications have been described since, including perfect lenses, superlenses, and hyperlenses, all functioning at deeply subwavelength level, invisibility cloaks, structures with extremely high, low (near-zero) and negative values of refractive index and many more (Cai and Shalaev 2010).

Metamaterials are not limited to optical waves, not even to all electromagnetic waves. They have been used for acoustic waves, mechanical waves, purely electrical and purely magnetic ones.

Of interest for MEMS and NEMS are ultrathin (quasi-2D) optical metamaterials—the metasurfaces (Bukhari et al. 2019; Jakšić et al. 2010). They ensure practically unlimited wavefront control and manipulation with optical space, including beam

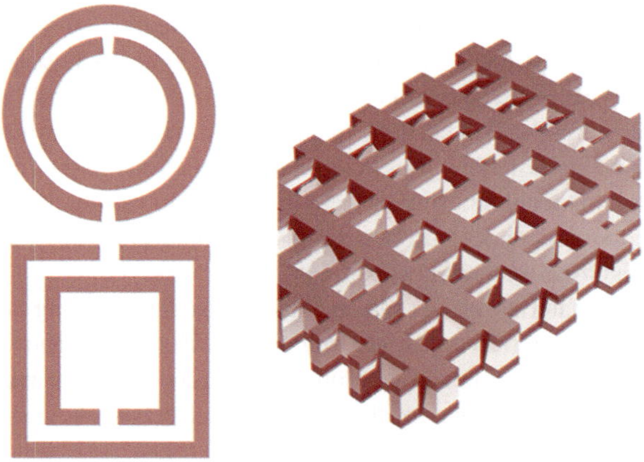

Fig. 8.5 Two examples of metamaterial structures. Left: split ring resonators (SRR) made of conductor (usually noble metal) as meta-atoms defining metamaterial unit cell. Top left: circular structure SRR. Bottom left: square structure SRR. Right: fishnet metamaterial with square array of rectangular subwavelength holes in a metal-dielectric-metal sandwich with nanoscale thickness

shaping and directing, near-perfect absorption (superabsorbers), full spectral dispersion tuning and control over values of optical parameters, and many more. All of these are achieved by using extremely compact structures with subwavelength (nanoscale) thickness. Optical metasurfaces represent one of the most important paradigms in modern applied optics and this particular branch of optics has been called metaoptics (Qiu et al. 2021). The number of practical applications is enormous and growing exponentially. Novel types of metasurfaces with drastically higher level of control and novel applications are being made with the help of machine learning and generally artificial intelligence (e.g., Jakšić et al. (2023), Nugroho et al. (2022)).

Nanostructured Biomaterials

To avoid misinterpretations, at the very beginning of this subsection, we make a distinction between a biological material that is a product of a living organism (including e.g. plant parts and their biomass, animal tissue including bones, blood, muscle, grease, etc.) and a biomaterial. The later is a wider term and may represent an artificial, biological or hybrid artificial-biological structure used with a goal of incorporating it in biological application and interacting with a living tissue.

Biomaterials may be fabricated from an organic or an inorganic material or a combination of both. They can be tailored for medical purposes which may be diagnostic, therapeutic or both simultaneously (theranostics). They may also include biomimetic nanomaterials, i.e. (nano)materials mimicking biological multifunctionalities. They may be utilized as building blocks of MEMS or NEMS for biological applications (BioMEMS and BioNEMS).

Nanostructured materials can be utilized to manufacture biomaterials with extremely high performance and with multifunctionalities that surpass the natural ones. Their chemical, mechanical, and general physical properties as well as their biological behavior can be arbitrarily tailored to suit particular needs. This makes them irreplaceable for a wide range of applications, including regenerative medicine (Khalilov 2023), tissue scaffolds (Wei and Ma 2008), antimicrobial function (Jampilek and Kralova 2022), implants like dental ones (Rasouli et al. 2018) or those for cardiologic purposes (Pryjmaková et al. 2020) (from stents to artificial hearts), drug delivery agents (Goldberg et al. 2007; Cai et al. 2020) with tailorable speed of release, etc. Some nanomaterials are used to destroy cancer, for instance metal (most often gold) nanoparticles with core-shell structure that are introduced into the bloodstream, are collected within the cancerous growth because of its accelerated metabolic processes, and are then irradiated by infrared laser which heats them up much more than the normal tissue and thus directly destroys malignant neoplasm (Pradhan et al. 2019). This process is known as photothermal therapy.

Nanostructured biomaterials are especially important because they allow for the introduction of multifunctionalities that do not exist in natural tissues. For instance, they allow for controlled incorporation of radionuclides into living tissue, thus facilitating theranostics (Farzin et al. 2019).

Nanostructured biomaterials for neural interfaces including neural prosthetics, neural recording, growth, and general neural stimulation are considered in Fattahi et al. (2014). Many nanomaterials are convenient for this purpose, including carbon allotropes (CNT, graphene), semiconductor and metal nanowires, electrically conductive polymers, metal-organic frameworks, etc. The main challenges in this field are related to the biocompatibility of these materials, their durability over time, and mechanical properties. The final goal is to develop organic or inorganic nanomaterials that will mimic the behavior of neural tissue and serve as interfaces between electrodes and neurons.

Quantum Materials

Quantum materials (Basov et al. 2017) are a broad term describing all different materials in which quantum effects are manifest. All of them stem from the spatial confinement of electrons in their nanoscale level building blocks i.e. the reduced dimensionality. Especially marked in 2D structures like graphene and generally 2D van der Waals crystals, they exhibit a host of novel effects or fundamentally modified effects that appear due to that confinement. Among those effects is the fact that often electrons cannot be regarded as independent. Actually, they strongly interact and form collective quasiparticles.

The quantum mechanical mechanisms like entanglement of electrons give rise to a plethora of fundamentally novel effects, and the situation is similar with quantum topological properties in systems with marked spin–orbit coupling. Paradoxically, the quantum material itself may be macroscopic, as long as its properties are caused by quantum effects.

Some of the quantum materials include topological insulators, superconductors, Weyl semimetals, graphene, and spin ices. Thus very diverse materials with a wide range of properties belong to this class, and one could consider the term "quantum materials" an umbrella term covering quite disparate systems. All of such systems have a behavior defined by quantum effects (Keimer and Moore 2017).

One must stress that the field of quantum materials is a lively developing one, with new discoveries appearing every day. The quoted materials and effects enrich the toolbox for MEMS and NEMS and extend their functionalities far beyond the conventional ones.

Functional Nanomaterials

Functional nanomaterials are artificial, natural (including biological) nanostructures that exhibit diverse targeted physical, chemical or biological functionalities. These functionalities are desirable from a certain applicative point of view. Two or more such functionalities can be combined within the same nanostructure, or it can offer a single desirable trait. Their properties are tailorable through variations of their geometry, dimensions, and structural materials.

A possible list of engineered functionalities of nanomaterials, by no means exhaustive, can be seen in Table 8.2. Short descriptions of pertinent mechanisms and applications are also presented.

The following subsections in this section present in some more details several nanomaterials that ensure targeted functionalities. They are not given in a particular order and are presented as an illustration only.

Table 8.2 Possible list of tailorable functionalities of nanomaterial

Type of functionality	Short description
Mechanical	An increase of mechanical strength, abrasion resistance, stiffness, hardness and toughness
Electrical	Conductivity increase up to superconductivity level (e.g. using CNT or graphene) Particle size-dependent electron confinement in quantum dots (this also reflects to optical properties)
Magnetic	2D monolayer magnetic nanomaterials ensure strongly enhanced magnetic behavior; similar is valid for magnetic nanoparticles and 1D magnetic structures. One of the examples is superparamagnetic behavior
Optical	Possibility to reach extreme light localizations of several orders of magnitude through the use of plasmonic nanoparticles and metamaterials/metasurfaces and to drastically modify spectral dispersions
Thermal conductivity	Some nanomaterials can be tailored to have very high thermal conductivity, which is useful for thermal management in MEMS and NEMS
Photothermal	Conversion of light into heat by means of plasmonic nanoparticles, used for killing off malignant neoplasms
Catalytic	Catalytic and photocatalytic nanomaterials ensure novel functionalities in microreactors for in situ hydrogen production, neutralization of hazardous pollutants, and fabrication of a new generation of enhanced fuel cells
Nanoporosity	Nanomaterials ensure production of microporous, mesoporous and macroporous structures with highly enlarged effective surfaces, which augments filtration/separation, adsorption, and catalysis
Biocompatibility	Possibility to use nanomaterial enhanced structures for scaffolds in tissue engineering (regenerative medicine), for drug delivery with controlled targeting and release, for in situ biomedical imaging, and theranostics

(continued)

Table 8.2 (continued)

Type of functionality	Short description
Antibacterial properties	Nanomaterials and nanocomposites can be used in wound sterilization and accelerated healing, but more generally for nanoprocessing of textiles making them applicable in biomedicine
Stimuli-responsive nanomaterials	All nanomaterials that can respond to some kind of external stimulus, e.g. light intensity, environmental pH factor change, electric or magnetic field, temperature change ensure smart behavior of devices based on them
Selectivity enhancement of sensors	Nanomaterial receptors can be optimized for given ligands, thus improving selectivity of adsorption-based chemical, biochemical or biological sensors
Energy conversion	Nanomaterials can be used in energy harvester MEMS and NEMS (e.g. solar cells, piezo-based nanogenerators, etc.
Self-healing	Some nanocomposites can perform reverse self repair of their mechanical damages through molecular motion and rearrangement in nanoscale structures

Ferroelectric Nanomaterials

Some materials exhibit a spontaneous electric polarization even when the external electric field is zero, i.e. in the natural state of their crystals their centers of the positive and negative charges do not coincide. Such behavior reminds of ferromagnetic materials which show a permanent magnetic moment, regardless of an external magnetic field. A ferroelectric is such a material whose spontaneous electric polarization can be reversed by applying a strong external electric field in the opposite direction. Thus ferroelectricity is in no way connected with ferromagnetic materials or indeed with magnetism, the only reason for "ferro" in its name being the mentioned electric analogy with spontaneous magnetic polarization. The ferroelectric effect and bulk materials that exhibit it have been known since 1920 (Anonymous 1920). Well known examples of bulk ferroelectrics include lead zirconium titanate (PZT), barium titanate ($BaTiO_3$), aluminum nitride (AlN), and lead titanate $PbTiO_3$ (Fig. 8.6).

Nanoscale ferroelectrics are of large practical interest for MEMS and NEMS devices but require the effect to appear in nanostructures. However, it turned out that a majority of bulk ferroelectrics lose that feature with a dimension decrease entering the nanoscale.

An emerging group of materials retains its ferroelectricity even in the form of 2D nanostructures. Notable examples include hafnium oxide (HfO_2) and 2D ferroelectrics (Qi et al. 2021) like graphene and molybdenum disulfide (MoS_2). Other examples are tin sulfide (SnS) semiconductor, CIPS ($CuInP_2S_6$), and indium selenide (α-In_2Se_3) (Dragoman et al. 2021). In most of the 2D ferroelectrics their strata are connected by van der Waals forces, which means that they can be exfoliated down to 2D monolayers. A use of ultrathin PZT ferroelectric has been described in Polcawich and Pulskamp (2017). Additionally, the known bulk perovskite or perovskite-like

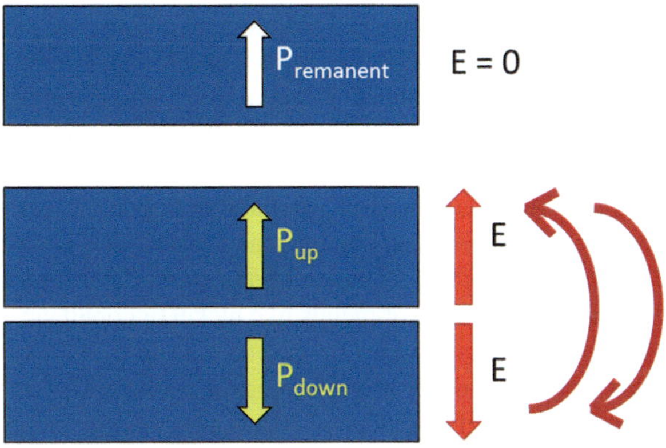

Fig. 8.6 Behavior of a ferroelectric material. Top: no external electric field; remanent polarization is exhibited. Middle: "Up" reversible polarization is switched due to external field. Bottom: "Down" reversible polarization is switched due to external field

ferroelectrics like barium titanate ($BaTiO_3$), strontium titanate ($SrTiO_3$), $SrRuO_3$, $PbTiO_3$ and $BiFeO_3$ have been proven to retain their ferroelectric properties even to thickness levels of a few unit cells.

Besides in ultrathin sheet 2D form, ferroelectric nanomaterials have been synthesized as nanoparticles, as well as 1D structures (nanowires and nanotubes) (Varghese et al. 2013). Ferroelectric nanocomposites have also been produced.

Besides their use in MEMS/NEMS as their sensing or actuating parts, ferroelectric nanomaterials find increasing use as advanced nanoelectronic devices like memristors, switches, negative capacitance transistors, tunnel junctions, etc., but also as different sensing devices. In addition, nanometer-thick tetragonal phase hafnium oxide produced by atomic layer deposition and doped with zirconium oxide has exhibited extremely large near-room temperature dielectric permittivity of 70 (Böscke et al. 2009). It is worth noting that all ferroelectrics also exhibit piezoelectric and pyroelectric behavior. Finally, since nanoscale ferroelectrics are one of the "hot topics" in applied science, their research is ongoing at the very moment of writing this text.

Magnetic Nanomaterials

Albeit 2D monolayer magnetic materials have been theoretically anticipated since 2011, mostly in transition metal halides and transition metal chalcogenides, their first experimental demonstration came as late as in 2017 (Gong et al. 2017; Huang et al. 2017). The experimental monolayer materials in which intrinsic ferromagnetism was discovered in the same year were $Cr_2Ge_2Te_6$ and CrI_3.

Compared to thin films of conventional magnetic materials, 2D monolayers have a host of advantages (Mak et al. 2019). These include, among others, higher surface to volume ratios (a property typical to a majority of nanomaterials), absence of problems with crystal lattice matching between the monolayer and other monolayers or bulk substrate, better flexibility and better resistance to mechanical damages (lower brittleness), lower transition temperatures, and ability to use critical fluctuations to manipulate spin states in spintronics (Ivanov et al. 2004).

Nowadays, magnetic nanomaterials include the quoted 2D structures, but also nanoparticles (0D structures) made of metals like nickel or cobalt, and of oxides like magnetite (Fe_3O_4) or maghemite (γ-Fe_2O_3). Nanocomposites of these nanoparticles, e.g., polymer matrix are also used, as well as agglomerates (clusters) of magnetic nanoparticles. 1D ferromagnetic structures like nanowires or nanotubes, but also similar 1D structures made from non-magnetic materials which are further functionalized by ferromagnetics, e.g. a combination of carbon nanotube and magnetic nanoparticle functionalization are applied as well.

Magnetic nanomaterials find applications in numerous fields. Besides their use in rewritable information storage (hard disk drives) they are also applied in electronics, sensors, scanning magnetic microscopy with atomic resolution, spintronics, and many more.

Piezoelectric Nanomaterials

Same as their macroscopic counterparts, piezoelectric nanomaterials convert mechanical energy into electricity and electrical energy into movement, only at the nanoscale (Yan and Jiang 2017). This conversion mechanism is strongly size-dependent throughout the achievable scale range. The same materials that are used as ferroelectrics at nanoscale can be utilized as piezoelectric nanomaterials as well. They include some perovskite materials like lead zirconate titanate (PZT), lead titanate ($PbTiO_3$), barium titanate ($BaTiO_3$), bismuth ferrite ($BiFeO_3$), but also wurtzite materials like zinc oxide (ZnO), zinc sulfide (ZnS), gallium nitride (GaN) and many more. Piezoelectric effect in crystalline inorganic materials appears as a consequence of the ions/atoms displacement of thin non-centrosymmetric unit cells of the crystals.

Many polymers are also used to fabricate piezoelectric nanomaterials (Smith and Kar-Narayan 2022). They include, for instance, polyvinylidene fluoride, also denoted as polyvinylidene difluoride—PVDF and related copolymers, polyacrylonitrile—PAN, poly (L-lactic acid)—PLLA, Poly (3-hydroxybutyrate-co-3-hydroxyvalerate)—PHBV, odd-numbered Nylons, etc. Generally speaking, piezoelectricity is common for whole groups of polymers, including polyamides, polyesters, polysaccharides, fluoropolymers, and polyureaspolypeptides. A number of ubiquitous natural materials found in living organisms exhibit piezoelectricity as well. This has brought to a theory that piezoelectricity is fundamental to many forms of life (Shamos and Lavine 1967). Some piezoelectric biopolymers include collagen (found in all bones), silk, cellulose (the primary constituent of all kinds of wood and plants generally), polyurea, diphenylalanine, chitin (i.e. N-acetylglucosamine), glycine, chitosan, even DNA itself (Fukada 1995).

Regarding the mechanism of piezoelectricity in polymers, its universal theory does not exist yet, and the current opinion is that it cannot even be created (Smith and Kar-Narayan 2022). The quoted reference also presents some methods to enhance the piezoelectric response of polymers since it is typically weaker than in e.g. piezoelectric ceramics.

Piezoelectric nanomaterials have been produced from the above mentioned materials in basically all nanomaterial geometries, including 0D (nanoparticles), 1D (nanofibers, nanorods, nanotubes), and 2D (sheets, ribbons, nanofilms).

Similar to ferroelectricity, piezoelectricity in nanomaterials is a hot topic in research and the works are ongoing. A number of applications of piezoelectric nanomaterials have been reported, including mechanical sensors (measuring pressure, mass, strain), energy harvesters (Pertin et al. 2021, 2022), flexible electronics, flexible MEMS/NEMS, various biomedical applications (Roy et al. 2022), and many more.

Thermal Management Materials

Thermal management nanomaterials are nanomaterials with low values of their interfacial thermal resistance and an inherently high thermal conductivity (Xing et al. 2022b). They ensure improved heat dissipation from small devices with high power densities like advanced electronic integrated chips, but also MEMS and NEMS.

Traditional solutions like thermal grease—a mixture of carrier matrix made of polyethylene glycol, paraffin wax or, earlier still, silicone or hydrocarbon oils with fillers like metals (silver, copper) or inorganic compounds (silicon carbide, oxides of aluminum or zinc) (Fletcher 1990) are not convenient for modern high power dissipation devices because they show relatively low values of thermal conduction. Albeit new solutions are still being proposed for thermal greases (e.g., Naghibi et al. (2020)) where mineral oil was mixed with graphene filler) this does not solve another problem with such approaches, i.e. the deterioration of the performance of the paste at elevated temperatures.

More recent approaches incorporate solutions with, e.g., metal matrix mixed with highly thermally conductive fillers. Sintered pastes containing two sizes of silver nanoparticles where the larger ones were used as the matrix while the smaller ones were fillers were described in Li et al. (2015). More recent approach to metal thermal management materials is to use a low melting point liquid metal (typically gallium or its alloys) with particles of another metal (e.g. copper) as fillers (Xing et al. 2022a) or a non-metal (an obvious idea is ball-milled and mixed graphene flakes, although other highly thermally conductive 2D materials like boron nitride can be used (Wang et al. 2021b).

Another large group of emerging thermal management materials is carbon-based ones. They include the already mentioned carbon allotrope and 2D nanomaterial graphene, but also 1D allotropes—carbon nanotubes and 0D—fullerenes (buckyballs). Other convenient carbon-based materials include diamond and diamondoids, graphene oxide, even graphite. As an example, a graphene monolith in the form of a honeycomb-aligned panel was described by Dai et al. (2019). A graphene–carbon nanotube aerogel was proposed in Lv et al. (2016).

Finally, a group of approaches to thermal management utilizes a polymer-based matrix. Since polymers are as a rule poor thermal conductors, it is crucial to make thermally conductive pathways through them. A concrete example is PDMS polymer matrix that includes both ball-milled graphene nanoflakes and silver nanowires (Chang et al. 2020). Solutions simultaneously including two or even more above-mentioned approaches have also been proposed.

Phase Change Materials

Phase-change materials (PCM) are compounds able to repeatedly switch between at least two different phases. These phase transitions may occur between liquid and solid state, but also between solid and solid, i.e. between different conformal states of the same material. Each of these phases will be related by distinctly different values of some of their physical properties, like thermal energy levels, optical scattering (reflectivity or transmissivity), electrical or thermal conductivity, mass density, etc.

In the case of thermal transition between phases, the PCM releases or absorbs sufficient amounts of thermal energy at phase transitions to deliver useful heating or cooling, where these conformations differ by their higher or lower energy states. In other words, some of these materials serve as efficient storage units for thermal energy of latent heat (Pielichowska and Pielichowski 2014; Goel et al. 2023). Materials used for this purpose include organic ones (typically paraffins and lipids, sometimes sugar alcohols) and inorganic (salt hydrates). Most recently, some innovative PCM were introduced, an example being calcium chloride hexahydrate–based composite (Zeng et al. 2023).

In the case of phase-change nanomaterials, their repeatibly switchable optical, electrical or thermal properties can be used to obtain high quality rewritable data storage (Raoux 2009). The switching of the optical properties is the basic mechanism of rewritable compact discs, digital versatile discs, and Blu-ray™ discs. The most often used PCM for optical data storage is $Ge_2Sb_2Te_5$.

Another use of PCM nanomaterials is for the enhancement of the effectiveness of thermal energy storage units through introducing PCM nanoparticles into the matrix made of a conventional thermal storage material—the so-called nanoparticles-enhanced PCM (Amidu et al. 2023).

Shape Memory Alloys

Shape memory alloys (SMA) are any metallic materials that can mechanically react to a stimulus like e.g. temperature change or mechanical stress of a specific value in such a way that they revert from any current shape to a previously defined ("trained") one. This is called the shape memory effect—these materials "remember" their previous shape and return to it. The effect can be one-way or two-way and it can be useful for various actuator devices (Kohl 2004). SMAs can also be mechanically deformed and return to the shape they had prior to that deformation as soon as the force causing the deformation is removed—this effect is called superelasticity and is also known as pseudoelasticity. Typically, the reversible deformation in superelasticity can be 5–10%.

The most often used SMA is nickel-titanium alloy NiTi (also known as nitinol), although some other metal combinations are used, e.g. Cu–Zn–Al and generally similar copper-based alloys or Fe–Mn–Si or similar iron-based alloys.

SMA materials are applicable in MEMS and NEMS (Stachiv et al. 2021). They are readily fabricated in the form of ultrathin films which retain their complete shape memory functionality, and can even augment it. Additionally, from the technological point of view, they are fabricated quite easily using standard planar procedures and the obtained quality is excellent—NiTi is deposited by simple sputtering.

The mechanism of SMA function is phase transition between their austenite (cubic lattice) and their martensite (monoclinic or orthorhombic) form. This transition can be triggered by heat and by stress. The heating approach is much more customary. The SMA has austenite form when cold and martensite when heated. They are "trained" by changing their shape to the desired one when cold: martensite can be deformed when cold but returns to its pre-deformed ("remembered") shape when heated (Fig. 8.7).

The desired basic shape can be "trained" by forming it while in martensite phase (cold), constraining it, e.g., by an external fixture, heating to the austenite phase and keeping it about 8–10 min at that temperature, and then quickly cooling it down to the martensite temperature. The new desired shape will be permanently retained in

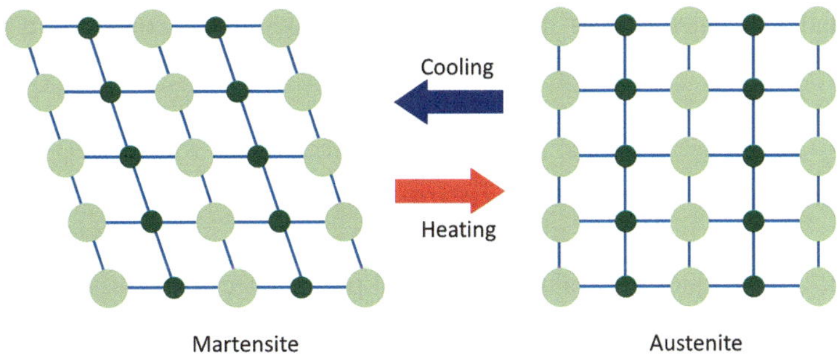

Fig. 8.7 Schematic presentation of two-way memory effect in SMA

this manner. After deforming it next time, every heating to the austenite phase will revert to the newly trained form.

Besides metal-based shape memory structures, polymer-based materials with shape memory properties are also available (shape memory polymers, SMP). This kind of material is able to retain two shapes, in some cases even three. Similarly to SMA, temperature changes are most often used to cause SMP shape transitions. Other shape shifting mechanisms applicable to SMP are electromagnetic field (optical signal) (Lendlein et al. 2005), external magnetic field, external electric field or even a liquid solution (Leng et al. 2005). Both SMA and SMP materials for MEMS and NEMS are considered in detail in the review paper (Stachiv et al. 2021).

Characterization and Testing of Novel Materials

The development of methods for the design and synthesis of novel materials and the development of methods for their characterization are intermixed. The development of materials also demands the development of characterization methods that are relevant for the known properties, but in micro/nano domain. The review (Stanimirović and Stanimirović 2009) delves into measuring mechanical properties in MEMS pointing out the lack of and necessity for the standardization of testing procedures. On the other hand, standards related to micro- and nano-scale surface metrology, ISO TC/213, have being developed since 2005 under the title "Geometrical Product Specifications (GPS)—Surface texture: Areal" within "Dimensional and geometrical product specifications and verification" (Blunt et al. 2005). There are many size effects that make the use of previously standardized methods for the sample preparation, handling, and interpretation of measured data difficult (Higo 2005). Accuracy changes with the scaling and homogeneity of low dimensional materials might become questionable, so the development of novel characterization methods, tailored for the novel (classes of) materials is an active field of research.

This section briefly lists analytical characterization techniques, electrical characterization methods, some computational methods for estimation of material parameters, and some methods for material assessment within the context of a specific application or a broader context (like the environmental impact).

Analytical Characterization Techniques

Analytical techniques for evaluating material properties can be classified based on the measurement objective into the groups like methods for surface profilometry (surface morphology and crystalline size), methods for measuring of bulk material properties (mechanical properties being the most important for many materials), and methods for specifying the material composition and the types of internal bonds.

Apart from the selection of methods based on the examined feature, the selection can be built on the measurement approach (Leng 2013), which can be based on the microscopic observation and analysis of electrons and ions (SEM, TEM, SIM...), can be based on mass spectroscopy of gasses, liquids and ions (GC, LC, IC, SIMS, TDS...), based on scanning probe microscopy (SPM, AFM), on the use of X-rays (XRD, XRR, XRF), based on vibration spectroscopy (FTIR, Raman) or on the absorption spectra (UV-Vis), to name a few (Zhang et al. 2008). Analytical methodology online wizard at (Smith et al. 2000) helps with the method choice with respect to a variety of analysis objectives related to shape and structure, composition and bonding states, abnormalities at the surface and in the bulk, composition of gases and liquids. Many novel characterization methods are being developed for a specific material or platform like for instance on-chip testing of polysilicon strength (Vicentini Ferreira do Valle et al. 2023), or a MEMS platform for in-situ testing of mechanical properties (Rajagopalan 2019). Such methods and devices are yet to be standardized and commercialized. Table 8.3 lists some of the best basic techniques, basic in the sense that they are often combined or upgraded to provide more complex information.

AES, Auger electron spectroscopy is used for the determination of the elemental composition of surfaces. It is based on the Auger effect (discovered independently by both Lise Meitner and Pierre Auger in the 1920s) and the tracking of energies during

Table 8.3 Some analytical techniques for the evaluation of novel materials

Method	Principle	Features
Scanning electron microscopy (SEM)	Scanning the sample with a focused electron beam	High depth of field, excellent resolution, and the ability to analyze conductive and non-conductive materials (the latter with a conductive coating)
Atomic force microscopy (AFM)	Creating a topographic map by using a sharp tip to scan the sample surface, measuring forces between the tip and the surface	Quantitative information on surface roughness with high spatial resolution and a capability to work in air, liquid, and vacuum
Transmission electron microscopy (TEM)	High-resolution images of internal and surface structures based on the analysis of transmitted electrons through the sample	Ultra-high resolution, capable of imaging at the atomic scale, and provides information on crystal structure
Confocal laser scanning microscopy (CLSM)	Uses laser light to scan the sample, capturing optical sections at different depths to reconstruct three-dimensional images	Non-destructive, and suitable for fluorescently labeled samples

(continued)

Table 8.3 (continued)

Method	Principle	Features
White light interferometry (WLI)	Uses interferometry to measure the height variations of the surface by analyzing the interference patterns of white light	High vertical resolution, non-contact, suitable for transparent and reflective surfaces
X-ray photoelectron spectroscopy (XPS)	Analyzing and measuring binding energies of electrons emitted upon X-ray irradiation	Provides information on surface chemistry and elemental composition and chemical state of the surface
Fourier transform infrared spectroscopy (FTIR)	Analyzing the scattering parameters of monochromatic light	Non-destructive, provides information on chemical composition and crystallinity, suitable for small sample sizes
X-ray diffraction (XRD)	Measures the diffraction pattern of X-rays when they interact with a crystalline material	Non-destructive, provides crystallographic information, and is suitable for layered materials
Auger electron spectroscopy (AES)	Measures the kinetic energy of electrons emitted from a sample	Reveals elemental composition of layers with high surface sensitivity
UV-visible spectroscopy Raman spectroscopy	UV-vis analyzes the absorption or emission spectra of a solution Raman analyzes the inelastic scattering of light	Non-destructive, provides chemical information on molecular vibrations and the composition of layers
Secondary ion mass spectrometry (SIMS)	Measures the mass-to-charge ratio of secondary ions ejected from the sample surface due to primary ion bombardment	High surface sensitivity, elemental and isotopic analysis
Time-of-flight secondary ion mass spectrometry (ToF-SIMS)	Measures the time of flight of secondary ions generated from the sample surface	Molecular identification and elemental composition with depth profiling capabilities and high lateral resolution
Electrochemical impedance spectroscopy (EIS)	Measures the impedance of an electrode	Electrochemical characterization, detection of changes in electrical behavior
Ultrasonic testing	Measures change in the transmission of ultrasonic waves	Non-destructive, can detect changes in material thickness and depth information
Thermogravimetric analysis (TGA)	Measures changes in weight with temperature	Thermal characterization, detection of weight changes

(continued)

Table 8.3 (continued)

Method	Principle	Features
Hall effect measurements	Behavior of conductive elements in a magnetic field	Bulk electrical characterization, carrier type, concentration, and mobility
Electroluminescence imaging	Captures light emission during electrical operation	Visualization of electrical activity, identifying local variations
Gas chromatography (GC) Liquid chromatography (LC) Ion chromatography (IC)	Separation, analysis, and quantification of individual components based on their retention times	Identification of compounds in a gas mixture, liquid solution or studying the composition of electrolytes or charged species

the transitions of electrons of an excited atom. Standard spectra for identification and interpretation of AES data for often used materials are given in McGuire (1979).

AFM, Atomic force microscopy is a microscopy technique for scanning the surface with a probe with a sharp tip. It can have many operation modes (contact, non-contact, tapping). The tip can also be used for scratching and profiling the surface i.e. for Scanning probe lithography (Nanoscience and nanotechnologies: opportunities and uncertainties | Royal Society 2004). It is a precise but slow method because it needs time to scan the surface and generate its software image based on the recorded interactions between the tip and all the locations along the scanned surface. Besides for scanning the surface it can also be used for investigating eigenmode resonance frequencies of various materials (even neural ones, diatom frustules (Cvjetinovic et al. 2023)) or for the determination of stiffness of the elastic modulus of micropillars (Cortelli et al. 2023). With the proper modifications of the apparatus, the method can be used for 3D imaging of the material (Zhang et al. 2023a).

THz-TDS is a technique, discovered more than fifty years ago but gaining importance today as being indestructible, the terahertz time-domain spectroscopy, technique where a material is probed with a femtosecond laser and the response in the form of the time-dependent electric field waveform is analyzed (Koch et al. 2023). Opposed to X-rays, THz radiation is non-ionizing, hence it is safe for biological tissues and structural health monitoring and the number of materials transparent in the THz domain makes it suitable for the measurements of microfluidic devices (Alfihed et al. 2018).

Thermal analyses of materials are based on the programmed heating and observing temperature-dependent phenomena of the sample (thermal stability, structure decomposition, phase transition...) by a variety of methods. According to the Handbook of Differential Scanning Calorimetry, DSC, Differential scanning calorimetry is the most common thermal analysis instrument, widely present in laboratories, yet often under-utilized in spite of the possibilities for Simultaneous DSC Techniques: DSC measurements are performed simultaneously with thermogravimetry (DSC-TGA),

DSC-dielectric analysis (DEA), DSC-thermomechanical analysis (TMA), and DSC-rheology (Rheo-DSC) (Szécsényi and Holló 2023). MEMS devices with micro-cantilevers (Zhou et al. 2022) and thermopiles (Zhang et al. 2023b) enable these measurements on a chip.

Mechanical characterization of MEMS materials encompasses numerous different tests and measurements of the response to various mechanical stimuli: strain measurements, tensile tests, bend tests (in-plane, out-of-plane, fixed ends), indentation tests, fracture tests, fatigue tests, creep tests, membrane tests, as portrayed in a book chapter in 2002 (Sharpe 2002) when round robin choice of method between interested laboratories was used as a starting point for convergence toward the standardization. Today, after two decades of continuous contributions through publications and patents, the topic is still evolving, allowing for the determination of the plastic deformation (Namazu 2023) or investigations of combined Thermal–Mechanical-Stress-Creep (TMSC) effect by an Isolation-Test Method (ITM) or investigations of combined Thermal–Mechanical-Stress-Creep (TMSC) effect by (ITM) (Zhang et al. 2022) or nanointendation approach (Farzin et al. 2019), to name a few.

Electrical Characterization

Electrical characterization relies on studying the materials performances in electrical circuits. It may involve measuring parameters such as resistance, capacitance, inductance, dielectric properties, voltage-current characteristics, all in various frequency domains.

Methods used and well known are Four-Point Probe Measurement, Hall Effect Measurements, Electrical Resistivity Measurement, Conductive AFM, Capacitance Measurement, Impedance Spectroscopy, Dielectric Spectroscopy, Dielectric Breakdown Testing, Impulse Voltage Testing, Piezoelectric Force Microscopy (PFM), Piezoelectric Resonance Analysis, Pyroelectric Coefficient Measurement, Pyroelectric Current Measurement, Seebeck Coefficient Measurement, Thermoelectric Module Testing, Magnetoelectric Coupling Measurements, Magnetoelectric Response Testing.

The complexity of novel materials makes these measurements a necessity in spite of the known properties of the constituent material segments. For instance, the intrinsic conductivity of MXene fillers can impart conductivity to the polyurethane matrix, but its contribution amount needs to be quantified. Depending on the intended application (e.g., sensors, coatings, energy storage), specific tests are often developed in a bundle or a test suite.

Computational Evaluation of Effective Material Properties

There is a great body of contemporary scientific literature presenting the synergy of design, modeling, simulation, and characterization of novel materials and devices. The advancement of software algorithms and methods and on the other hand hardware platforms for multiscale modeling, ab-initio calculations, first principles calculations, enables reliable predictions of future materials, simultaneously saving laboratory time and laboratory consumables and equipment. In combination with machine learning techniques and artificial intelligence-assisted methods the design has all prerequisites to result in fully optimized components (Pertin et al. 2021; Devi et al. 2022).

A valuable use of the computational evaluation of effective material properties is the determination of the effective refractive index of fishnet structured metamaterials (Tanasković et al. 2014), there it is a tool for the discovery of regions with the negative refractive index change. The computational evaluation of effective material properties can be used as proof that aperture-based material scans also exhibit the negative refractive index (Tanasković et al. 2015, 2016), and can direct the experimental work to the related wavelength region of interest.

Computational evaluation of effective material properties in general are reliable, it has been documented that the computational evaluation of effective material properties of composites reinforced by randomly distributed spherical particles was in good agreement with the relevant analytical techniques (Kari et al. 2007b). The same can be said for the computational evaluation of optoelectronic properties for organic/carbon materials (Shuai et al. 2014), computational evaluation of the flexoelectric effect in dielectric solids (Abdollahi et al. 2014), computational evaluation of tensile properties of additively manufactured hexa- and tetrachiral auxetic cellular structures (Novak et al. 2021), computational evaluation of optoelectronic, thermodynamic and electron transport properties of $CuYZ_2$ ($Z = S$, Se and Te) chalcogenides semiconductors (Elamin Ketfi et al. 2022), the listing is literally endless.

Conclusion

This chapter presented a short compendium on exploring innovative materials with respect to their use for advanced MEMS/NEMS applications. It provided an introductory overview of such materials relevant to Micro-Electro-Mechanical Systems (MEMS) and Nano-Electro-Mechanical Systems (NEMS), systems that have gained prominence in diverse industries, ranging from aerospace to healthcare, due to their diminutive scale ranging from micrometers to nanometers. The focus of this chapter was on the materials with the potential to serve as the new fundamental elements for MEMS/NEMS and microsystems and nanosystem technologies.

The exploration presented here has been structured to cover various aspects of innovative materials in MEMS/NEMS, commencing with a fundamental understanding of their properties, categorized in accordance with the physical, chemical, mechanical, acoustical, electrical, magnetic, optical, radiological, and thermal attributes of materials.

The materials were examined in three main categories, with respect to their historical emergence, first mentioning the materials used in microelectronic and early MEMS-based devices, highlighting their role in enabling the evolution of the technology. Then, the core of this chapter centered on emerging materials that are poised to transform the landscape of MEMS/NEMS, such as nanomaterials (0D like quantum dots and general nanoparticles, 1D like nanowires and including CNT, 2D and quasi-2D like graphene or MXenes) nanocomposites, metamaterials, biomaterials (organic and inorganic), biomimetic materials, and quantum materials. In the third category, the focus was on the most promising materials for the demanding applications of the contemporary world and the near future, functional materials that are essential for the operation of MEMS/NEMS devices, such as ferroelectric materials, magnetic materials, phase-changing materials, shape memory alloys, etc.

In addition, the critical area of material characterization and testing, fundamental for understanding material properties and ensuring their reliability and durability, is also addressed. An array of chosen analytical methods employed to assess material properties is also presented.

This comprehensive overview aims to facilitate further research of novel materials optimized for applications within the MEMS/NEMS domain.

Acknowledgements This research was funded by the Ministry of Education, Science, and Technological Development of Republic of Serbia, grant number 451-03-47/2023-01/200026. The authors wish to express their thanks to David Jakšić, M.Sc in ME, for his help with 3D modeling in Blender computer graphics software.

Competing Interests The authors have no conflicts of interest to declare that are relevant to the content of this chapter.

References

Abdollahi A, Peco C, Millán D, Arroyo M, Arias I (2014) Computational evaluation of the flexoelectric effect in dielectric solids. J Appl Phys 116(9):093502. https://doi.org/10.1063/1.4893974

Alfihed S, Bergen MH, Ciocoiu A, Holzman JF, Foulds IG (2018) Characterization and integration of terahertz technology within microfluidic platforms. Micromachines 9(9). https://doi.org/10.3390/mi9090453

Alviz-Meza A, Orozco-Agamez J, Quinayá DCP, Alviz-Amador A (2023) Bibliometric analysis of fourth industrial revolution applied to material sciences based on web of science and scopus databases from 2017 to 2021. ChemEngineering 7(1):2. https://doi.org/10.3390/CHEMENGINEERING7010002

Amidu MA, Ali M, Alkaabi AK, Addad Y (2023) A critical assessment of nanoparticles enhanced phase change materials (NePCMs) for latent heat energy storage applications. Sci Rep 13(1). https://doi.org/10.1038/s41598-023-34907-0

Anonymous (1920) Minutes of the Washington Meeting, April 23 and 24, 1920. Am Phys Soc XV(6):505–567. https://doi.org/10.1103/physrev.15.505

Ashby MF (2011) Materials selection in mechanical design, 4th edn. Butterworth-Heinemann, Burlington, MA

Basov DN, Averitt RD, Hsieh D (2017) Towards properties on demand in quantum materials. Nat Mater 16(11). https://doi.org/10.1038/nmat5017

Beyeler D, Muntwyler S, Beyeler F (2018) EP3489654A1—Mems nanoindenter chip with indenter probe and reference probe https://patents.google.com/patent/EP3489654A1/en. Accessed 14 Nov 2023

Beygelzimer Y, Kulagin R, Fratzl P, Estrin Y (2021) The earth's lithosphere inspires materials design. Adv Mater 33(3):2005473. https://doi.org/10.1002/adma.202005473

Bhatnagar P, Vashaee D (2022) Development of MEMS process compatible $(Bi,Sb)_2(Se,Te)_3$-based thin films for scalable fabrication of planar micro-thermoelectric generators. Micromachines 13(9). https://doi.org/10.3390/mi13091459

Blunt L, Jiang X, Scott PJ (2005) Advances in micro and nano-scale surface metrology. Key Eng Mater 295–296:431–436. https://doi.org/10.4028/www.scientific.net/KEM.295-296.431

Böscke TS, Hung PY, Kirsch PD, Quevedo-Lopez MA, Ramírez-Bon R (2009) Increasing permittivity in HfZrO thin films by surface manipulation. Appl Phys Lett 95(5):052904. https://doi.org/10.1063/1.3195623

Bukhari SS, (Yiannis) Vardaxoglou J, Whittow W (2019) A metasurfaces review: definitions and applications. Appl Sci 9(13). https://doi.org/10.3390/app9132727

Byakodi M et al (2022) Emerging 0D, 1D, 2D, and 3D nanostructures for efficient point-of-care biosensing. Biosens Bioelectron X 12:100284. https://doi.org/10.1016/j.biosx.2022.100284

Cai W, Shalaev V (2010) Optical metamaterials. Springer, New York, NY. https://doi.org/10.1007/978-1-4419-1151-3

Cai A-Y, Zhu Y-J, Qi C (2020) Biodegradable inorganic nanostructured biomaterials for drug delivery. Adv Mater Interfaces 7(20):2000819. https://doi.org/10.1002/admi.202000819

Cao Y et al (2023) Attractive electrocaloric and flexoelectric properties of Bi4Ti3O12 films. Phys B Condens Matter 669:415310. https://doi.org/10.1016/J.PHYSB.2023.415310

Chang T-C, Kwan Y-K, Fuh Y-K (2020) A reduced percolation threshold of hybrid fillers of ball-milled exfoliated graphite nanoplatelets and AgNWs for enhanced thermal interface materials in high power electronics. Compos Part B Eng 191:107954. https://doi.org/10.1016/j.compositesb.2020.107954

Cheng Y, Zhou C, Yuan BG, Wu DJ, Wei Q, Liu XJ (2015) Ultra-sparse metasurface for high reflection of low-frequency sound based on artificial Mie resonances. Nat Mater 14(10):1013–1019. https://doi.org/10.1038/nmat4393

Chynoweth AG (1958) Pyromagnetic effect: a method for determining curie points. J Appl Phys 29:563–565. https://doi.org/10.1063/1.1723222

Coelho P, Bessa C, Landeck J, Silva C (2023) Industry 5.0: the arising of a concept. Procedia Comput Sci 217:1137–1144. https://doi.org/10.1016/j.procs.2022.12.312

Cormier R, Elliott M (2017) SMART marine goals, targets and management—Is SDG 14 operational or aspirational, is 'Life Below Water' sinking or swimming? Mar Pollut Bull 123(1–2):28–33. https://doi.org/10.1016/J.MARPOLBUL.2017.07.060

Cortelli G et al (2023) Determination of stiffness and the elastic modulus of 3D-printed micropillars with atomic force microscopy-force spectroscopy. ACS Appl Mater Interfaces 15(5):7602–7609. https://doi.org/10.1021/acsami.2c21921

Cvjetinovic J et al (2023) Probing vibrational eigenmodes in diatom frustules via combined in silico computational study and atomic force microscopy experimentation. Appl Phys Lett 123(18):184101. https://doi.org/10.1063/5.0171503

Dai W et al (2019) Metal-level thermally conductive yet soft graphene thermal interface materials. ACS Nano 13(10):11561–11571. https://doi.org/10.1021/acsnano.9b05163
Devi S, Guha K, Jakšić O, Baishnab KL, Jakšić Z (2022) Optimized design of a self-biased amplifier for seizure detection supplied by piezoelectric nanogenerator: metaheuristic algorithms versus ANN-assisted goal attainment method. Micromachines 13(7):1104.1–1104.23. https://doi.org/10.3390/mi13071104
de Ribeiro PO et al (2023) Anisotropic magneto-resistivity and magnetocaloric effect in DyAl2. J Alloys Compd 173149. https://doi.org/10.1016/j.jallcom.2023.173149
Dragoman M, Aldrigo M, Dragoman D, Iordanescu S, Dinescu A, Modreanu M (2021) The rise of ferroelectricity at nanoscale: nanoelectronics is rediscovering the ferroelectricity. IEEE Nanotechnol Mag 15(5):8–19. https://doi.org/10.1109/MNANO.2021.3098217
Estrin Y et al (2021) Architecturing materials at mesoscale: some current trends. Mater Res Lett 9(10):399–421. https://doi.org/10.1080/21663831.2021.1961908
Farzin L, Sheibani S, Moassesi ME, Shamsipur M (2019) An overview of nanoscale radionuclides and radiolabeled nanomaterials commonly used for nuclear molecular imaging and therapeutic functions. J Biomed Mater Res A 107(1):251–285. https://doi.org/10.1002/jbm.a.36550
Fattahi P, Yang G, Kim G, Abidian MR (2014) A review of organic and inorganic biomaterials for neural interfaces. Adv Mater 26(12):1846–1885. https://doi.org/10.1002/ADMA.201304496
Fletcher LS (1990) A review of thermal enhancement techniques for electronic systems. IEEE Trans Compon Hybrids Manuf Technol 13(4):1012–1021. https://doi.org/10.1109/33.62543
Fu Q, Sun W (2001) Mie theory for light scattering by a spherical particle in an absorbing medium. Appl Opt 40(9):1354–1361. https://doi.org/10.1364/AO.40.001354
Fukada E (1995) Piezoelectricity of biopolymers. Biorheology 32(6):593–609. https://doi.org/10.1016/0006-355X(95)00039-C
Ghosh SK et al (2017) Electrospun gelatin nanofiber based self-powered bio-e-skin for health care monitoring. Nano Energy 36:166–175. https://doi.org/10.1016/j.nanoen.2017.04.028
Goel V, Saxena A, Kumar M, Thakur A, Sharma A, Bianco V (2023) Potential of phase change materials and their effective use in solar thermal applications: a critical review. Appl Therm Eng 219:119417. https://doi.org/10.1016/j.applthermaleng.2022.119417
Goldberg M, Langer R, Jia X (2007) Nanostructured materials for applications in drug delivery and tissue engineering. J Biomater Sci Polym Ed 18(3):241–268. https://doi.org/10.1163/156856207779996931
Goncharenko AV, Venger EF, Pinchuk AO (2014) Homogenization of quasi-1d metamaterials and the problem of extended bandwidth. Opt Express 22(3):2429–2442. https://doi.org/10.1364/OE.22.002429
Gong C et al (2017) Discovery of intrinsic ferromagnetism in two-dimensional van der Waals crystals. Nature 546(7657). https://doi.org/10.1038/nature22060
González-Viñas W, Mancini HL (2004) An introduction to materials science. Princeton University Press. https://doi.org/10.1515/9781400880058
Higo Y (2005) Characterization of micro/nano electro-mechanical systems (MEMS/NEMS). Fatigue Fract Eng Mater Struct 28(8):655–655. https://doi.org/10.1111/J.1460-2695.2005.00933.X
Huang B, Clark G, Navarro-Moratalla E, Klein DR, Cheng R, Seyler KL, Zhong D, Schmidgall E, McGuire MA, Cobden DH, Yao W, Xiao D, Jarillo-Herrero P, Xu X (2017) Layer-dependent ferromagnetism in a van der Waals crystal down to the monolayer limit. Nature 546(7657). https://doi.org/10.1038/nature22391
Ivanov VA, Aminov TG, Novotortsev VM, Kalinnikov VT (2004) Spintronics and spintronics materials. Russ Chem Bull 53(11):2357–2405. https://doi.org/10.1007/s11172-005-0135-5
Jakšić Z, Vuković S, Matovic J, Tanasković D (2010) Negative refractive index metasurfaces for enhanced biosensing. Materials 4(1):1–36. https://doi.org/10.3390/ma4010001
Jakšić Z, Jakšić O (2020) Biomimetic nanomembranes: an overview. Biomimetics 5(2):24. https://doi.org/10.3390/BIOMIMETICS5020024

Jakšić Z, Obradov M, Jakšić O (2022a) Bio-inspired nanomembranes as building blocks for nanophotonics, plasmonics and metamaterials. Biomimetics 7(4):222. https://doi.org/10.3390/BIOMIMETICS7040222

Jakšić Z, Obradov M, Jakšić O, Tanasković D (2022b) A family of perforated submicrometer core–shell plasmonic particles bio-inspired by leafhopper brochosomes. Opt Quantum Electron 54(8):1–14. https://doi.org/10.1007/S11082-022-03767-9

Jakšić Z, Devi S, Jakšić O, Guha K (2023) A comprehensive review of bio-inspired optimization algorithms including applications in microelectronics and nanophotonics. Biomimetics 8(3):278. https://doi.org/10.3390/BIOMIMETICS8030278

Jampilek J, Kralova K (2022) Advances in nanostructures for antimicrobial therapy. Materials 15(7). https://doi.org/10.3390/ma15072388

Kabir M, Kazari H, Ozevin D (2018) Piezoelectric MEMS acoustic emission sensors. Sens Actuators Phys 279:53–64. https://doi.org/10.1016/J.SNA.2018.05.044

Kari S, Berger H, Gabbert U (2007a) Numerical evaluation of effective material properties of randomly distributed short cylindrical fibre composites. Comput Mater Sci 39(1):198–204. https://doi.org/10.1016/j.commatsci.2006.02.024

Kari S, Berger H, Rodriguez-Ramos R, Gabbert U (2007b) Computational evaluation of effective material properties of composites reinforced by randomly distributed spherical particles. Compos Struct 77(2):223–231. https://doi.org/10.1016/j.compstruct.2005.07.003

Kari S, Berger H, Gabbert U, Guinovart-Dıaz R, Bravo-Castillero J, Rodrıguez-Ramos R (2008) Evaluation of influence of interphase material parameters on effective material properties of three phase composites. Compos Sci Technol 68(3):684–691. https://doi.org/10.1016/j.compscitech.2007.09.009

Keimer B, Moore JE (2017) The physics of quantum materials. Nat Phys 13(11). https://doi.org/10.1038/nphys4302

Ketfi ME, Bennacer H, Essaoud SS, Ziane MI, Boukortt A (2022) Computational evaluation of optoelectronic, thermodynamic and electron transport properties of CuYZ2 (Z = S, Se and Te) chalcogenides semiconductors. Mater Chem Phys 277:125553. https://doi.org/10.1016/j.matchemphys.2021.125553

Khalilov R (2023) A comprehensive review of advanced nano-biomaterials in regenerative medicine and drug delivery. Adv Biol Earth Sci 8(1):5–18

Koch M, Mittleman DM, Ornik J, Castro-Camus E (2023) Terahertz time-domain spectroscopy. Nat Rev Methods Primer 3(1). https://doi.org/10.1038/s43586-023-00232-z

Kohl M (2004) Shape memory microactuators. In: Microtechnology and MEMS. Springer, Berlin, Heidelberg. https://doi.org/10.1007/978-3-662-09875-2

Kumar A, Kaur D (2023) Magnetic field tunable piezoelectric resonator for MEMS applications. Sens Actuators Phys 364:114803. https://doi.org/10.1016/j.sna.2023.114803

Kumar A et al (2024) MEMS-based piezoresistive and capacitive microphones: a review on materials and methods. Mater Sci Semicond Process 169:107879. https://doi.org/10.1016/J.MSSP.2023.107879

Lee H-Y, Yao T (2002) Design and evaluation of omnidirectional one-dimensional photonic crystals. J Appl Phys 93(2):819–830. https://doi.org/10.1063/1.1530726

Lendlein A, Jiang H, Jünger O, Langer R (2005) Light-induced shape-memory polymers. Nature 434(7035). https://doi.org/10.1038/nature03496

Leng J, Lv H, Liu Y, Du S (2008) Comment on 'Water-driven programable polyurethane shape memory polymer: demonstration and mechanism' [Appl Phys Lett 86:114105 (2005)]. Appl Phys Lett 92(20):206105. https://doi.org/10.1063/1.2936288

Leng Y (2013) Materials characterization: introduction to microscopic and spectroscopic methods. Wiley-VCH Verlag GmbH & Co. KGaA. https://books.google.rs/books?hl=en&lr=&id=kdc4AAAAQBAJ&oi=fnd&pg=PA1&ots=ajMGzVb2Dc&sig=uwonEes-KQd-GY6iOTPfk3QnKqk&redir_esc=y#v=onepage&q&f=false. Accessed 27 Dec 2023

Li Z, Herrmann K (2006) DE102006052153B4—Indentor and method for measuring mechanical properties of materials—Google Patents. https://patents.google.com/patent/DE102006052153B4/en. Accessed 14 Nov 2023

Li M, Xiao Y, Zhang Z, Yu J (2015) Bimodal sintered silver nanoparticle paste with ultrahigh thermal conductivity and shear strength for high temperature thermal interface material applications. ACS Appl Mater Interfaces 7(17):9157–9168. https://doi.org/10.1021/acsami.5b01341

Ling Y-C, Yoo SJB (2023) Review: tunable nanophotonic metastructures. Nanophotonics 12(20):3851–3870. https://doi.org/10.1515/nanoph-2023-0034

Liu H, Lalanne P (2008) Microscopic theory of the extraordinary optical transmission. Nature 452(7188). https://doi.org/10.1038/nature06762

Loevestam G et al (2010) Considerations on a definition of nanomaterial for regulatory purposes. JRC58726. https://doi.org/10.2788/98686

Lv P, Tan X-W, Yu K-H, Zheng R-L, Zheng J-J, Wei W (2016) Super-elastic graphene/carbon nanotube aerogel: a novel thermal interface material with highly thermal transport properties. Carbon 99:222–228. https://doi.org/10.1016/j.carbon.2015.12.026

Mak KF, Shan J, Ralph DC (2019) Probing and controlling magnetic states in 2D layered magnetic materials. Nat Rev Phys 1(11). https://doi.org/10.1038/s42254-019-0110-y

McGuire GE (1979) Auger electron spectroscopy reference manual. Springer US, Boston, MA. https://doi.org/10.1007/978-1-4757-1702-0

Mehmood Z, Haneef I, Udrea F (2018) Material selection for micro-electro-mechanical-systems (MEMS) using Ashby's approach. Mater des 157:412–430. https://doi.org/10.1016/j.matdes.2018.07.058

Mehmood M et al (2023) Investigation of structural, electronic, elastic, magnetic and thermodynamic properties of antiperovskites XCRh3 (X = Cd, Ta, W, Re, Os, Ir, Pt, Au, Hg, Ce, Pr, Nd, Pm, Sm, Eu, Tb). Phys B Condens Matter 649:414442. https://doi.org/10.1016/j.physb.2022.414442

Meng C, Thrane PCV, Ding F, Bozhevolnyi SI (2022) Full-range birefringence control with piezoelectric MEMS-based metasurfaces. Nat Commun13(1). https://doi.org/10.1038/s41467-022-29798-0

Mikolajick T et al (2021) Next generation ferroelectric materials for semiconductor process integration and their applications. J Appl Phys 129(10):100901. https://doi.org/10.1063/5.0037617

Minegishi T (1963) Magneto-thermoelectric power and magnetostriction of FeNi3 single crystals. Jpn J Appl Phys 2(3):151. https://doi.org/10.1143/JJAP.2.151

Mohapatra RK (2020) Recent advances on engineering materials: micro to nano scale. Micro Nanosyst 12(2):80. https://doi.org/10.2174/187640291202200618115118

MST|Analytical Methodology Wizard. https://www.mst.or.jp/Portals/0/en/en_wizard.html. Accessed 27 Dec 2023

Naghibi S et al (2020) Noncuring graphene thermal interface materials for advanced electronics. Adv Electron Mater 6(4):1901303. https://doi.org/10.1002/aelm.201901303

Namazu T (2023) Mechanical property measurement of micro/nanoscale materials for MEMS: a review. IEEJ Trans Electr Electron Eng 18(3):308–324. https://doi.org/10.1002/TEE.23747

Nanoscience and nanotechnologies: opportunities and uncertainties | Royal Society. The Royal Society (2004). https://royalsociety.org/topics-policy/publications/2004/nanoscience-nanotechnologies/. Accessed 02 Jan 2024

Novak N et al (2021) Experimental and computational evaluation of tensile properties of additively manufactured hexa- and tetrachiral auxetic cellular structures. Addit Manuf 45:102022. https://doi.org/10.1016/j.addma.2021.102022

Nugroho FAA et al (2022) Inverse designed plasmonic metasurface with parts per billion optical hydrogen detection. Nat Commun 13(1). https://doi.org/10.1038/s41467-022-33466-8

Ozevin D (2020) MEMS acoustic emission sensors. Appl Sci 10(24):8966. https://doi.org/10.3390/APP10248966

Patel S (2023) Flexoelectricity in ceramics and their application. Elsevier

Patil S, Lee DW (2019) Status review on the MEMS-based flexible supercapacitors. J. Micromech Microeng 29(9):093001. https://doi.org/10.1088/1361-6439/AB1C9B

Penumuru DP, Muthuswamy S, Karumbu P (2020) Identification and classification of materials using machine vision and machine learning in the context of industry 4.0. J Intell Manuf. https://doi.org/10.1007/s10845-019-01508-6

Pergal MV, Jakšić OM, Rašljić-Rafajilović MB, Vasiljević-Radović DG (2023) Chapter 1. Polymer-based microelectromechanical systems (MEMS) and microfluidic devices: engineering and applications. In: Wythers MC (ed) Advances in materials science research, vol 63. Nova Science Publishers, pp 1–118. https://novapublishers.com/shop/advances-in-materials-science-research-volume-63/. Accessed 04 July 2023

Perkin E, Gubala V (2017) Nanomaterials—On the brink of revolution? Or the endless pursuit of something unattainable? Chem Int 39(2):10–13. https://doi.org/10.1515/ci-2017-0206

Pertin O, Guha K, Jakšić O (2021) Artificial intelligence-based optimization of a bimorph-segmented tapered piezoelectric MEMS energy harvester for multimode operation. Computation 9(8):84. https://doi.org/10.3390/COMPUTATION9080084

Pertin O, Guha K, Jakšić O, Jakšić Z, Iannacci J (2022) Investigation of nonlinear piezoelectric energy harvester for low-frequency and wideband applications. Micromachines 13(9):1399. https://doi.org/10.3390/mi13091399

Petelin N, Kalin M, Kitanovski A (2023) A conceptual design of a thermal switch capacitor in a magnetocaloric device: experimental characterization of properties and simulations of operating characteristics. J Phys Energy 5(3):035007. https://doi.org/10.1088/2515-7655/ace288

Pielichowska K, Pielichowski K (2014) Phase change materials for thermal energy storage. Prog Mater Sci 65:67–123. https://doi.org/10.1016/j.pmatsci.2014.03.005

Polcawich RG, Pulskamp JS (2017) Lead Zirconate Titanate (PZT) for M/NEMS. In: Bhugra H, Piazza G (eds) Piezoelectric MEMS resonators, in microsystems and nanosystems. Springer International Publishing, Cham, pp 39–71. https://doi.org/10.1007/978-3-319-28688-4_2

Poongodi A, Thangaraj N (2023) Synthesis and characterization of ferrous manganese tungsten thin films for magnetic MEMS devices. J Ovonic Res 19(3):253–258. https://doi.org/10.15251/JOR.2023.193.253

Pradhan A, Kumari A, Srivastava R, Panda D (2019) Quercetin encapsulated biodegradable plasmonic nanoparticles for photothermal therapy of hepatocellular carcinoma cells. ACS Appl Bio Mater 2(12):5727–5738. https://doi.org/10.1021/acsabm.9b00764

Pradhan P (2023) A threefold approach to rescue the 2030 Agenda from failing, Special Topic: Coupling Human and Natural Systems. Natl Sci Rev 10(7):nwad015. https://doi.org/10.1093/nsr/nwad015

Pryjmaková J, Kaimlová M, Hubáček T, Švorčík V, Siegel J (2020) Nanostructured materials for artificial tissue replacements. Int J Mol Sci 21(7). https://doi.org/10.3390/ijms21072521

Qi L, Ruan S, Zeng Y-J (2021) Review on recent developments in 2D ferroelectrics: theories and applications. Adv Mater 33(13):2005098. https://doi.org/10.1002/adma.202005098

Qiu C-W, Zhang T, Hu G, Kivshar Y (2021) Quo vadis, metasurfaces? Nano Lett 21(13):5461–5474. https://doi.org/10.1021/acs.nanolett.1c00828

Rajagopalan J (2019) Microelectromechanical systems (MEMS)-based testing of materials. In: Schmauder S, Chen C-S, Chawla KK, Chawla N, Chen W, Kagawa Y (eds) Handbook of mechanics of materials. Springer, Singapore, pp 1955–1979. https://doi.org/10.1007/978-981-10-6884-3_45

Rajamehala M, Phiri CK, Singh M VP (2022) Approaches, attributes and applications of matrix nanocomposites—A review. Cogent Eng 9(1):2152650. https://doi.org/10.1080/23311916.2022.2152650

Raoux S (2009) Phase change materials. Annu Rev Mater Res 39(1):25–48. https://doi.org/10.1146/annurev-matsci-082908-145405

Rasouli R, Barhoum A, Uludag H (2018) A review of nanostructured surfaces and materials for dental implants: surface coating, patterning and functionalization for improved performance. Biomater Sci 6(6):1312–1338. https://doi.org/10.1039/C8BM00021B

Rizkin BA, Popovic FG, Hartman RL (2019) Review article: spectroscopic microreactors for heterogeneous catalysis. J Vac Sci Technol 37(5):50801. https://doi.org/10.1116/1.5108901
Rogolino A, Savio G (2021) Trends in additively manufactured microfluidics, microreactors and catalytic materials. Mater Adv 2(3):845–855. https://doi.org/10.1039/D0MA00704H
Roy A, Dwari D, Ram MK, Datta P (2022) Chapter 13—Piezoelectric nanomaterials for biomedical applications. In: Pal K, Sarkar A, Sarkar P, Bandara N, Jegatheesan V (eds) Food, medical, and environmental applications of nanomaterials, in Micro and Nano Technologies. Elsevier, pp 355–377. https://doi.org/10.1016/B978-0-12-822858-6.00012-1
Rumpf R (2022) Electromagnetic and photonic simulation for the beginner: finite-difference frequency-domain in MATLAB. Artech House, Boston. https://empossible.net/fdfdbook/. Accessed 25 Dec 2023
Sannino D (2021) Types and classification of nanomaterials. In: Tahir MB, Rafique M, Sagir M (eds) Nanotechnology: trends and future applications. Springer, Singapore, pp 15–38. https://doi.org/10.1007/978-981-15-9437-3_2
Sekhar MC, Veena E, Kumar NS, Naidu KCB, Mallikarjuna A, Basha DB (2023) A review on piezoelectric materials and their applications. Cryst Res Technol 58(2):2200130. https://doi.org/10.1002/CRAT.202200130
Shamos MH, Lavine LS (1967) Piezoelectricity as a fundamental property of biological tissues. Nature 213(5073). https://doi.org/10.1038/213267a0
Sharpe WN (2002) 3 Mechanical properties of MEMS materials. In: The MEMS handbook, in Mechanical engineering series. CRC Press, pp 1–33
Shuai Z, Wang D, Peng Q, Geng H (2014) Computational evaluation of optoelectronic properties for organic/carbon materials. Acc Chem Res 47(11):3301–3309. https://doi.org/10.1021/ar400306k
Smith M, Kar-Narayan S (2022) Piezoelectric polymers: theory, challenges and opportunities. Int Mater Rev 67(1):65–88. https://doi.org/10.1080/09506608.2021.1915935
Smith DR, Padilla WJ, Vier DC, Nemat-Nasser SC, Schultz S (2000) Composite medium with simultaneously negative permeability and permittivity. Phys Rev Lett 84(18):4184–4187. https://doi.org/10.1103/PhysRevLett.84.4184
Smith DR, Schultz S, Markoš P, Soukoulis CM (2002) Determination of effective permittivity and permeability of metamaterials from reflection and transmission coefficients. Phys Rev B 65(19):195104. https://doi.org/10.1103/PhysRevB.65.195104
Stachiv I, Alarcon E, Lamac M (2021) Shape memory alloys and polymers for MEMS/NEMS applications: review on recent findings and challenges in design, preparation, and characterization. Metals 11(3). https://doi.org/10.3390/met11030415
Stanimirović Z, Stanimirović I (2009) Mechanical properties of MEMS materials. In: Takahata K (ed) Micro electronic and mechanical systems. IntechOpen, pp 3-1–3-32. https://doi.org/10.5772/7010
Stiharu I, Andronenko S, Zinnatullin A, Vagizov F (2023) SiCNFe ceramics as soft magnetic material for MEMS magnetic devices: a Mössbauer study. Micromachines 14(5). https://doi.org/10.3390/mi14050925
Sumithra P, Thiripurasundari D (2017) Review on computational electromagnetics. Adv Electromagn 6(1). https://doi.org/10.7716/aem.v6i1.407
Szécsényi KM, Holló BB (2023) Chapter 6—Simultaneous DSC techniques. In: Menczel JD, Grebowicz J (eds) The handbook of differential scanning calorimetry. Butterworth-Heinemann, pp 659–791. https://doi.org/10.1016/B978-0-12-811347-9.00007-2
Tanasković D, Obradov M, Jakšić O, Jakšić Z (2014) A low-loss double-fishnet metamaterial based on transparent conductive oxide. Phys Scr 2014(T162):014048.1–014048.5. https://doi.org/10.1088/0031-8949/2014/T162/014048
Tanasković D, Jakšić Z, Obradov M, Jakšić O (2015) Super unit cells in aperture-based metamaterials. J Nanomater 2015:312064.1–312064.9. https://doi.org/10.1155/2015/312064
Tanasković D, Obradov M, Jakšić O, Jakšić Z (2016) Nonlocal effects in double fishnet metasurfaces nanostructured at deep subwavelength level as a path toward simultaneous sensing of multiple

chemical analytes. Photon Nanostruct Fundam Appl 18:36–42. https://doi.org/10.1016/j.photonics.2015.12.003
Tripathy A, Saravanakumar B, Mohanty S, Nayak SK, Ramadoss A (2021) Comprehensive review on flexoelectric energy harvesting technology: mechanisms, device configurations, and potential applications. ACS Appl Electron Mater 3(7):2898–2924. https://doi.org/10.1021/ACSAELM.1C00267/ASSET/IMAGES/MEDIUM/EL1C00267_0023.GIF
Transforming our world: the 2030 Agenda for Sustainable Development | Department of Economic and Social Affairs. https://sdgs.un.org/2030agenda. Accessed 24 Sep 2023
Varghese J, Whatmore RW, Holmes JD (2013) Ferroelectric nanoparticles, wires and tubes: synthesis, characterisation and applications. J Mater Chem C 1(15):2618–2638. https://doi.org/10.1039/C3TC00597F
Veselago VG (1968) The electrodynamics of substances with simultaneously negative values of and μ. Sov Phys Uspekhi 10(4):509. https://doi.org/10.1070/PU1968v010n04ABEH003699
Vicentini Ferreira do Valle T et al (2023) MEMS reliability: on-chip testing for the characterization of the out-of-plane polysilicon strength. Micromachines 14(2). https://doi.org/10.3390/mi14020443
Voiculescu I, Nurashikin Nordin A (2012) Acoustic wave based MEMS devices for biosensing applications. Biosens Bioelectron 33:1–9. https://doi.org/10.1016/j.bios.2011.12.041
Wang H, Ma Y, Zheng Q, Cao K, Lu Y, Xie H (2021a) Review of recent development of MEMS speakers. Micromachines 12(10):1257. https://doi.org/10.3390/MI12101257
Wang C et al (2021b) A general approach to composites containing nonmetallic fillers and liquid gallium. Sci Adv 7(1):eabe3767. https://doi.org/10.1126/sciadv.abe3767
Wei G, Ma PX (2008) Nanostructured biomaterials for regeneration. Adv Funct Mater 18(22):3568–3582. https://doi.org/10.1002/adfm.200800662
Will-Cole AR et al (2022) Tutorial: Piezoelectric and magnetoelectric N/MEMS—Materials, devices, and applications. J Appl Phys 131(24):241101. https://doi.org/10.1063/5.0094364/16508697/241101_1_ACCEPTED_MANUSCRIPT.PDF
Xing W et al (2022a) Gallium-based liquid metal composites with enhanced thermal and electrical performance enabled by structural engineering of filler. Adv Eng Mater 24(9):2101678. https://doi.org/10.1002/adem.202101678
Xing W, Xu Y, Song C, Deng T (2022b) Recent advances in thermal interface materials for thermal management of high-power electronics. Nanomaterials 12(19). https://doi.org/10.3390/nano12193365
Yan Z, Jiang L (2017) Modified continuum mechanics modeling on size-dependent properties of piezoelectric nanomaterials: a review. Nanomaterials 7(2). https://doi.org/10.3390/nano7020027
Yuan M, Cao Z, Luo J, Chou X (2019) Recent developments of acoustic energy harvesting: a review. Micromachines 10(1):48. https://doi.org/10.3390/MI10010048
Yuan X et al (2021) Recent progress in the design and fabrication of multifunctional structures based on metamaterials. Curr Opin Solid State Mater Sci 25(1):100883. https://doi.org/10.1016/J.COSSMS.2020.100883
Yurkov AS, Dejneka A, Yudin PV (2019) Flexoelectric polarization induced by inhomogeneous heating and implications for energy harvesting. Int J Solids Struct 162:96–104. https://doi.org/10.1016/J.IJSOLSTR.2018.12.003
Zeng Z, Huang D, Zhang L, Sheng X, Chen Y (2023) An innovative modified calcium chloride hexahydrate–based composite phase change material for thermal energy storage and indoor temperature regulation. Adv Compos Hybrid Mater 6(2):80. https://doi.org/10.1007/s42114-023-00654-z
Zhang S, Li L, Kumar A (2008) Materials characterization techniques. CRC Press. https://doi.org/10.1201/9781420042955
Zhang Y, Sun J, Liu H, Liu Z (2022) Modeling and measurement of thermal–mechanical-stress-creep effect for RF MEMS switch up to 200 °C. Micromachines 13(2):166. https://doi.org/10.3390/MI13020166

Zhang H, Geng J, Gao H, Rong W, Xie H (2023a) The development of 3D atomic force microscopy with magnetically driven-orthogonal cantilever probes. Engineering 24:84–93. https://doi.org/10.1016/j.eng.2022.06.010
Zhang H, Jia H, Feng W, Ni Z, Xu P, Li X (2023b) Ultra-responsive MEMS sensing chip for differential thermal analysis (DTA). Sensors 23(3). https://doi.org/10.3390/s23031362
Zhou W et al (2022) Chip-based MEMS platform for thermogravimetric/differential thermal analysis (TG/DTA) joint characterization of materials. Micromachines 13(3):445. https://doi.org/10.3390/mi13030445

GPSR Compliance

The European Union's (EU) General Product Safety Regulation (GPSR) is a set of rules that requires consumer products to be safe and our obligations to ensure this.

If you have any concerns about our products, you can contact us on ProductSafety@springernature.com

In case Publisher is established outside the EU, the EU authorized representative is:

Springer Nature Customer Service Center GmbH
Europaplatz 3
69115 Heidelberg, Germany

Batch number: 10704307

Printed by Printforce, the Netherlands